Introduction to
Calculus

WOLFRAM ✸ eTEXTBOOK SERIES

Introduction to
Calculus

A COMPUTATIONAL APPROACH

John Clark | Devendra Kapadia

Introduction to Calculus: A Computational Approach
by John Clark and Devendra Kapadia
Copyright © 2024 by Wolfram Media, Inc.

Wolfram Media, Inc. | wolfram-media.com

ISBN 978-1-57955-091-2 (paperback)
ISBN 978-1-57955-080-6 (digital online)
ISBN 978-1-57955-092-9 (Kindle)

Library of Congress Control Number: 2024944758

This book is based on the Wolfram U interactive course Introduction to Calculus:
wolfr.am/WolframU-IntroToCalc

For information about permission to reproduce selections from this book, write to
permissions@wolfram.com.

Typeset with Wolfram Notebooks: wolfram.com/notebooks

Access the interactive Wolfram
Notebook edition of this textbook:

wolfr.am/eTextbook-IntroToCalc

Opening these notebooks will require
Mathematica, Wolfram|One or Wolfram Player.

Most organizations have a site license for Mathematica.
To find out if you have access, visit wolfr.am/siteinfo.

If you don't have access through your organization,
visit wolfr.am/downloads.

Contents

To the Student

Calculus has provided stunningly accurate models for a wide range of systems in science, engineering, economics and other fields during the last three centuries. Hence, the study of this subject is an essential part of the curriculum at high schools and colleges everywhere.

This book will give you a comprehensive introduction to the basic concepts of calculus along with their history and applications. It begins with functions and limits, followed by differential calculus and its applications, and then moves on to integral calculus and its applications. You will learn each concept in a traditional manner but you will also learn how to build intuition and solve real-world problems with blinding speed using the world-class functionality for doing calculus computations and visualizations in Mathematica.

Perhaps the most important feature of this approach is the ease with which you can vary any problem to gain insight into its essential features while Mathematica does the tedious calculations for you. Also, when you have finished studying this book, you will have acquired considerable expertise with using the powerful and versatile Mathematica system.

We recommend that you use this book along with the popular interactive Wolfram U course on "Introduction to Calculus" which has already helped thousands of students all over the world including those preparing for the AP Calculus exam in the United States and others in seemingly distant places such as Nasarawa in Nigeria.

The mastery of the fundamental concepts of calculus is a major milestone in your academic career. We hope that our book will help you to achieve this milestone!

The Wolfram Calculus and Algebra Team
October 2023

1 | What Is Calculus?

The Science of Change

Calculus is, in short, the science of change. Examples of change are all around us.

For instance, the second hand on a clock changes every second:

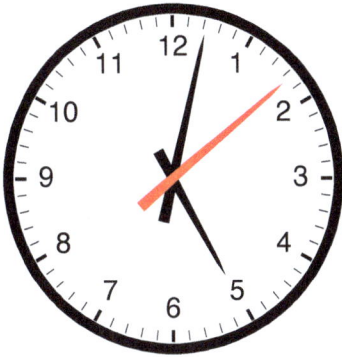

Also, the planets in the solar system change their position every instant as they revolve around the Sun:

orbital motion ▶ ⌃ ⌄ →

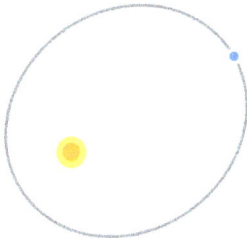

If your body became static (no longer pumping blood, no longer burning calories, etc.), you would die. Calculus is an essential tool for deeply understanding change in a mathematical way.

The Four Main Problems

Calculus was originally developed to solve four main problems that have interested humanity since time immemorial. These problems were studied considerably by the Greeks and later investigated by mathematical scholars of the seventeenth and eighteenth centuries.

They are the problems of:

1. Finding a tangent line to a curve

2. Calculating the area under a curve

3. Investigating the velocity of a particle given either its position or acceleration

4. Optimizing a process by finding the corresponding function's maxima and minima

Tangent to a Curve

Finding the tangent to a curve was a problem that had applications in many different fields. The problem was inherently geometrical, but physicists in optics and mechanics also wanted to find its solution.

In optics, it was crucial in investigating how light entered a lens. In mechanics, it was known that moving bodies on a path went in the direction of the tangent line to the path.

Today, point-slope form is used to find the equations of lines, but calculating the slope at a single point on a curve is difficult. The best method is to use two points that are very close to each other to approximate the slope.

Consider the following interactive example:

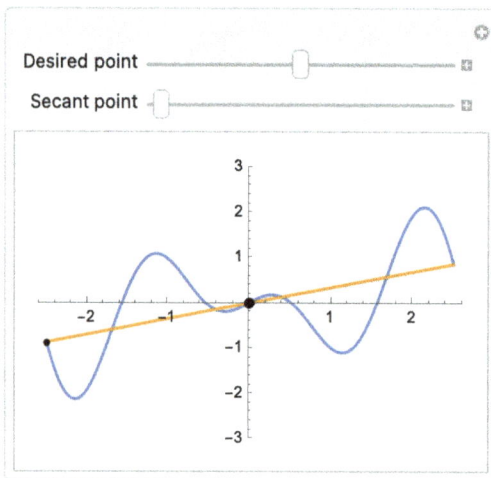

As the secant point gets closer to the desired point, the result is a line with slope closer to the desired instantaneous slope. However, letting the points overlap will result in an error.

Area and Length

Finding the area and length of curves was a very important problem in celestial mechanics. The goal was to find the distance a planet travels in a given length of time.

The problem of area and length was also extended to finding the area between two curves, calculating the volumes of solids, finding the center of mass for an object, and even calculating the gravitational force that something like a planet exerted on other objects.

The Greeks made some progress in this area. Archimedes used the method of exhaustion to find the area of a circle.

Here is an example that gives the area of a regular polygon with n sides as n goes from 3 to 50:

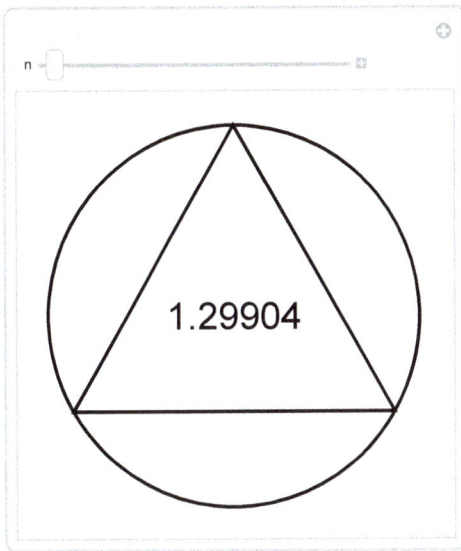

As the number of sides goes up, the polygons look more and more like the circumscribed circle. Likewise, the area for a regular polygon with apothem 1 approaches $\pi \sim 3.14$. The circle with radius 1 has area π.

The Velocity Problem

Given a particle's position, scientists wanted to calculate its velocity at various instants. The standard formula for velocity could not be used, because that calculated the average velocity of the particle over a given time period. To calculate instantaneous velocity, they had to use a time period that was essentially 0, so that no time actually transpired.

Scientists also wanted to calculate a particle's velocity given its acceleration, or the rate at which the velocity was changing. For particles with constant acceleration, this was not difficult, but for a particle with a variable acceleration, the problem was more challenging.

In the next example, a particle is following a path with its unit velocity (orange) and unit acceleration (purple) shown:

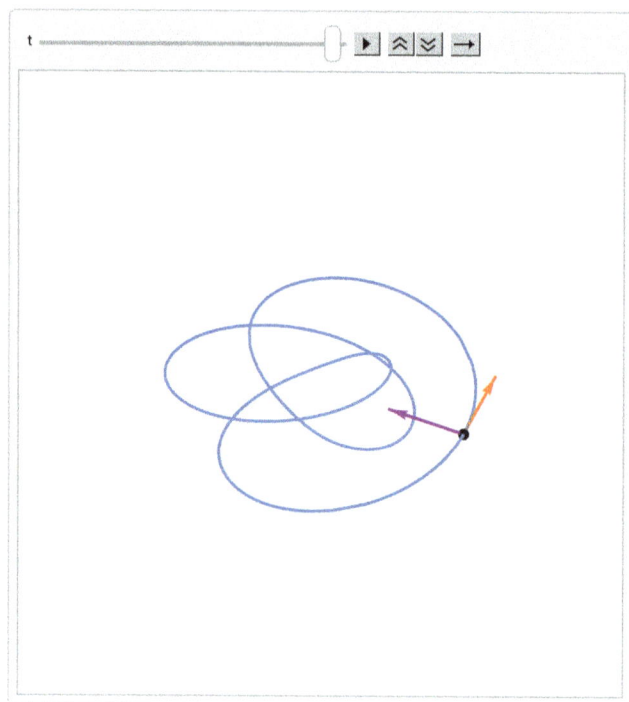

Optimization

Humanity has always been interested in performing tasks in the best way possible. Warriors wanted to shoot their weapons so that they covered the most distance, in addition to hitting their target. Scientists wanted to calculate the greatest and least distances that a planet was from the Sun. Merchants wanted to know how many items to produce in order to minimize costs and maximize revenue.

Consider the following (relatively simple) problem: a farmer has only 500 ft. of fencing and wants to fence off a rectangle with the maximum amount of area. What should the dimensions of the rectangle be?

Here is a model to illustrate:

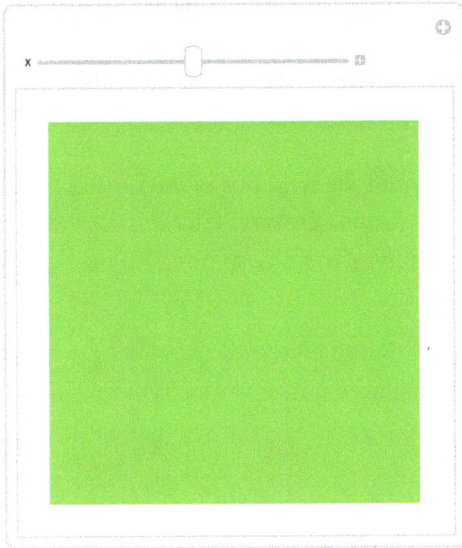

It is clear that the rectangle should be a 125 ft. by 125 ft. square, but how could this solution have been arrived at without trial and error?

Early Development

Besides the Greeks, mathematicians in the seventeenth century were trying to figure how to solve these four problems.

One of the first pioneers was Pierre de Fermat, who in his *Methodus ad Disquirendam Maximam et Minimam* (1637) provided the basic notion of how to calculate the tangent line to a curve using geometrical methods.

One of the most influential pioneers of calculus was John Wallis, who in his *Arithmetica Infinitorum* (1655) solved many problems using analysis and the method of indivisibles. He even came up with a way to calculate $\pi / 2$!

Johannes Kepler, the great astronomer, used the methods of the Greeks to calculate the volume of kegs for wine dealers in *Stereometria Doliorum* (1615). In his work, the beginning trend can be seen of approximating areas by splitting up regions into smaller and more manageable shapes.

Galileo in *Two New Sciences* (1638) argued that the distance traveled by a point could be calculated by finding the corresponding area under a velocity-time curve.

Bonaventura Francesco Cavalieri, a student of Galileo's, was one of the first to suppose that dividing a region into infinitely many pieces would be monumental in calculating area. Indeed, this can be seen in the principle that bears his name.

Isaac Barrow in his *Lectiones Geometricae* (1669) used what is now known as the differential triangle to help calculate the tangent to a curve through purely geometrical means and was one of the first to connect the tangent line problem to the area problem.

Discovery by Newton and Leibniz

With all the progress made by previous scientists, one may wonder what Newton and Leibniz had to offer. In short, both summarized and expanded upon previous work by giving the general methods of the differential and integral calculus. Since the work involved in calculating areas and tangents was such an involved process, many mathematicians could not see the general method they were utilizing.

Isaac Newton had his idea for calculus as early as 1665, and only published his work on it in passing roughly 20 years later in his masterful *Philosophiae Naturalis Principia Mathematica* (1687). He later published an official paper on the calculus in 1693. Newton was a scientist first, and a mathematician second. This was clearly seen in the *Principia*, where he developed the laws of classical mechanics and planetary motion.

Gottfried Wilhelm Leibniz was a diplomat, "philosopher, lawyer, historian, philologist, and pioneer geologist" (*Mathematical Thought from Ancient to Modern Times*, Morris Kline, 370). He published his work on calculus in 1684 in the *Acta Eruditorum*, providing the notation still used today.

The question soon arose as to who should receive initial credit for the discovery. Although Leibniz did publish his paper first, Newton was always collaborating with other scientists and mathematicians and sharing his calculus techniques with them in the years between 1665 and 1687.

Newton accused Leibniz of plagiarism, and a now-famous argument over priority erupted, leaving scientists in England to advocate for Newton, while scientists in continental Europe supported Leibniz.

It has been said that English mathematics as a whole was "deprived of contributions that some of the ablest minds might have made" because of this struggle (Kline, 381).

Later Developments

Even though Leibniz and Newton provided papers on calculus, both were rather unclear in their explanations. Newton with his fluxions and Leibniz with his dx's and dy's ultimately had their works prepared for a larger audience through the work of the Bernoulli brothers (Jacob and Johann).

Jacob and Johann were both mathematicians in their own right, and Johann was an early adopter of the differential calculus.

Gillaume de l'Hôpital (the person l'Hôpital's rule is named for) used the notes of Johann Bernoulli to create the first differential calculus book, *Analyse des infiniment petits* (1696). His clear diction and style were instrumental in popularizing the calculus for wider audiences.

Maria Gaetana Agnesi wrote the first calculus book that covered both differential and integral calculus and was the first woman to be appointed a mathematics professor at a university. The witches of Agnesi is a family of curves named in her honor.

Augustin-Louis Cauchy and Karl Weierstrass later made the reasoning behind calculus more precise and provided the modern foundation for calculus that is used today.

Calculus in Today's World

Nowadays, calculus is used in very diverse fields all over the world.

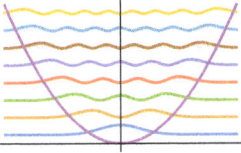

Quantum mechanics uses the Schrödinger equation, which is defined using concepts that come from calculus.

Financial experts use the famous Black-Scholes equation to find the price for options in a stock portfolio. It also uses advanced concepts from calculus.

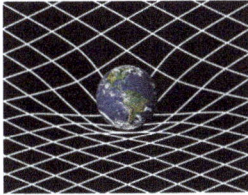

Albert Einstein developed his general theory of relativity using math based on calculus.

Economists use calculus to analyze markets and see the current trends.

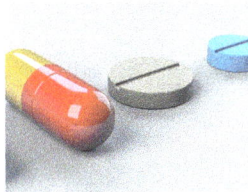

In medicine, calculus is used to figure out the proper dosing strategy for a new drug that has been developed.

Scientists at NASA use calculus to help direct their rockets during spaceflight.

Why Study Calculus?

This lesson has already covered the main advantages of calculus in fields such as economics, physics, engineering and mathematics. However, not every student wants to be an economist, physicist, engineer or mathematician!

Ultimately, calculus is very useful because its applications are so broad. It is useful for companies trying to maximize their profits; it is useful for scientists wanting to analyze natural phenomena; it is useful when calculating estimations.

For those wanting to take higher-level STEM courses at the college level, calculus is usually a prerequisite. To not know calculus as a senior undergraduate in STEM would be very detrimental!

Above all else though, it is important to study calculus because it is a major milestone in any student's career. Many people already cower at the word "calculus," so successfully completing a calculus course can provide a huge boost to one's self-confidence.

To know how to solve problems that baffled even the Greeks, as well as the greatest minds of the seventeenth and eighteenth centuries, is sure to boost one's ego.

About This Course

This 40-part lecture series will roughly cover the entire subject matter of a typical AP Calculus AB course. It will first start with functions and limits, followed by differential calculus and its applications, and then move on to integral calculus and its applications. Differential calculus studies how quantities change. Integral calculus studies how quantities accumulate. Near the end, there will also be a lesson on how to make programs that use calculus in the Wolfram Language™.

Even though it is important to understand the reasoning behind the theory, for the most part the lessons refrain from providing formal proofs for the various theorems presented.

Who Is This Course For?

The Wolfram U course on calculus is useful for:

Beginners wanting to learn calculus for the first time.

Professionals who already know calculus but may want a refresher.

Anyone who wants to know how the Wolfram Language is useful for studying calculus.

Summary

As a mathematical subject, calculus is an essential concept that shows what humanity is capable of.

As a scientific tool, calculus is a very useful apparatus that has changed how people view the world.

As an educational hurdle, calculus is one of many tasks that a student should face and be proud to master.

In the next lesson, the course will formally begin with an introduction to functions.

2 | Functions

Overview

Calculus is all about change. Changing speeds, changing areas, changing volumes, change, change, change. In general, quantities change **with respect to some other quantity (or quantities)**.

A moving car's **distance** changes with respect to **time**. A cylinder's **surface area** changes with respect to its **radius and height**. A mathematical way of **precisely** defining such change is needed.

This lesson endeavors to do so, and to show how to do it in the Wolfram Language.

Functions

Mathematics gives a way of precisely defining the way one quantity changes with respect to some other quantities.

First, here are some definitions:

A **set** is a collection of objects (for example, the set of even positive numbers is 2, 4, 6, 8, 10, …).

An **element** is an object in a set. So 2 is an element of the set of even positive numbers.

Now there is enough information to define the special tool:

A **function** f is a rule that assigns to each element x in a set A *exactly one* element, called $f[x]$, in a set B.

The value $f[x]$ is called the **value of f at x** and is usually read as "f of x."

The set A is called the **domain** of f. The set B is called the **codomain** of f.

The set of all possible values of f given elements in the domain of f is called the **range** of f.

Note that the range is contained in the codomain of f, but they do not have to be equal.

Symbols (like x) in the domain are called **independent variables**, while symbols (like y) in the range of f are called **dependent variables**.

Here is an example of a function. Its domain is all real numbers, its codomain is all real numbers, and its range is the interval $[-1, 1]$:

In[ ]:= **Plot[Sin[x], {x, −10, 10}]**

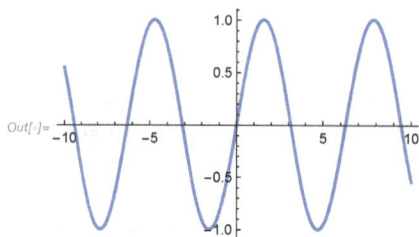

Making Functions with the Wolfram Language

To make functions in the Wolfram Language, the following notation is used:

In[ ]:= **f[x_] := x^2**

This function, called f, takes the independent variable x and produces the dependent variable x^2.

Now use it on the value 2:

In[ ]:= **f[2]**

Out[ ]= **4**

Be sure to copy the left-hand side **exactly**, otherwise it will not work. **To make a function f you need f[x_]:=**.

Do not forget the underscore _ or the colon : , and be sure to use x in the definition to make it vary with x.

Here is a function that depends on variables x and t:

In[ ]:= **g[x_, t_] := x t + t^3**

Plug in the values $x = 2$ and $t = 3$:

In[ ]:= **g[2, 3]**

Out[ ]= **33**

Notice that the right-hand side **x t+t^3** has the same independent variables (**x** and **t**) as the left-hand side (**g[x_, t_]**).

Sketching Graphs

You can also sketch the **graph of a function**. For a function f with domain A, its graph is the set of ordered pairs $\{(x, f[x]) \mid x \in A\}$.

The notation $\{x \mid x \in A\}$ is used to indicate sets. $\{x \mid x \in A\}$ says the set consists of the values x where x is an element of A. In this case, $\{(x, f[x]) \mid x \in A\}$ means there is a set of ordered pairs $(x, f[x])$ where x is an element of A.

To sketch 2D graphs with the Wolfram Language, use the built-in Plot function (to make 3D graphs, use Plot3D).

The following plot makes the graph of the function $f[x] = x^2$ from 0 to 10:

In[ ]:= **f[x_] := x^2**

You can also use the expression x^2:

In[ ]:= **Plot[f[x], {x, 0, 10}]**

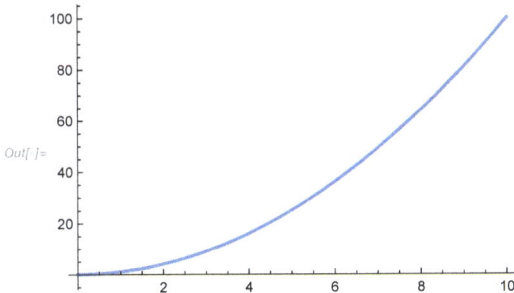

Evaluating Functions

You can also plug math expressions into functions.

Plug the expression $x + 1$ into the function $f[x] = x^2$:

In[ ]:= **f[x_] := x^2**

In[ ]:= **f[x + 1]**

Out[ ]= $(1 + x)^2$

If you want to expand the given expression, use Expand:

In[ ]:= **Expand[f[x + 1]]**

Out[ ]= $1 + 2x + x^2$

In the future, `//` will be used at the end of a function to also use **Expand**:

In[ ]:= **f[x + 1] // Expand**

Out[ ]= $1 + 2x + x^2$

Plug the expression $x + y$ into the expression $g[x] = 3\,x^3 - 2\,x^2 + 6\,x - 4$:

In[ ]:= **g[x_] := 3 x ^ 3 − 2 x ^ 2 + 6 x − 4**

In[ ]:= **g[x + y] // Expand**

Out[ ]= $-4 + 6x - 2x^2 + 3x^3 + 6y - 4xy + 9x^2y - 2y^2 + 9xy^2 + 3y^3$

Function Representations

Even though the Wolfram Language has a simple way of representing functions, in real life there are many ways to do so.

A function can be described **verbally**: a square's area is given by squaring the length of one of its sides.

A function can be described using a **table of values**:

side length	1	2	3	4	5	6	7	8	9	10
square area	1	4	9	16	25	36	49	64	81	100

A function can be described with its **graph**:

In[ ]:= **Plot[x ^ 2, {x, 0, 10}]**

Out[ ]=

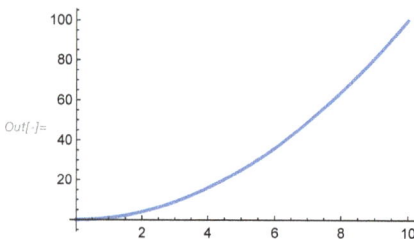

A function can be described **algebraically**: squarearea[sidelength] = sidelength^2.

The Wolfram Language works best with algebraically described functions:

In[ ]:= **squarearea[*sidelength_*] := *sidelength* ^ 2**

In[ ]:= **squarearea[55]**

Out[ ]= 3025

All of these represent the same function.

Cost of a Box

Consider the following example:

A metal box has volume 100 cm^3. It height is twice its length. If the material for the top and bottom costs \$5 per square centimeter, and the material for the sides costs \$8 per square centimeter, write a function that gives the total cost of the material needed to make the box in terms of the box's length.

The surface area of a rectangular prism [box figure labeled height, width, length] is given by the following expression:

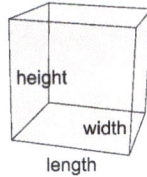

In[ ]:= **surfacearea = 2 length * width + 2 length * height + 2 width * height;**

In this case, the cost of the material can be written as:

In[ ]:= **cost = 5 (2 length * width) + 8 (2 length * height + 2 width * height);**

You know the height is two times the length: height = $2 *$ length.

You can now express the width of the box in terms of the length using the volume with Solve, since volume = length $*$ width $*$ height:

In[ ]:= **Solve[100 == length * width * height /. height → 2 length, width]**

Out[ ]= $\left\{\left\{\text{width} \rightarrow \dfrac{50}{\text{length}^2}\right\}\right\}$

Now you can make a function for the cost of the material:

In[ ]:= **boxcost[*length_*] :=**
    5 (2 *length* * 50 / (*length* ^ 2)) + 8 (2 *length* * (2 *length*) + 2 * 50 / (*length* ^ 2) * (2 *length*))

For a box with length 10 cm, the cost is \$3410:

In[ ]:= **boxcost[10]**

Out[ ]= **3410**

Function Domains

Most of the functions used here have **all real numbers** as their domain. In other words, any real number can be plugged into the function to get a result. However, the previous function **boxcost** has a more restricted domain.

Plug a negative number into the function:

In[·]:= **boxcost[−1]**

Out[·]= −2068

Even though you got an answer, the answer does not make sense. The cost cannot be negative; you do not spend negative money (earn money?) to make the box. The length cannot be negative either; that does not make sense.

Plugging in 0 also makes you run into trouble, since the function is not defined there:

In[·]:= **boxcost[0] // Quiet**

Out[·]= Indeterminate

If you input an invalid value, you will get errors. Quiet silences these errors (so you cannot see the error messages that would show up otherwise). So the domain of the function is just positive real numbers.

FunctionDomain gives the values where a function is defined:

In[·]:= **FunctionDomain[boxcost[length], length]**

Out[·]= length < 0 || length > 0

In this case, the function is not defined at 0, but the domain should only be values greater than 0. It is important to use **common sense** when finding the domain of a function. Context is everything!

Vertical Line Test

The definition of a function says a function f assigns to each element x in a set A *exactly one* element, called $f[x]$, in a set B.

The following graph is therefore **not** the graph of a function:

In[·]:= **ParametricPlot[{Cos[t], Sin[t]}, {t, 0, 2 π}]**

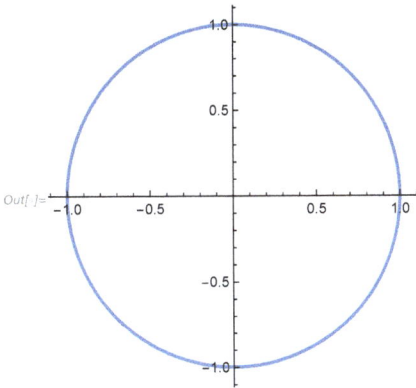

Out[·]:=

It assigns each element in the interval $(-1, 1)$ to *two* elements in the codomain.

In general, you can determine if a graph is a graph of a function by using the **vertical line test**. The vertical line test states that a curve in the coordinate plane is the graph of a function **if and only if** no vertical line passes through the curve *more than once*. So by the vertical line test, the preceding graph is not the graph of a function, since vertical lines pass through it *twice* in the interval $(-1, 1)$.

Piecewise Functions

Sometimes a function can be described by different formulas in different parts of its domain. Such functions are called **piecewise defined** (shortened here to **piecewise**) **functions**.

The built-in function RealAbs is an example of a piecewise function. It is x when $x \geq 0$ and $-x$ when $x < 0$.

Here is its graph in the interval $[-5, 5]$:

In[·]:= **Plot[RealAbs[x], {x, −5, 5}]**

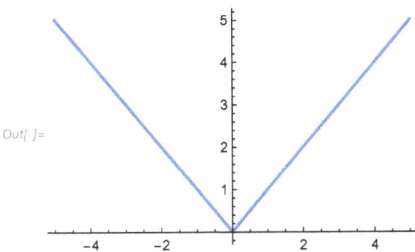

Out[·]:=

You can also recreate it using the built-in Piecewise:

In[·]:= **realabs[*x*_] := Piecewise[{{−*x*, *x* < 0}}, *x*]**

In[·]:= **Plot[realabs[x], {x, −5, 5}]**

Out[·]=

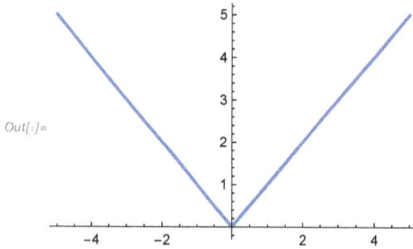

Symmetry

Functions can also have innate symmetry. An **even function** f is a function with the property that $f[-x] = f[x]$. Visually, its graph is symmetric with respect to the y axis. For example, $f[x] = x^2$ is an even function, since $f[-x] = (-x)^2 = x^2 = f[x]$.

This can also be seen in its graph:

In[·]:= **f[*x*_] := *x*^2**

In[·]:= **Plot[f[x], {x, −2, 2}]**

Out[ ]=

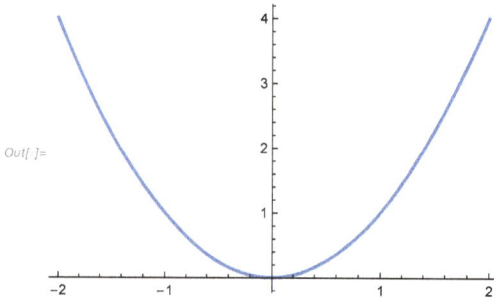

An **odd function** f is a function with the property that $f[-x] = -f[x]$. Visually, its graph is symmetric with respect to the origin. In other words, if you take your function's graph and rotate it 180 degrees about the origin, you will get the same graph as before.

For example, $f[x] = x^3$ is an odd function, since $f[-x] = (-x)^3 = -x^3 = -f[x]$.

This can also be seen in its graph:

In[·]:= **f[x_] := x^3**

In[·]:= **Plot[f[x], {x, −2, 2}]**

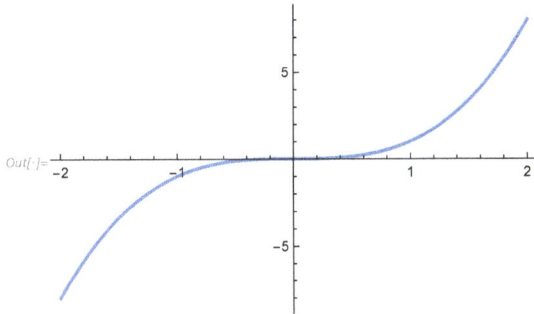

Increasing and Decreasing Functions

Look again at the function $f[x] = x^2$:

In[·]:= **f[x_] := x^2**

In[·]:= **Plot[f[x], {x, −1, 1}]**

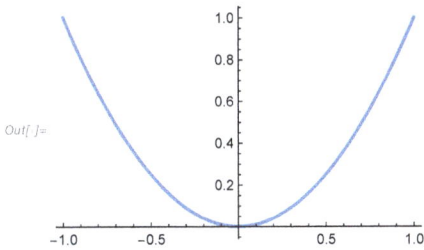

Notice that the function gets smaller as you go from −1 to 0, and gets larger as you go from 0 to 1. This can be expressed by saying the function is decreasing on [−1, 0] and increasing on [0, 1].

In general, a function is **increasing** on the interval [a, b] if $f[x_1] < f[x_2]$ for x_1, x_2 in [a, b] and $x_1 < x_2$. A function is **decreasing** on the interval [a, b] if $f[x_1] > f[x_2]$ for x_1, x_2 in [a, b] and $x_1 < x_2$.

The following function is neither increasing nor decreasing:

In[·]:= **g[x_] := 4**

In[·]:= **Plot[g[x], {x, −2, 2}]**

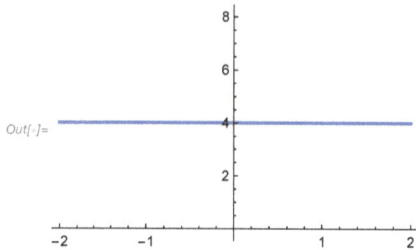

Out[·]=

Summary

To say that a quantity changes with respect to some other quantities precisely, functions are used.

Functions and their graphs are easy to make in the Wolfram Language.

Mathematical expressions can also be plugged into functions.

The domain of a function is the set of all possible inputs for the function.

The range of a function is the set of all possible outputs for the function.

Functions can represented using words, tables, graphs and formulas.

The vertical line test can be used to determine whether a graph is the graph of a function.

Piecewise-defined functions use different formulas for different parts of their domains.

Functions can have even or odd symmetry.

Functions can be increasing, decreasing or neither.

The next lesson will go over the basic functions covered in this course.

Exercises

Exercise 1—Sketch a Polynomial

Sketch the graph of the function $f[x] = 2\,x^6 - 3\,x^3 + 1$ in the interval $[-1,\ 1.5]$.

Solution

First, define the function in the Wolfram Language:

In[ ]:= **f[x_] := 2 x^6 − 3 x^3 + 1**

Then graph it with Plot:

In[ ]:= **Plot[f[x], {x, −1, 1.5}]**

Out[ ]=

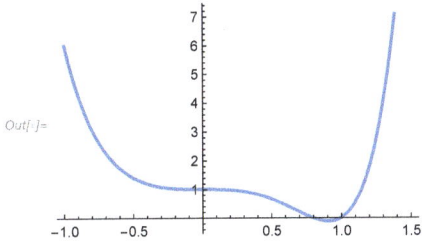

Exercise 2—Difference Quotient

The expression $(f[x + h] - f[x])/h$ is known as a **difference quotient**. Find the difference quotient for the function $f[x] = x^2 - 4\,x$.

Solution

Define the function:

In[ ]:= **f[x_] := x^2 − 4 x**

Then evaluate the expression $(f[x + h] - f[x])/h$:

In[ ]:= **(f[x + h] − f[x])**

Out[ ]= $4\,x - x^2 - 4\,(h + x) + (h + x)^2$

Simplify it with Simplify:

In[ ]:= **Simplify[(f[x + h] − f[x])/h]**

Out[ ]= $-4 + h + 2\,x$

You could also invoke it with //:

In[ ]:= **(f[x + h] – f[x])/h // Simplify**

Out[ ]= $-4 + h + 2x$

Exercise 3—Step Functions

The following table represents a function that gives the cost of mailing a packages with weight w:

0.75	if $0 \leq w < 2$
1	if $2 \leq w < 4$
1.25	if $4 \leq w < 6$
1.5	if $6 \leq w < 8$
1.75	if $8 \leq w < 10$
2	if $w > 10$

Graph the function using Piecewise and Plot in the interval [0, 15].

Solution

In[ ]:= **f[w_] := Piecewise[{{0.75, 0 ≤ w < 2}, {1, 2 ≤ w < 4},**
{1.25, 4 ≤ w < 6}, {1.5, 6 ≤ w < 8}, {1.75, 8 ≤ w < 10}, {2, w > 10}}, Null]

In[ ]:= **Plot[f[w], {w, 0, 15}]**

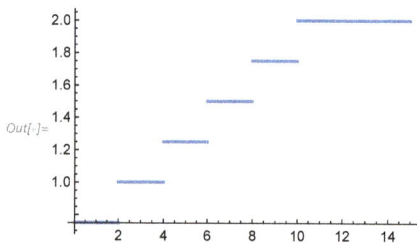

Notice the function is constant on every interval. Functions like this are called **step functions**.

Exercise 4—Even or Odd?

Determine if the following functions are even, odd or neither:

$$f[x] = x^5$$
$$g[x] = 2^x$$
$$h[x] = x^4 - 3x^2$$

Solution

The first function is odd, since $f[-x] = (-x)^5 = -x^5 = -f[x]$.

The second function is neither even nor odd, since
$g[-x] = 2^{-x} \neq 2^x = g[x]$ and $g[-x] = 2^{-x} \neq -2^x = -g[x]$.

The third function is even, since $h[-x] = (-x)^4 - 3(-x)^2 = x^4 - 3x^2 = h[x]$.

You could also see this in their graphs:

```
In[ ]:= f[x_] := x^5;
       g[x_] := 2^x;
       h[x_] := x^4 - 3 x^2;
```

```
In[ ]:= Plot[{f[x], g[x], h[x]}, {x, -2, 2}, PlotLegends → "Expressions"]
```

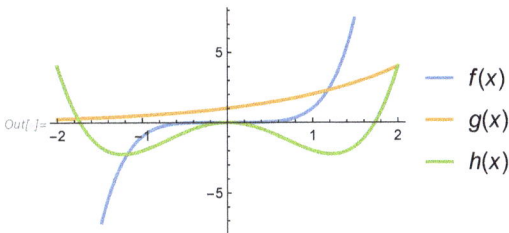

Exercise 5—Increasing and Decreasing Function

The graph of a function is given:

```
In[ ]:= Plot[Piecewise[{{-2 x^2 + (5 x^4)/4 - (x^6)/6, -2.5 ≤ x ≤ 2}, {-4.36, x < -2.5}}, 4/3], {x, -4, 4}]
```

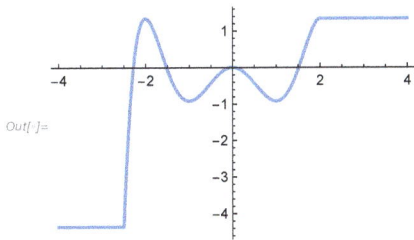

Where is it increasing and decreasing?

Solution

From the graph, you can see the function is increasing on $[-2.5, 2]$, $[-1, 0]$ and $[1, 2]$.

From the graph, you can see the function is increasing on $[-2.5, 2]$, $[-1, 0]$ and $[1, 2]$.

It is decreasing on $[-2, -1]$ and $[0, 1]$.

It is neither increasing nor decreasing on $(-\infty, -2.5]$ and $[2, \infty)$.

3 | The Elementary Functions

Overview

So far, you have made functions to express how one quantity changes with respect to others. However, there are so many different kinds of functions that one can get lost among them all.

Here is a plot of several functions:

```
In[ ]:= f[x_] := x + 2;
        g[x_] := 6 x^3 - x - 3;
        h[x_] := 3^x;
        r[x_] := 15 Log[x];
        s[x_] := 10 Sin[x];
        t[x_] := 2/(x^2);
        Plot[{f[x], g[x], h[x], r[x], s[x], t[x]}, {x, -10, 10}, PlotLegends → "Expressions"]
```

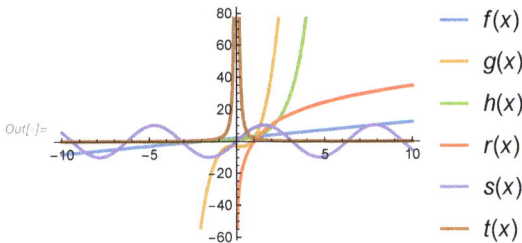

What can you assume about these functions based solely on their graphs?

This lesson categorizes many of the types of functions you will encounter later on in the course.

Linear

Consider the following function:

```
In[ ]:= f[x_] := x + 3
```

If you plot it, you can see that it makes a line:

In[·]:= **Plot[f[x], {x, −1, 1}]**

Out[·]=

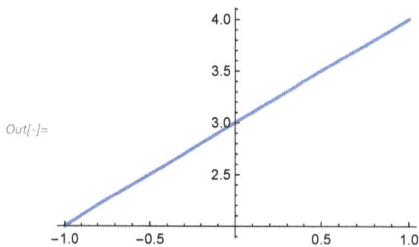

The general function $f[x] = m\,x + b$ is called a **linear function** and has **slope** m and **y-intercept** b. The function crosses the y axis at b and increases at the constant rate of m units for every 1 unit increase of x.

Polynomials

The following function is known as a **polynomial**:

In[·]:= **f[x_] := 3 x^4 − x^2 + 7 x + 6**

Here is its plot:

In[·]:= **Plot[f[x], {x, −2, 2}]**

Out[·]=

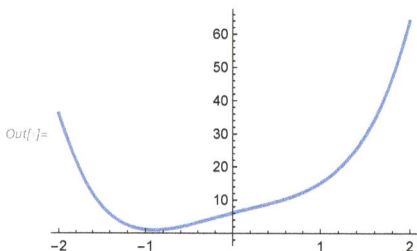

Polynomials generally have the form $f[x] = a_n x^n + a_{n-1} x^{n-1} + \ldots + a_1 x + a_0$, where n is non-negative and the constants $a_n, \ldots, a_0$ are called its **coefficients**.

If $a_n \neq 0$, then the **degree** of the polynomial is n.

The function shown has degree 4.

Linear functions have degree 1.

A function with degree 2 is called a **quadratic function**.

A function with degree 3 is called a **cubic function**.

Power Functions

A **power function** is a function of the form $f[x] = x^a$, where a is a constant. Power functions can take several forms. When a is a positive integer, the result is polynomials.

These polynomials have special patterns, as seen in the following graph:

In[ ]:= **Plot[{x, x^2, x^3, x^4, x^5, x^6}, {x, −1, 1}, PlotLegends → "Expressions"]**

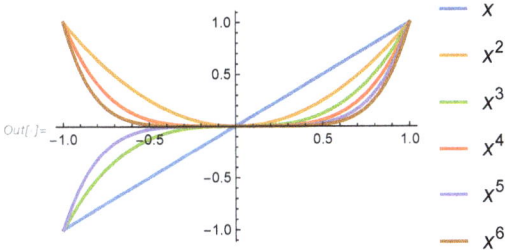

The odd degree functions are odd and the even degree functions are even: for example, $x^3 = -(-x)^3$, so the cube function is an odd function, and $x^4 = (-x)^4$, so the fourth power is an even function.

When $a = 1/n$ for n a positive integer, the result is **root functions**. For example, the square root function $f[x] = \sqrt{x}$ is a root function, with exponent $1/2$.

The root functions also have special patterns, as seen in this graph:

In[ ]:= **Plot[{x^(1/2), Surd[x, 3], x^(1/4), Surd[x, 5]}, {x, −1, 1}, PlotLegends → "Expressions"]**

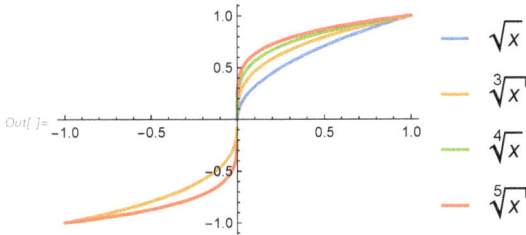

For n odd, the function is defined everywhere. For n even, the function is only defined on $[0, \infty)$.

Rational

Let $P[x]$ and $Q[x]$ be two polynomials. A **rational function** f is a ratio of the polynomials P and Q. In other words, $f[x] = \frac{P[x]}{Q[x]}$. When Q is constant, the result is the polynomials as before.

The reciprocal function $f[x] = x^{-1} = \frac{1}{x}$ is a rational function.

Here is its graph:

In[ ]:= **Plot[1 / x, {x, −1, 1}]**

Out[ ]=

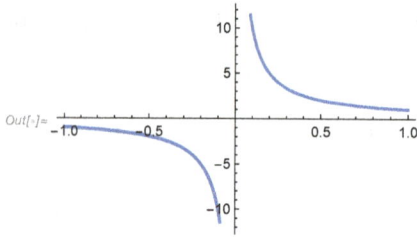

Notice how the graph does not seem to take on a value at $x = 0$. That is because the function is undefined there.

The function evaluates to $\frac{1}{0}$, which is undefined for real numbers. In general, rational functions are defined everywhere, except where their denominators are 0.

Algebraic

All the functions covered so far are called **algebraic functions**. That is because they can be made by adding, subtracting, multiplying, dividing or taking roots of polynomial functions and other algebraic functions.

The following function is an algebraic function:

In[ ]:= **f[x_] := Sqrt[x + Sqrt[3 x ^ 3 + 1]] / CubeRoot[x ^ 2]**

Here is its graph:

In[ ]:= **Plot[f[x], {x, −1, 1}]**

Out[ ]=

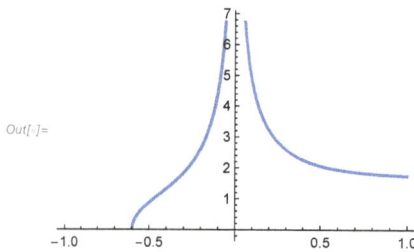

Algebraic functions come in a variety of shapes.

Trigonometric

The sine, cosine and tangent functions are **trigonometric** functions and are not algebraic. They can be defined based on the ratios of the sides of a triangle, but the general functions go beyond that. When inputting values into trigonometric functions in the Wolfram Language, it is assumed that radian measure is used.

So Sin[30] ≠ 1 / 2, but Sin[π / 6] = 1 / 2. To make it evaluate as degrees, add Degree at the end:

In[ ]:= **{Sin[30], Sin[π/6], Sin[30 Degree]}**

Out[ ]= $\left\{ \text{Sin[30]}, \frac{1}{2}, \frac{1}{2} \right\}$

Here are the graphs of the three trigonometric functions sine, cosine and tangent in the range $[-2\,\pi,\ 2\,\pi]$:

In[ ]:= **Table[Plot[i, {x, −2 π, 2 π}, PlotLabel → Style[i, Large]], {i, {Sin[x], Cos[x], Tan[x]}}]**

Out[ ]=

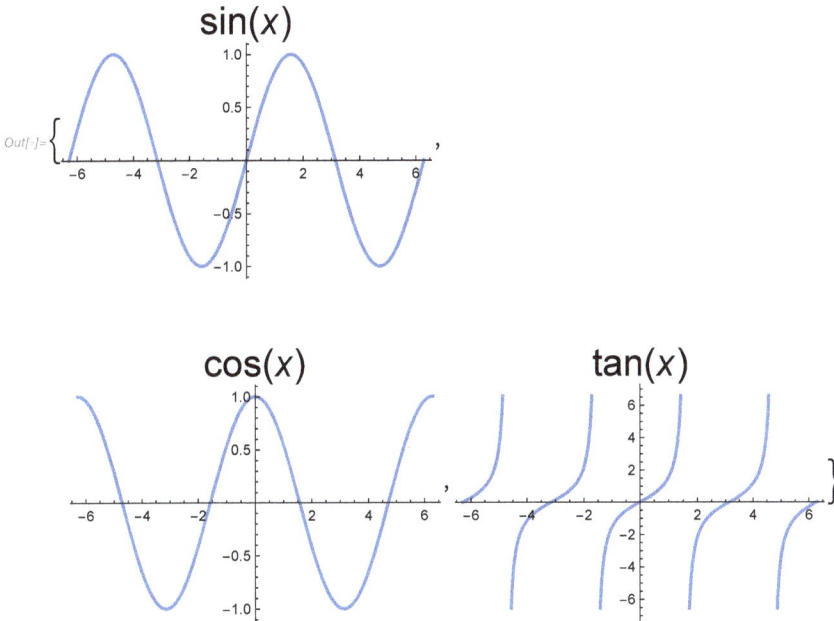

Notice that each trigonometric function is very repetitive; they take on the same value every $2\,\pi$ interval.

When a function f has the property $f[x + t] = f[x]$ for all x, it is **periodic with period t**.

Sine and cosine have period $2\,\pi$.

Tangent has period π.

Reciprocal Trigonometric

The reciprocals of sine, cosine and tangent are **cosecant**, **secant** and **cotangent**, respectively. In other words, Cosecant = $\frac{1}{\text{Sine}}$, Secant = $\frac{1}{\text{Cosine}}$, and Cotangent = $\frac{1}{\text{Tangent}}$.

Secant, cosecant and cotangent are also trigonometric functions. Here are their graphs and reciprocals in the range $[-2\pi, 2\pi]$:

In[ ]:= `Table[Plot[{i, 1/i}, {x, -2 π, 2 π}, PlotLabel → Style[i, Large]], {i, {Sec[x], Csc[x], Cot[x]}}]`

Out[ ]:=

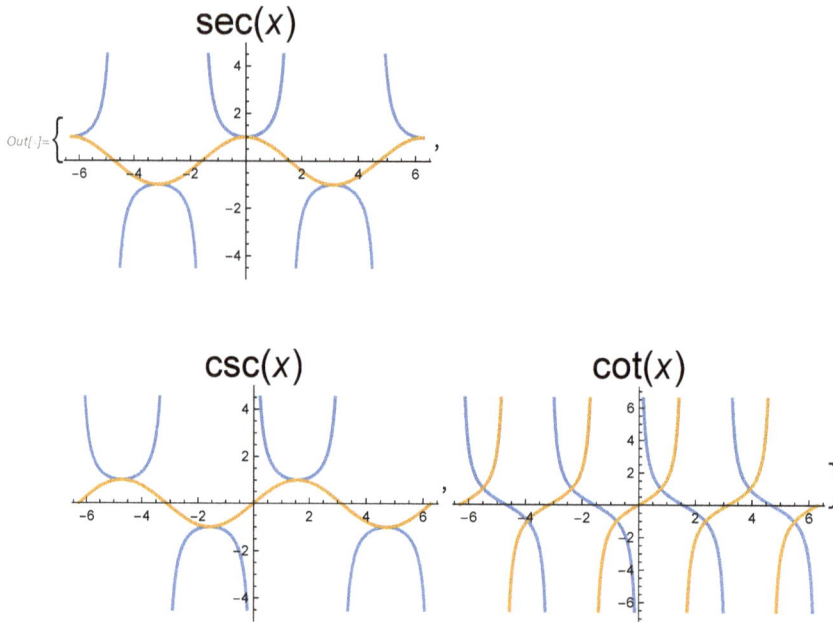

Notice that secant, cosecant and cotangent (in red) have multiple places where they are undefined. These places correspond to places where their reciprocals cosine, sine and tangent (in blue) are 0.

Secant and cosecant have period 2π.

Cotangent has period π.

Exponential

Whereas power functions have the form x^a, exponential functions have the form a^x. When $a > 1$, there is an **exponential growth function**.

Here is a plot of the exponential growth function $f[x] = 2^x$:

In[ ]:= **Plot[2^x, {x, −1, 1}]**

Out[ ]=

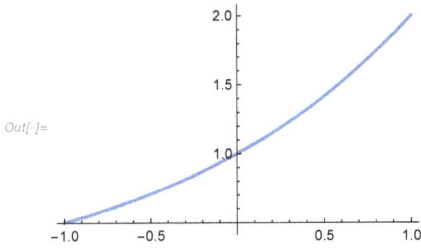

When $0 < a < 1$, there is an **exponential decay function**.

Here is a plot of the exponential decay function $g[x] = 0.5^x$:

In[ ]:= **Plot[0.5^x, {x, −1, 1}]**

Out[ ]=

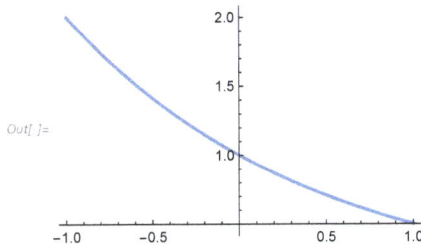

All exponential functions of the form a^x pass the y axis at the point (0, 1), because $a^0 = 1$ for any number a.

Logarithmic

Logarithmic functions have the form $f[x] = \log_a x$ and can be defined as the inverses to the exponential functions. So $a^{f[x]} = x$. Here a is known as the base.

It is assumed here that whenever the base is left out in the expression $\mathsf{Log}_a x$, the base is the natural number $e \sim 2.718$.

The Wolfram Language represents logarithmic functions with the **Log** function.

Here is the plot of the logarithmic function with base e (and a general sketch for base greater than 1):

In[·]:= **Plot[Log[x], {x, −1, E}]**

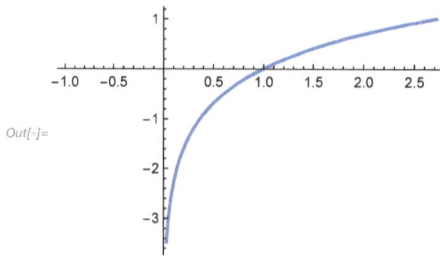

Out[·]=

Notice that the domain is all positive real numbers (i.e. $(0, \infty)$). Log[e] = 1 since $e^1 = e$.

Here is the plot of the logarithmic function with base 1 / 4 (and a general sketch for base between 0 and 1):

In[·]:= **Plot[Log[1 / 4, x], {x, 0, 4}]**

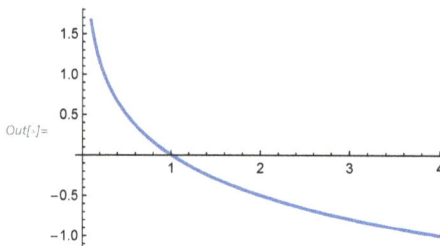

Out[·]=

Summary

Functions have many different forms and can have a variety of graphs.

Linear functions have graphs that look like lines.

Polynomial functions include linear functions and have all real numbers as their domain.

Power function have their inputs taken to a constant power.

Rational functions include polynomial functions and are defined wherever their denominators are not 0.

Algebraic functions include rational functions and are created by adding, subtracting, multiplying, dividing and taking roots.

Trigonometric functions have their basis in geometry and are periodic.

Exponential functions and their inverses, logarithmic functions, have positive base a and can either grow or decay.

The next lesson will cover limits, the starting point of any beginning course into calculus.

Exercises

Exercise 1—Properties of a Line

A line goes through the points $(4, 1)$ and $(-3, 7)$. Find its slope and y-intercept, and plot the function in the range $[-5, 5]$.

Solution

Since there is a line, its function should be linear and have the form $f[x] = m x + b$, where m is the slope and b is the y-intercept.

The slope can be calculated by dividing the change in y by the change in x for the two points:

In[ ]:= **m = (7 − 1)/(−3 − 4)**

Out[ ]= $-\dfrac{6}{7}$

Now plug one of the points into the function to solve for b:

In[ ]:= **Solve[1 == m * 4 + b, b]**

Out[ ]= $\left\{\left\{b \to \dfrac{31}{7}\right\}\right\}$

Since the function has equation $f[x] = \dfrac{-6}{7} x + \dfrac{31}{7}$, here is the plot of the function with the two points:

In[ ]:= **f[x_] := −6 x / 7 + 31 / 7**

The plot in the range $[-5, 5]$ is:

In[ ]:= **Plot[f[x], {x, −5, 5}]**

Out[ ]=

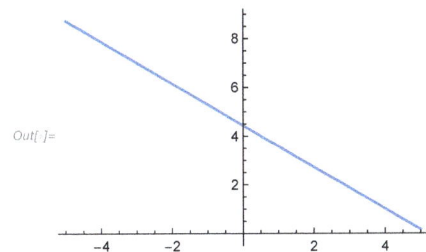

Exercise 2—Polynomials

A polynomial has degree 6 and has coefficients $a_6 = 2$, $a_5 = 6$, $a_4 = 0$, $a_3 = 1$, $a_2 = -7$, $a_1 = 3$ and $a_0 = 14$. Plot the polynomial on a graph. What is its value when you plug in the number 45?

Solution

Based on the given information, the polynomial has the form $2\,x^6 + 6\,x^5 + x^3 - 7\,x^2 + 3\,x + 14$.

Make it a function:

In[·]:= **f[x_] := 2 x^6 + 6 x^5 + x^3 − 7 x^2 + 3 x + 14**

Plot the polynomial:

In[·]:= **Plot[f[x], {x, −4, 3}]**

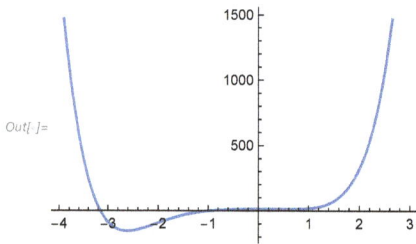

Evaluate the function at 45:

In[·]:= **f[45]**

Out[·]= 17 714 777 099

The function (and polynomial) has value 17,714,777,099 when $x = 45$.

Exercise 3—Rational

Plot the following rational function to find its domain:

In[·]:= **f[x_] := (x^3 + 6 x − 7) / (20 x^4 + 4 x^3 − 359 x^2 − 130 x + 888)**

Solution

Here is the function's plot in the range $[-5, 5]$:

In[ ]:= **Plot[f[x], {x, −5, 5}]**

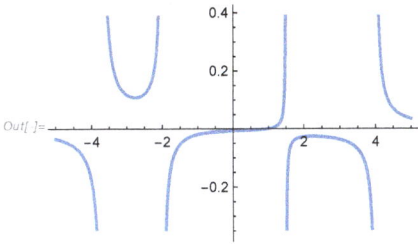

Out[ ]=

To find the function's domain, set its denominator to 0 and solve for x:

In[ ]:= **Solve[20 x^4 + 4 x^3 − 359 x^2 − 130 x + 888 == 0, x]**

Out[ ]= $\left\{\left\{x \to -\frac{37}{10}\right\}, \{x \to -2\}, \left\{x \to \frac{3}{2}\right\}, \{x \to 4\}\right\}$

The function has domain $(-\infty, -3.7) \cup (-3.7, -2) \cup (-2, 1.5) \cup (1.5, 4) \cup (4, \infty)$.

You could have also calculated it with FunctionDomain:

In[ ]:= **FunctionDomain[f[x], x]**

Out[ ]= $x < -\frac{37}{10} \,||\, -\frac{37}{10} < x < -2 \,||\, -2 < x < -\frac{3}{2} \,||\, \frac{3}{2} < x < 4 \,||\, x > 4$

Exercise 4—Planet Aurora

The fictional planet Aurora takes 501 days to revolve around its star. The number of hours of daylight can be modeled by the following function:

In[ ]:= **daylight[*t*_] := 15 + 4.7 Cos[2 π / 501 (*t* − 240)]**

where t is the number of days after its new year. Plot the function and find how long a day lasts on the 397th day of the Auroran year.

Solution

The daylight function is plotted with Plot:

In[·]:= **Plot[daylight[t], {t, −501, 501}]**

Out[·]=

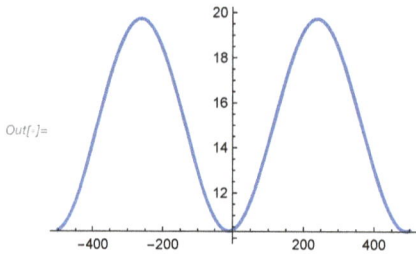

Plug 397 into the function:

In[·]:= **daylight[397]**

Out[·]= 13.1776

On the 397th day of the year, the Auroran day lasts about 13.18 hours (or 13 hours, 10 minutes and about 36 seconds).

Exercise 5—Bacteria Colony

A bacteria colony grows according to the following exponential growth model, where t is in hours:

In[·]:= **bacterialgrowth[t_] := 1.97 ^ t**

If the colony gets too big (say, reaches an amount M), antibiotics are used to cause the colony to die according to the following exponential decay model, where t is in hours:

In[·]:= **bacterialdecay[m_, t_] := m * 0.67 ^ t**

If the bacteria colony is allowed to grow for 30 hours, and is then hit by antibiotics, how long will it take the colony to contain only 1000 members?

Solution

First, evaluate the size of the colony at 30 hours:

In[·]:= **bacterialgrowth[30]**

Out[·]= 6.82318×10^8

Then, plug this number into the exponential decay function and solve for t when the function equals 1000:

```
In[·]:= Solve[1000 == bacterialdecay[bacterialgrowth[30], t], t] // Quiet
```

```
Out[·]= {{t → 33.5431}}
```

It will take about 33.542 hours (or 33 hours, 32 minutes and about 36 seconds) to reduce the bacterial population back to 1000.

Here are the two graphs together:

```
In[·]:= Plot[{Piecewise[{{bacterialgrowth[t], 0 ≤ t ≤ 30}}, Null],
        Piecewise[{{bacterialdecay[bacterialgrowth[30], t − 30], 30 ≤ t ≤ 54}}, Null]},
      {t, 0, 54}, PlotRange → All, PlotLegends → {"Growth", "Decay"}]
```

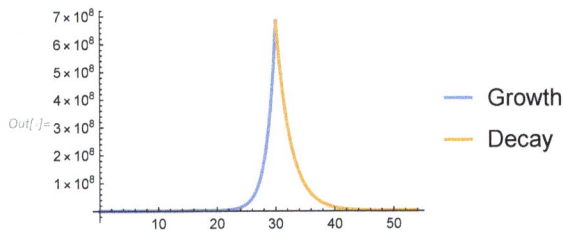

4 | The Limit of a Function

Overview

The functions covered so far compute desired quantities, given some input.

Here is the plot of the function $f[x] = x$:

In[ ]:= **Plot[x, {x, −1, 1}]**

Out[ ]=

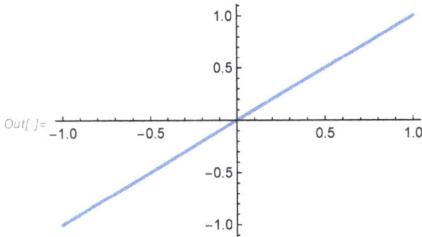

However, sometimes the functions are not defined at certain points, even though it looks like they are from their graphs.

Here is the graph of the function $g[x] = x^2 / x$:

In[ ]:= **Plot[x^2/x, {x, −1, 1}]**

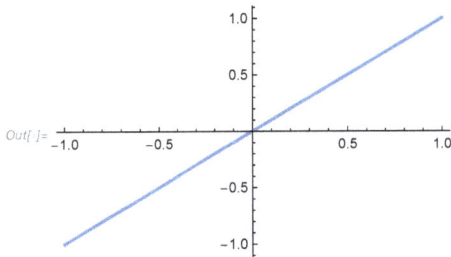

Out[ ]=

It looks just like the previous graph, but it is not defined at 0:

In[ ]:= **g[x_] := x^2/x**

In[ ]:= **g[0] // Quiet**

Out[ ]= Indeterminate

Even though it is not defined there, you might still like to know the behavior of the function near the point. This lesson will explore this notion mathematically with the concept known as a **limit**.

Notion of a Limit

Limits arise naturally in many areas of applied and theoretical mathematics. To find a limit, you see what the function approaches as x approaches a certain value.

For example, consider the following function:

In[·]:= **f[x_] := x^2 − x + 2**

The following tables give values of f for values of x that get closer and closer to 1 (but not including 1!):

$$\left\{ \begin{array}{|c|c|} \hline x & x^2 - x + 2 \\ \hline 0 & 2. \\ 0.5 & 1.75 \\ 0.9 & 1.91 \\ 0.95 & 1.9525 \\ 0.99 & 1.9901 \\ 0.995 & 1.99503 \\ 0.999 & 1.999 \\ \hline \end{array} , \begin{array}{|c|c|} \hline x & x^2 - x + 2 \\ \hline 2 & 4. \\ 1.5 & 2.75 \\ 1.1 & 2.11 \\ 1.05 & 2.0525 \\ 1.01 & 2.0101 \\ 1.005 & 2.00503 \\ 1.001 & 2.001 \\ \hline \end{array} \right\}$$

It appears the function values get closer and closer to 2 as the values get closer and closer to 1 from either side.

Here is a graph that does the same thing.

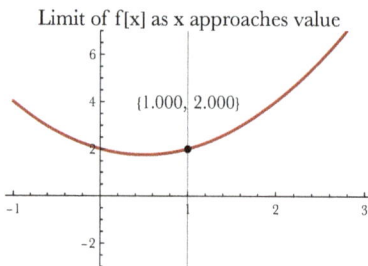

Limit of f[x] as x approaches value

{1.000, 2.000}

Trigonometric Rational Function

Compute the following limit as x approaches 0:

In[·]:= **f[x_] := Sin[x] / x**

Make a table of values for the function:

x	$\frac{\sin(x)}{x}$	x	$\frac{\sin(x)}{x}$
−1	0.841471	1	0.841471
−0.5	0.958851	0.5	0.958851
−0.1	0.998334	0.1	0.998334
−0.05	0.999583	0.05	0.999583
−0.01	0.999983	0.01	0.999983
−0.005	0.999996	0.005	0.999996
−0.001	1.	0.001	1.

Here is a plot of the function:

In[ ]:= **Plot[f[x], {x, −1, 1}]**

Out[ ]=

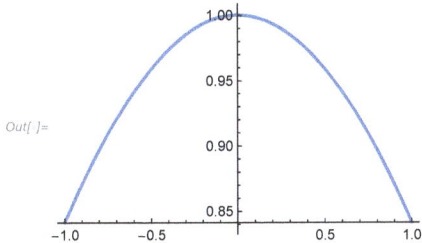

It is clear that the function approaches 1 as x approaches 0.

With Limit, you can calculate the limit of any function at any point. The preceding limit is:

In[ ]:= **Limit[f[x], x → 0]**

Out[ ]= **1**

Rational Function with a Removable Discontinuity

Compute the following limit as x approaches − 1:

In[ ]:= **f[x_] := (x + 1)/(x^2 − 1)**

Begin by making a table of values for the function:

x	$\frac{x+1}{x^2-1}$	x	$\frac{x+1}{x^2-1}$
−2	−0.333333	0	−1.
−1.5	−0.4	−0.5	−0.666667
−1.1	−0.47619	−0.9	−0.526316
−1.05	−0.487805	−0.95	−0.512821
−1.01	−0.497512	−0.99	−0.502513
−1.005	−0.498753	−0.995	−0.501253
−1.001	−0.49975	−0.999	−0.50025

Here is a plot of the function:

In[·]:= **Plot[f[x], {x, −2, 0}, GridLines → {{−1}, {−1/2}}]**

Out[·]=

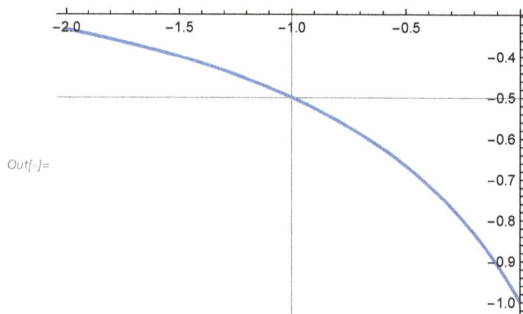

Grid lines are added with GridLines to help see the function value. It is clear that the function approaches −0.5 as x approaches −1.

You can confirm this answer using Limit:

In[·]:= **Limit[f[x], x → −1]**

Out[·]= $-\dfrac{1}{2}$

Piecewise Function

Compute the following limit as x approaches −1:

In[·]:= **g[x_] := Piecewise[{{−0.75, x == −1}}, (x + 1)/(x^2 − 1)]**

Begin by making a table of values for the function:

x	$\dfrac{x+1}{x^2-1}$	x	$\dfrac{x+1}{x^2-1}$
−2	−0.333333	0	−1.
−1.5	−0.4	−0.5	−0.666667
−1.1	−0.47619	−0.9	−0.526316
−1.05	−0.487805	−0.95	−0.512821
−1.01	−0.497512	−0.99	−0.502513
−1.005	−0.498753	−0.995	−0.501253
−1.001	−0.49975	−0.999	−0.50025

Here is a plot of the function:

In[ ]:= **Plot[g[x], {x, −2, 0}, GridLines → {{−1}, {−1/2}},**
  Epilog → {PointSize[Large], Point[{−1, g[−1]}]}]

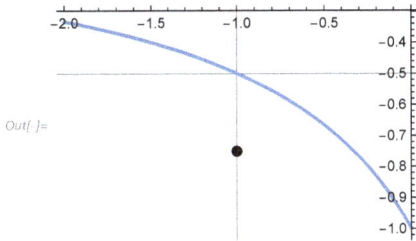

Out[ ]=

The function still approaches − 0.5 as *x* approaches − 1.

You can confirm this answer using Limit:

In[ ]:= **Limit[g[x], x → 0]**

Out[ ]= **−1**

Algebraic Function with a Removable Discontinuity

Compute the following limit as *x* approaches 0:

In[ ]:= **f[x_] := (Sqrt[x^2 + 4] − 2)/x^2**

Begin by making a table of values for the function:

x	$\frac{\sqrt{x^2+4}-2}{x^2}$	x	$\frac{\sqrt{x^2+4}-2}{x^2}$
−2	0.207107	0	Indeterminate
−1.5	0.222222	−0.5	0.246211
−1.1	0.233506	−0.9	0.238483
−1.05	0.234804	−0.95	0.237295
−1.01	0.235818	−0.99	0.236316
−1.005	0.235943	−0.995	0.236192
−1.001	0.236043	−0.999	0.236093

Here is a plot of the function:

In[·]:= **Plot[f[x], {x, −1, 1}]**

Out[·]=

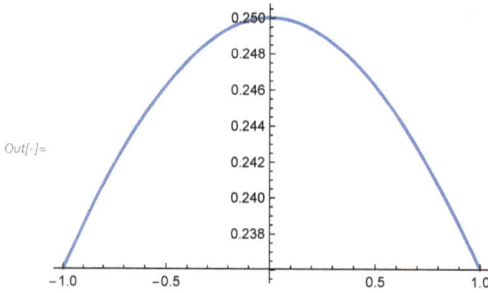

It is clear that the function approaches 0.25 as x approaches 0.

You can confirm this answer using Limit:

In[·]:= **Limit[f[x], x → 0]**

Out[·]= $\dfrac{1}{4}$

Trigonometric Function

Compute the following limit as x approaches 0:

In[·]:= **f[x_] := Cos[π / x]**

Begin by making a table of values for the function:

x	$\cos\left(\frac{\pi}{x}\right)$	x	$\cos\left(\frac{\pi}{x}\right)$
−1	−1.	1	−1.
−0.5	1.	0.5	1.
−0.1	1.	0.1	1.
−0.05	1.	0.05	1.
−0.01	1.	0.01	1.
−0.005	1.	0.005	1.
−0.001	1.	0.001	1.

It appears the function approaches 1 as x approaches 0, but it is much more complicated than it appears.

Here is an interactive example to illustrate.

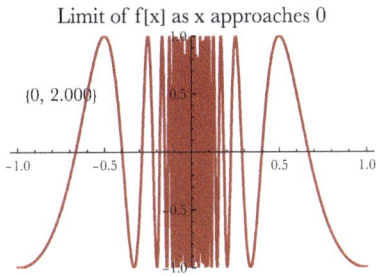

Limit of f[x] as x approaches 0

The function does not appear to approach any particular value as x approaches 0, as shown by Limit:

In[ ]:= **Limit[f[x], x → 0]**

Out[ ]= Indeterminate

Sum of Trigonometric and Polynomial Functions

Compute the following limit as x approaches 0:

In[ ]:= **f[x_] := 4 x^5 + 3 Cos[100 x]/200**

Begin by making a table of values for the function:

x	$4x^5 + \frac{3}{200}\cos(100x)$	x	$4x^5 + \frac{3}{200}\cos(100x)$
−1	−3.98707	1	4.01293
−0.5	−0.110526	0.5	0.139474
−0.1	−0.0126261	0.1	−0.0125461
−0.05	0.00425368	0.05	0.00425618
−0.01	0.00810453	0.01	0.00810453
−0.005	0.0131637	0.005	0.0131637
−0.001	0.0149251	0.001	0.0149251

It appears the function approaches $3/200$ (0.015) as x approaches 0.

Here is a plot of the function:

In[ ]:= **Plot[f[x], {x, –0.1, 0.1}, PlotRange → All]**

Out[ ]=

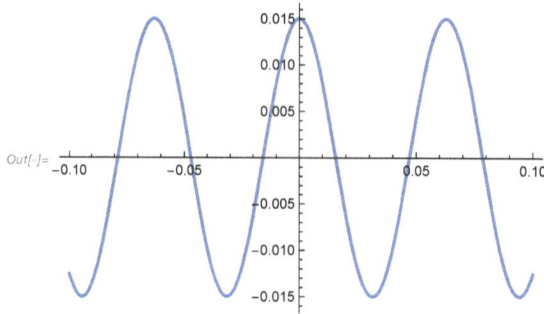

You can confirm this answer using Limit:

In[ ]:= **Limit[f[x], x → 0]**

Out[ ]= $\dfrac{3}{200}$

One-Sided Limits

The **left-hand limit** is the value a function approaches as it goes from the left.

The **right-hand limit** is the value a function approaches as it goes from the right.

The limit for a function **exists** if the left-hand limit and right-hand limit are **equal**.

Compute the following limit as x approaches 0:

In[ ]:= **f[x_] := Piecewise[{{–1, x < 0}}, 1]**

Given the nature of f, a table is not necessary. Here is a plot of f instead:

In[ ]:= **Plot[f[x], {x, –1, 1}]**

Out[ ]=

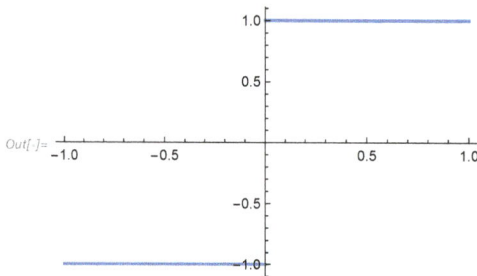

It is clear that left-hand limit at 0 is – 1, and the right-hand limit at 0 is 1.

Confirm this answer using Limit and the option Direction. "FromBelow" means a left-hand limit. "FromAbove" means a right-hand limit:

In[ ]:= **{Limit[f[x], x → 0, Direction → "FromBelow"],**
       Limit[f[x], x → 0, Direction → "FromAbove"]}

Out[ ]= **{−1, 1}**

The limit therefore does not exist:

In[ ]:= **Limit[f[x], x → 0]**

Out[ ]= Indeterminate

General Piecewise Function

Compute the limit of the following function as x approaches 1, 4 and 6:

In[ ]:= **f[x_] := Piecewise[{{4, x == 1}, {3 Cos[x − 1], x < 1}, {3 Sqrt[x], x > 1 && x < 6}}, 3 x − 13]**

Begin by plotting the function:

In[ ]:= **Plot[f[x], {x, 0, 7}, GridLines → {{1, 4, 6}, {3, 6}},**
       Epilog → {PointSize[Large], Point[{1, f[1]}]}]

Out[ ]=

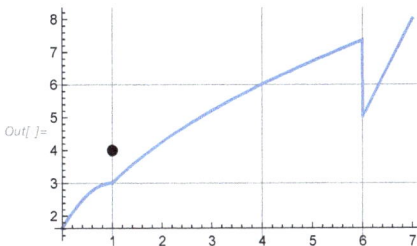

It is clear that the function approaches 3 as x approaches 1, approaches 6 for $x = 4$, and does not exist for $x = 6$.

Confirm this answer using Limit:

In[ ]:= **Limit[f[x], x → #] & /@ {1, 4, 6}**

Out[ ]= {3, 6, Indeterminate}

You can use the symbols #, & and /@ to find the limit for each of the values in the list.

Again, the limit does not exist at 6, since the left-hand and right-hand limits at 6 are different:

In[ ]:= **{Limit[f[x], x → 6, Direction → "FromBelow"], Limit[f[x], x → 6, Direction → "FromAbove"]}**

Out[ ]= $\left\{3 \sqrt{6}, 5\right\}$

Table of Values

Compute the limit for the following function as x approaches -1, given the values $x = -2, -1.5, -1.1, -1.01, -1.001, -0.999, -0.99, -0.95, -0.9, -0.5$ and 0.

In[·]:= **f[x_] := (x^2 + 2 x)/(x^2 - 4 x - 5)**

Make a chart for the function using Grid, Join and Table:

x	$\frac{2x+x^2}{-5-4x+x^2}$
-2	0
-1.5	-0.230769
-1.1	-1.62295
-1.01	-16.6373
-1.001	-166.639
-0.999	166.694
-0.99	16.6928
-0.95	3.35294
-0.9	1.67797
-0.5	0.272727
0	0

It appears the function approaches ∞ from the left and $-\infty$ from the right as x approaches -1. So the limit should not exist.

Confirm with Limit:

In[·]:= **{Limit[f[x], x → −1, Direction → "FromBelow"],**
        Limit[f[x], x → −1, Direction → "FromAbove"]}

Out[·]= **{−∞, ∞}**

In[·]:= **Limit[f[x], x → −1]**

Out[·]= **Indeterminate**

Vertical Asymptote 1

Compute the following limit as x approaches 1:

In[·]:= **f[x_] := 1/(x − 1)^2**

Begin by making a table of values for the function:

x	$\frac{1}{(x-1)^2}$		x	$\frac{1}{(x-1)^2}$
0	1.		2	1.
0.5	4.		1.5	4.
0.9	100.		1.1	100.
0.95	400.		1.05	400.
0.99	10 000.		1.01	10 000.
0.995	40 000.		1.005	40 000.
0.999	$1. \times 10^6$		1.001	$1. \times 10^6$

Here is a plot of the function:

In[ ]:= **Plot[f[x], {x, 0, 2}]**

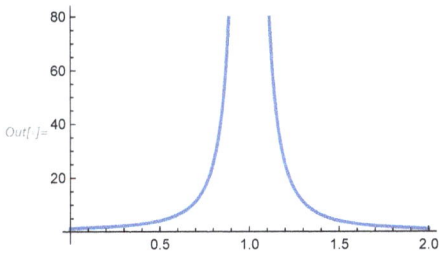

Out[ ]=

It is clear that the function approaches ∞ from both sides as x approaches 1.

Confirm with Limit:

In[ ]:= **Limit[f[x], x → 1]**

Out[ ]= ∞

Vertical Asymptote 2

Compute the following limit as x approaches 0:

In[ ]:= **f[x_] := 1 / x**

Begin by making a table of values for the function:

x	$\frac{1}{x}$		x	$\frac{1}{x}$
−1	−1.		1	1.
−0.5	−2.		0.5	2.
−0.1	−10.		0.1	10.
−0.05	−20.		0.05	20.
−0.01	−100.		0.01	100.
−0.005	−200.		0.005	200.
−0.001	−1000.		0.001	1000.

Add a plot to make it clearer:

In[·]:= **Plot[f[x], {x, −1, 1}]**

Out[·]=

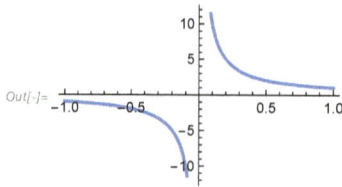

It is clear that the function approaches −∞ from the left and ∞ from the right as x approaches 0; therefore the limit does not exist.

Confirm with Limit:

In[·]:= **{Limit[f[x], x → 0, Direction → "FromBelow"], Limit[f[x], x → 0, Direction → "FromAbove"]}**

Out[·]= **{−∞, ∞}**

In[·]:= **Limit[f[x], x → 0]**

Out[·]= **Indeterminate**

Finding the Asymptotes of a Function

Find the asymptotes of the following function in the range -2π to 2π:

In[·]:= **f[x_] := Cot[x]**

Begin by plotting the function:

In[·]:= **Plot[f[x], {x, −2 π, 2 π}, Frame → True,**
    FrameTicks → {{Automatic, None}, {{−2 π, −π, 0, π, 2 π}, None}}]

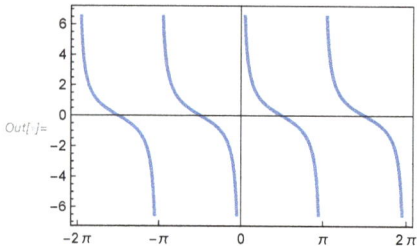

Out[·]=

The function approaches −∞ from the left and ∞ from the right as x approaches integer multiples of π:

In[·]:= **list = Table[n π, {n, −2, 2}]**

Out[·]= **{−2 π, −π, 0, π, 2 π}**

Limit shows the left- and right-hand limits are not the same at 0:

$_{In[\cdot]:=}$ **{Limit[f[x], x → 0, Direction → "FromBelow"], Limit[f[x], x → 0, Direction → "FromAbove"]}**

$_{Out[\cdot]:=}${$-\infty$, ∞}

For all integer multiples of π, the limit does not exist:

$_{In[\cdot]:=}$ **Limit[f[x], x → list]**

$_{Out[\cdot]:=}${Indeterminate, Indeterminate, Indeterminate, Indeterminate, Indeterminate}

Difference Quotient and Table of Values

Compute the limit of the following function as h approaches 0:

$_{In[\cdot]:=}$ **f[h_] := ((x + h)^2 − x^2)/h**

Graphing this will get you nowhere, so make a table of values for h on the left and right of 0:

-1	$-1 + 2x$
-0.5	$-0.5 + 2.x$
-0.1	$-0.1 + 2.x$
-0.01	$-0.01 + 2.x$
-0.001	$-0.001 + 2.x$
0.001	$0.001 + 2.x$
0.01	$0.01 + 2.x$
0.05	$0.05 + 2.x$
0.1	$0.1 + 2.x$
0.5	$0.5 + 2.x$
1	$1 + 2x$

It is clear that the function approaches $2x$ as h approaches 0.

Confirm with Limit:

$_{In[\cdot]:=}$ **Limit[f[h], h → 0]**

$_{Out[\cdot]:=}$ $2x$

Summary

Limits give the function value a function approaches as its input approaches some value.

The function does not need to be defined at the value to have a limit there.

As long as the right-hand limit and left-hand limit are equal at a point, the limit exists at the point.

Tables are useful for finding limits.

The next lesson will cover the limit laws, which give ways to calculate limits without using tables.

Exercises

Exercise 1—Rational Function

Compute the limit of the following function as x approaches 1:

In[ ]:= **f[x_] := (x^5 − 1)/(x^10 − 1)**

Solution

Begin by making a table of values for the function:

x	$\frac{x^5-1}{x^{10}-1}$	x	$\frac{x^5-1}{x^{10}-1}$
0	1.	2	0.030303
0.5	0.969697	1.5	0.116364
0.9	0.628737	1.1	0.383067
0.95	0.563767	1.05	0.439313
0.99	0.51256	1.01	0.487565
0.995	0.506265	1.005	0.493766
0.999	0.501251	1.001	0.498751

Here is a plot of the function:

In[ ]:= **Plot[f[x], {x, 0, 2}, GridLines → {{1}, None}]**

Out[ ]=

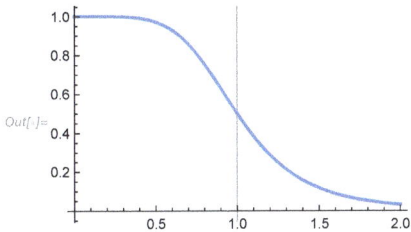

It is clear that the function approaches 0.5 as x approaches 1.

Confirm with Limit:

In[ ]:= **Limit[f[x], x → 1]**

Out[ ]= $\dfrac{1}{2}$

Exercise 2—Trig Function Given a Table of Values

Compute the limit for the following function as x approaches 0, given the values $x = -1, -0.5, -0.1, -0.01, -0.001, 0.001, 0.01, 0.05, 0.1, 0.5$ and 1.

In[ ]:= **f[x_] := Sin[x]/(x − Tan[x])**

Solution

Make a chart for the function:

−1.	−1.50961
−0.5	−10.3542
−0.1	−298.302
−0.01	−29 998.3
−0.001	$-3. \times 10^6$
0.001	$-3. \times 10^6$
0.01	−29 998.3
0.05	−1198.3
0.1	−298.302
0.5	−10.3542
1.	−1.50961

It appears the function approaches $-\infty$ from the left and right as x approaches 0. So the limit should be $-\infty$.

Confirm with Limit:

In[ ]:= **Limit[f[x], x → 0]**

Out[ ]= $-\infty$

Exercise 3—Difference Quotient

Compute the limit of the following function as h approaches 0:

In[ ]:= **f[h_] := ((2 + h)^5 − 32)/h**

Solution

Begin by making a table of values for the function:

x	$\frac{(x+2)^5-32}{x}$	x	$\frac{(x+2)^5-32}{x}$
−1	31.	1	211.
−0.5	48.8125	0.5	131.313
−0.1	72.3901	0.1	88.4101
−0.05	76.0988	0.05	84.1013
−0.01	79.204	0.01	80.804
−0.005	79.601	0.005	80.401
−0.001	79.92	0.001	80.08

Here is a plot of the function:

In[ ]:= **Plot[f[h], {h, −1, 1}]**

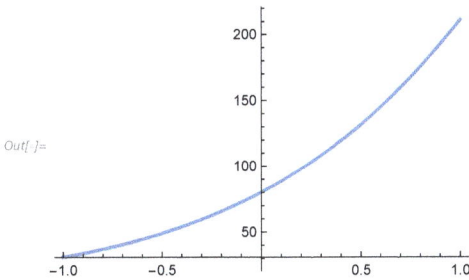

Out[ ]=

It looks like the function approaches 80 as h approaches 0.

Confirm with Limit:

In[ ]:= **Limit[f[h], h → 0]**

Out[ ]= 80

Exercise 4—Where Does the Limit Not Exist?

Find the places where the limit does not exist for the following function:

In[ ]:= **f[x_] := Piecewise[{{1 + x, x < −1}, {2 x^2, −1 ≤ x ≤ 1}}, 3 − x]**

Solution

Plot the function:

In[∘]:= **Plot[f[x], {x, −2, 4}]**

Out[∘]=

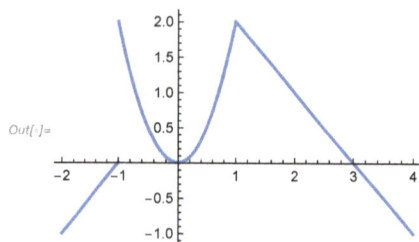

It appears the limit does not exist at − 1.

Confirm with Limit:

In[∘]:= **{Limit[f[x], x → −1, Direction → "FromBelow"],**
       Limit[f[x], x → −1, Direction → "FromAbove"]}

Out[∘]= **{0, 2}**

At $x = 1$, the limit also exists even though it is not smooth there:

In[∘]:= **{Limit[f[x], x → 1, Direction → "FromBelow"], Limit[f[x], x → 1, Direction → "FromAbove"]}**

Out[∘]= **{2, 2}**

Exercise 5—An Application from Special Relativity

The theory of special relativity was developed by Albert Einstein.

In special relativity, the mass of a particle with rest mass m_0 and velocity v is given by the following function:

In[∘]:= **mass[v_] := m0 / Sqrt[1 − v^2 / c^2]**
     c = 299 792 458;
     m0 = 1;

Here c is the speed of light in a vacuum. Let the rest mass be 1.

What is the limit of the mass as v approaches c?

Solution

See what happens as v approaches c from the left:

2.99792×10^8	$12\,243.2$
2.99792×10^8	$17\,314.5$
2.99792×10^8	$38\,716.4$
2.99792×10^8	$122\,432.$
2.99792×10^8	$387\,169.$

Because of the size of c, there is some error, but overall it appears to approach ∞ from the left.

Confirm with Limit:

```
In[ ]:= Limit[mass[v], v → c, Direction → "FromBelow", Assumptions → m0 > 0 && c > 0]
```

```
Out[ ]= ∞
```

5 | The Laws of Limits

Overview

Finding limits with a table can be tedious work. With the **limit laws**, you can do this much more easily.

This lesson will go over the limit laws and use the example functions f and g to illustrate them:

In[ ]:= **f[x_] := 2 x + 2**
g[x_] := 3 Cos[x]/2

Here are their plots:

In[ ]:= **Plot[{f[x], g[x]}, {x, −1, 1}, PlotLegends → "Expressions"]**

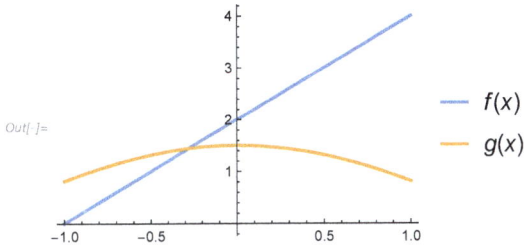

Out[ ]=

Their limits at 0 are 2 and $\frac{3}{2}$, respectively:

In[ ]:= **Limit[f[x], x → 0]**

Out[ ]= **2**

In[ ]:= **Limit[g[x], x → 0]**

Out[ ]= $\frac{3}{2}$

Sum Law

The limit of a sum is the sum of the limits. So the limit for $f[x] + g[x]$ as x approaches 0 is $2 + 3/2 = 7/2$.

Limit agrees:

In[ ]:= **Limit[f[x] + g[x], x → 0]**

Out[ ]= $\frac{7}{2}$

Here is a plot of all three functions:

In[·]:= **Plot[{f[x], g[x], f[x] + g[x]}, {x, −1, 1}, PlotLegends → "Expressions"]**

Out[·]=

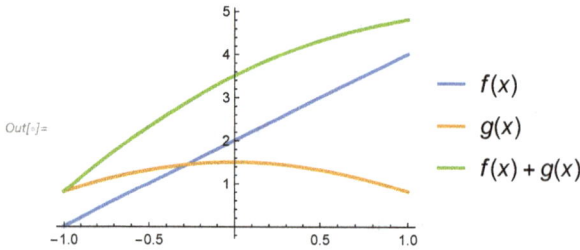

— $f(x)$
— $g(x)$
— $f(x) + g(x)$

From the sum law follow the difference law and the scalar multiplication law.

Difference law: The limit of a difference is the difference of the limits.

Scalar multiplication law: The limit of a constant times a function is the constant times the limit of the function.

The limit for $f[x] - g[x]$ as x approaches 0 is $2 - 3/2 = 1/2$, and the limit for $3 f[x]$ as x approaches 0 is $3 * 2 = 6$:

In[·]:= **{Limit[f[x] − g[x], x → 0], Limit[3 f[x], x → 0]}**

Out[·]= $\left\{\dfrac{1}{2}, 6\right\}$

Product Law

The limit of a product is the product of the limits. So the limit for $f[x] * g[x]$ as x approaches 0 is $2 * 3/2 = 3$.

Limit agrees:

In[·]:= **Limit[f[x] ∗ g[x], x → 0]**

Out[·]= 3

Here is a plot of all three functions:

In[·]:= **Plot[{f[x], g[x], f[x] ∗ g[x]}, {x, −1, 1}, PlotLegends → "Expressions"]**

Out[·]=

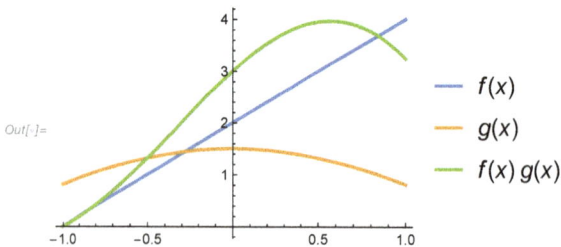

— $f(x)$
— $g(x)$
— $f(x) \, g(x)$

The product law naturally extends to powers of a function.

Power law: The limit of a power of a function is the power of the limit of the function.

The limit of $f[x]^3$ as x approaches 0 is $2^3 = 8$:

In[ ]:= **Limit[f[x]^3, x → 0]**

Out[ ]= 8

Quotient Law

The limit of a quotient is the quotient of the limits, so long as the limit of the denominator is not 0. So the limit for $\frac{f[x]}{g[x]}$ as x approaches 0 is $2/(3/2) = 4/3$.

Limit agrees:

In[ ]:= **Limit[f[x]/g[x], x → 0]**

Out[ ]= $\dfrac{4}{3}$

Here is a plot of all three functions:

In[ ]:= **Plot[{f[x], g[x], f[x]/g[x]}, {x, −2, 2}, PlotLegends → "Expressions"]**

Out[ ]=

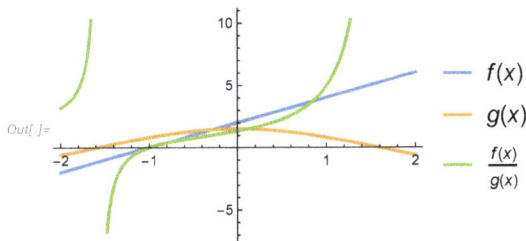

Note that $\dfrac{f}{g}$ cannot have its limit calculated at $\pi/2$ with the quotient law because g is 0 there:

In[ ]:= **g[π/2]**

Out[ ]= 0

In[ ]:= **Limit[f[x]/g[x], x → π/2]**

Out[ ]= Indeterminate

Limit of a Polynomial

The limit of a polynomial can be calculated if a few things are accepted.

First, accept that the limit of a constant function $f[x] = c$ as x approaches **any** value is the constant c:

In[ ]:= **Limit[c, x → a] == c**

Out[ ]= **True**

Also accept that the limit of the function $g[x] = x$ at any value a is just a:

In[ ]:= **Limit[x, x → a] == a**

Out[ ]= **True**

Since polynomials can be constructed by taking g and multiplying it by itself as well as adding and multiplying various constant functions f, the limit of any polynomial at any point a is calculated by the sum law and product law to just be the value of the polynomial at a.

The limit of the polynomial $f[x] = 2x^2 - 4x + 3$ as x approaches 4 is just $2(4)^2 - 4(4) + 3 = 19$.

Limit agrees:

In[ ]:= **Limit[2 x^2 − 4 x + 3, x → 4]**

Out[ ]= **19**

Limit of a Rational Function

Compute the limit for the following function as x approaches -2:

In[ ]:= **f[x_] := (x^3 − x^2 + 2)/(5 x − 3)**

A rational function is a ratio of two polynomial functions. Since you have the quotient law and know how to compute limits for polynomials, this is rather easy.

Its constituent polynomials are $g[x] = x^3 - x^2 + 2$ and $h[x] = 5x - 3$:

In[ ]:= **g[x_] := x^3 − x^2 + 2**

In[ ]:= **h[x_] := −3 + 5 x**

In[ ]:= **f[x] == g[x]/h[x]**

Out[ ]= **True**

Limit shows that you get the same limit if you divide the limits of the polynomials:

In[ ]:= **{Limit[g[x]/h[x], x → −2], Limit[g[x]/h[x], x → −2] == f[−2]}**

Out[ ]:= $\left\{\dfrac{10}{13}, \text{True}\right\}$

Here are their plots; they overlap because they are the same:

In[ ]:= **Plot[{f[x], g[x]/h[x]}, {x, −3, −1}, PlotLegends → "Expressions"]**

Out[ ]=

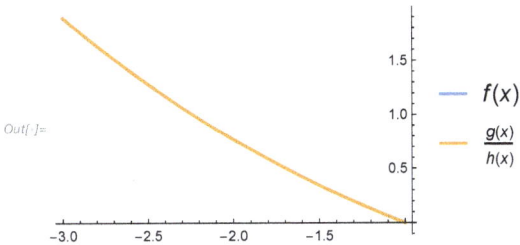

Limit of a Rational Function: Special Example

Compute the limit for the following function as x approaches -1.

In[ ]:= **f[x_] := (x^2 − 1)/(x + 1)**

The denominator equals 0 at -1, so you cannot use the quotient law:

In[ ]:= **x + 1 /. x → −1**

Out[ ]= 0

To find the limit, factor the numerator and see if it has a factor that cancels out the denominator. Use Factor:

In[ ]:= **Factor[x^2 − 1]**

Out[ ]= (−1 + x) (1 + x)

You can see a $(x + 1)$ also appears in the numerator, so simplifying the expression will give $(x - 1)$. Use Simplify:

In[ ]:= **Simplify[f[x]]**

Out[ ]= −1 + x

Therefore the limit is -2, and you can see this in the graph of f:

In[ ]:= **Limit[f[x], x → −1]**

Out[ ]= −2

In[·]:= **Plot[f[x], {x, −2, 0}]**

Out[·]=

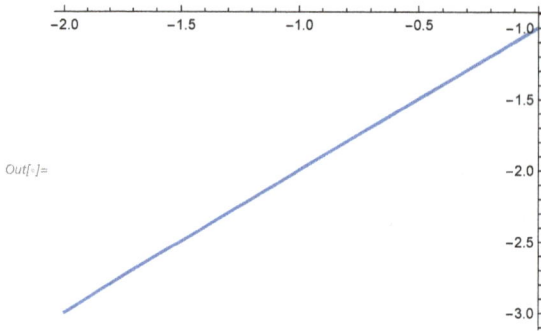

Difference Quotient

Find the limit for the following function as h approaches 0:

In[·]:= **f[$h\_$] := ((2 + h)^2 − 4)/h**

Plot the function:

In[·]:= **Plot[f[h], {h, −1, 1}]**

Out[·]=

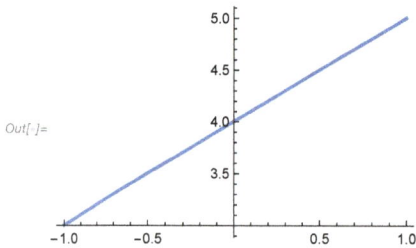

Expand the numerator and see if anything cancels out with the denominator. Use Expand:

In[·]:= **Expand[((2 + h)^2 − 4)]**

Out[·]= $4 h + h^2$

Since h is also a factor of the denominator, cancel it out:

In[·]:= **Factor[f[h]]**

Out[·]= $4 + h$

It is clear now that the limit as h approaches 0 should be 4.

Confirm with Limit:

In[·]:= **Limit[f[h], h → 0]**

Out[·]= 4

One-Sided Limits: Absolute Value Function

Sometimes finding the left- and right-hand limits of a function is the easiest way to find the overall limit. Recall that the limit of a function exists if and only if the left-hand and right-hand limits exist and are equal.

Compute the limit for the absolute value function $|x|$ as x approaches 0:

In[ ]:= **f[x_] := RealAbs[x]**

Begin by plotting the function:

In[ ]:= **Plot[f[x], {x, −1, 1}]**

Out[ ]=

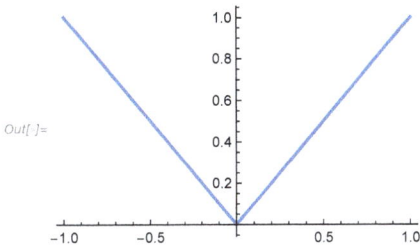

It is clear that the left-hand limit and right-hand limit are both 0 at 0.

Now confirm this answer using Limit:

In[ ]:= **{Limit[f[x], x → 0, Direction → 1], Limit[f[x], x → 0, Direction → −1]}**

Out[ ]= **{0, 0}**

Therefore the limit is 0, and Limit agrees:

In[ ]:= **Limit[f[x], x → 0]**

Out[ ]= **0**

Nonexistent Limit

Compute the limit for the following function as x approaches 0:

In[ ]:= **f[x_] := RealAbs[x] / x**

Begin by plotting the function:

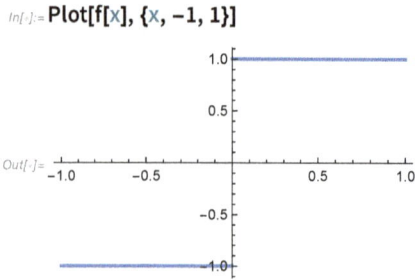

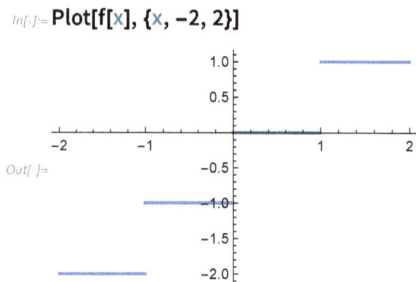

In[ ]:= **Plot[f[x], {x, −1, 1}]**

Out[ ]=

It is clear that the left-hand limit at 0 is − 1, and the right-hand limit at 0 is 1.

In[ ]:= **{Limit[f[x], x → 0, Direction → 1], Limit[f[x], x → 0, Direction → −1]}**

Out[ ]= **{−1, 1}**

Confirm with Limit:

In[ ]:= **Limit[f[x], x → 0]**

Out[ ]= **Indeterminate**

Therefore the limit is nonexistent.

Floor Function

The **floor function** calculates the greatest integer less than or equal to x.

Find the places where the limit does not exist for the floor function in the range − 2 to 2:

In[ ]:= **f[x_] := Floor[x]**

Begin by plotting the function:

In[ ]:= **Plot[f[x], {x, −2, 2}]**

Out[ ]=

It appears the limit does not exist at any of the integers.

For example, at 0 the left- and right-hand limits are − 1 and 0 respectively:

In[ ]:= **{Limit[f[x], x → 0, Direction → 1], Limit[f[x], x → 0, Direction → −1]}**

Out[ ]= **{−1, 0}**

Limit confirms this at each of the integers − 2, − 1, 0, 1 and 2. Also use #, &, /@ and Range:

In[ ]:= **Limit[f[x], x → #] & /@ Range[−2, 2]**

Out[ ]= {Indeterminate, Indeterminate, Indeterminate, Indeterminate, Indeterminate}

Squeeze Theorem

If $f[x]$ is **greater** than $g[x]$ near a, then the limit of f at a is **greater** than the limit of g at a. With this, you can extend to three functions and use what is called the **squeeze theorem**: If $f[x] \geq g[x]$ near a, and $g[x] \geq h[x]$ near a, and the limits of f and h at a are **both** L, then the limit of g at a is also L.

Use the squeeze theorem to find the limit for the following function as x approaches 0:

In[ ]:= **g[x_] := x^2 Cos[1 / x]**

Since the cosine lies strictly between 1 and − 1, the functions that bound g should be x^2 and $-x^2$. You can see this in a graph:

In[ ]:= **f[x_] = x^2;**
h[x_] = −x^2;
Plot[{f[x], g[x], h[x]}, {x, −0.1, 0.1}, PlotLegends → "Expressions"]

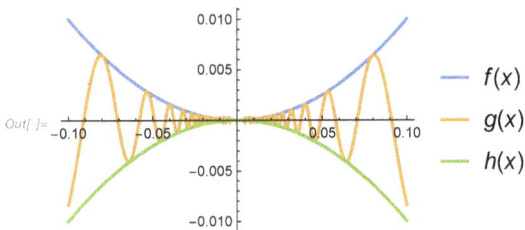

You already know the limits as x approaches 0 for f and h, since they are polynomial functions:

In[ ]:= **{Limit[g[x], x → 0], Limit[h[x], x → 0]}**

Out[ ]= **{0, 0}**

Therefore the limit should be 0 for g as well, as shown by Limit:

In[·]:= **Limit[f[x], x → 0]**

Out[·]= 0

Summary

Instead of having to look at tables to determine limits, the limit laws give a way to find the limits of functions mathematically.

The laws include situations for sums, differences, products and quotients.

From these laws, one can easily find the limit of any polynomial.

For rational and general algebraic functions, sometimes it is best to try factoring the function to calculate the limit.

For piecewise functions, sometimes it is best to calculate the left and right limits of the function to calculate the limit.

For a function that lies between two other functions, the squeeze theorem can be helpful when calculating the limit.

The next lesson will cover continuous functions, which will make calculating the limits of functions even easier.

Exercises

Exercise 1—Polynomial

Compute the limit of the following function as x approaches 1:

In[ ]:= **f[x_] := x^6 + 3 x^2 + 9**

Solution

Since f is a polynomial, you can just plug in 1 to get the answer:

In[ ]:= **f[1]**

Out[ ]= 13

Limit agrees:

In[ ]:= **Limit[f[x], x → 1] == 13**

Out[ ]= True

Here is its graph:

In[ ]:= **Plot[f[x], {x, 0, 2}]**

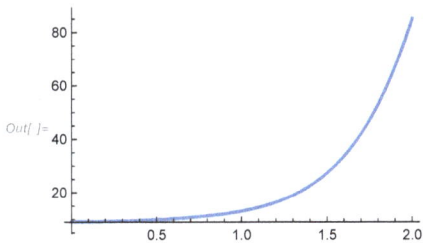

Exercise 2—Limit of Sum

Compute the limit of the following function as x approaches -4:

In[ ]:= **f[x_] := 3 x + Abs[x + 4]**

Solution

Use the sum law to solve this:

In[ ]:= **g[x_] := 3 x**
    h[x_] := Abs[x + 4]

Add the limits of the two functions together:

In[ ]:= **Limit[g[x], x → −4] + Limit[h[x], x → −4]**

Out[ ]= **−12**

Limit agrees:

In[ ]:= **Limit[f[x], x → −4] == −12**

Out[ ]= **True**

Here is its graph:

In[ ]:= **Plot[f[x], {x, −5, −3}]**

Out[ ]=

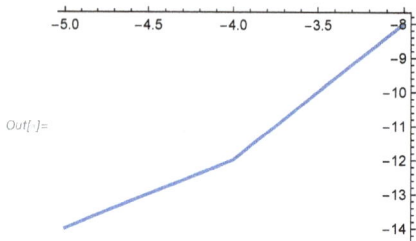

Exercise 3—Limit of a Product

Compute the limit of the following function as *x* approaches 2:

In[ ]:= **f[x_] := (x^4 − 5 x) (3 − Sqrt[x])**

Solution

Use the product law, since it is a product of two polynomials:

In[ ]:= **g[x_] := (x^4 − 5 x)**
　　h[x_] := (3 − Sqrt[x])

Multiply the limits of the two functions together:

In[ ]:= **{Limit[g[x], x → 2] * Limit[h[x], x → 2], Limit[g[x], x → 2] * Limit[h[x], x → 2] // N}**

Out[ ]= $\left\{ 6\left(3 - \sqrt{2}\right), 9.51472 \right\}$

Limit agrees:

In[ ]:= **Limit[f[x], x → 2] == 6 (3 − Sqrt[2])**

Out[ ]= **True**

Here is its graph:

In[ ]:= **Plot[f[x], {x, 1.5, 2.2}, PlotRange → {−5, 15}, GridLines → {{2}, {6 (3 − Sqrt[2])}}}]**

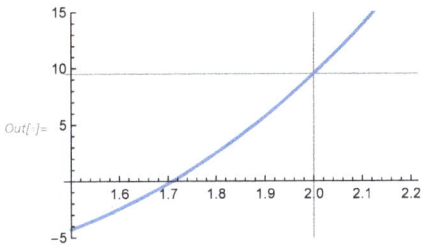

Out[ ]=

Exercise 4—Limit of a Quotient

Compute the limit of the following function as *x* approaches 0:

In[ ]:= **f[x_] := (2 x^4 + 5) / (7 x + 9)**

Solution

This is a rational function, so try to plug in 0:

In[ ]:= **f[0]**

Out[ ]= $\dfrac{5}{9}$

Limit agrees:

In[ ]:= **Limit[f[x], x → 0] == 5/9**

Out[ ]= **True**

Here is its graph:

In[ ]:= **Plot[f[x], {x, −1, 1}]**

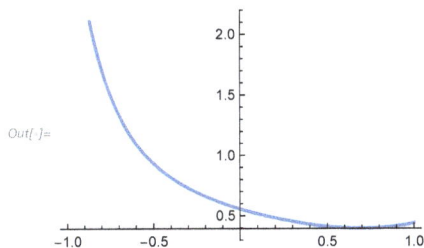

Out[ ]=

Exercise 5— Another Limit of a Quotient

Compute the limit of the following function as x approaches 0:

In[·]:= **f[x_] := (Sqrt[5 + x] – Sqrt[5]) / x**

Solution

See if simplifying it makes it easier to deal with:

In[·]:= **FullSimplify[f[x]]**

Out[·]= $\dfrac{1}{\sqrt{5} + \sqrt{5+x}}$

It is now clear that the limit should be $\dfrac{1}{2\sqrt{5}}$:

Confirm with the original function:

In[·]:= **Limit[f[x], x → 0]**

Out[·]= $\dfrac{1}{2\sqrt{5}}$

Here is its graph with a horizontal grid line at $1 / \left(2\sqrt{5} \right)$:

In[·]:= **Plot[f[x], {x, –1, 1}, GridLines → {None, {1 / (2 Sqrt[5])}}]**

Out[·]=

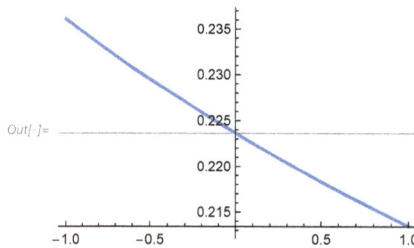

Exercise 6—Squeeze Theorem

Compute the limit for the following function:

In[·]:= **f[x_] := x Sin[1 / x]**

Solution

Using the product rule will not work, since the limit does not exist for $\mathrm{Sin}\left[\frac{1}{x}\right]$:

In[ ]:= **Limit[Sin[1/x], x → 0]**

Out[ ]= Indeterminate

Since the sine lies between 1 and – 1, use the squeeze theorem with boundary functions $|x|$ and $-|x|$ (you cannot use x and $-x$ because $x \geq -x$ for $x \geq 0$, but $-x > x$ for $x < 0$):

In[ ]:= **g[x_] := RealAbs[x]**

In[ ]:= **h[x_] := –RealAbs[x]**

The limit as x approaches 0 for g and h is 0. Therefore the limit for f should also be 0:

In[ ]:= **{{Limit[g[x], x → 0],**
Limit[h[x], x → 0]}, Limit[f[x], x → 0] == 0}

Out[ ]= {{0, 0}, True}

You can also see this in their graphs:

In[ ]:= **Plot[{f[x], g[x], h[x]}, {x, –2, 2}, PlotLegends → "Expressions"]**

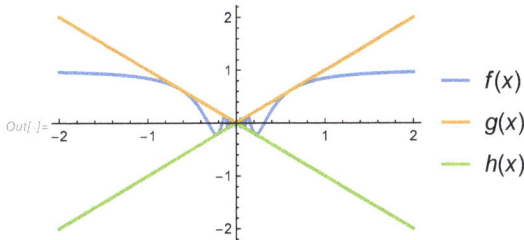

Final Exercise—Length Contraction

In the theory of special relativity, the Lorentz contraction formula states that an object with rest length L_0 and velocity v with respect to a stationary observer has new length with respect to the observer given by:

In[ ]:= **Lorentz[v_] := L0 Sqrt[1 – v^2 /c^2]**
L0 = 1;
c = 299 792 458;

Here c is the speed of light in a vacuum. Set the rest length to 1.

What is the limit of the object's length as v approaches c?

Solution

See what happens as v approaches c from the left:

$In[\cdot]:=$ **Limit[Lorentz[v], v → c, Direction → "FromBelow"]**

$Out[\cdot]=$ 0

The length shrinks to zero!

Here is an interactive diagram to illustrate. The slider controls the velocity and alters the length of the line segment with respect to a stationary observer:

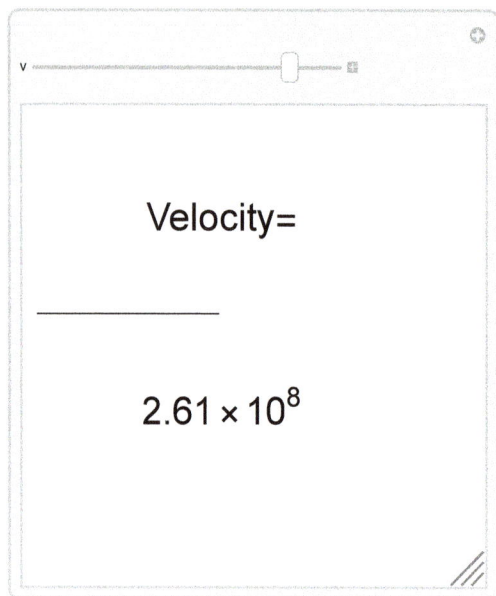

v ——————————⬜——▣

Velocity=

_____

2.61×10^8

6 | Continuity

Overview

Previously, it was noted that the limit of a polynomial is just the value of the polynomial:

$In[\cdot]:=$ **f[x_] := 2 x^4 − 3 x − 9**

$In[\cdot]:=$ **Limit[f[x], x → a]**

$Out[\cdot]=$ $-9 - 3a + 2a^4$

A function whose limit at any point is just the value of the function at that point is given a special name. Such a function is called **continuous**.

This lesson will give examples of continuous functions, show when a function is not continuous, and include a practical application of continuity.

Continuous Functions

A function f is continuous at a point a if $f[a]$ equals the limit of f at a.

It is important to note that for a function to be continuous at a point:

1. $f[a]$ must be defined.
2. The limit of f at a must exist.
3. The limit of f at a must equal $f[a]$.

Visually speaking, a function that is continuous at every point in an interval should have no gaps or breaks in its graph.

For example, the following function is continuous at 1:

$In[\cdot]:=$ **f[x_] := x^2 − x + 1**

$In[\cdot]:=$ **f[1] == Limit[f[x], x → 1]**

$Out[\cdot]=$ **True**

Here is its graph:

In[·]:= **Plot[f[x], {x, 0, 2}]**

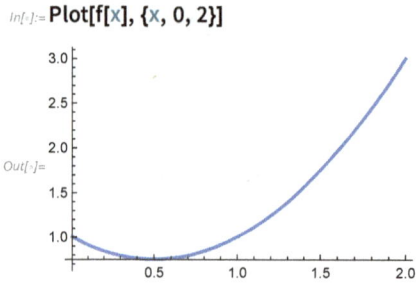

Out[·]=

Polynomials and Rational Functions

Polynomials and rational functions are continuous everywhere they are defined.

Here is an example of a polynomial function and a rational function:

In[·]:= **f[x_] := 5 x^3**

In[·]:= **g[x_] := x / (x + 1)**

f is continuous at every point, and g is continuous at every point except for the point $x = -1$ (where it is undefined).

Look at their graphs:

In[·]:= **{Plot[f[x], {x, -2, 2}], Plot[g[x], {x, -2, 2}]}**

Out[·]=

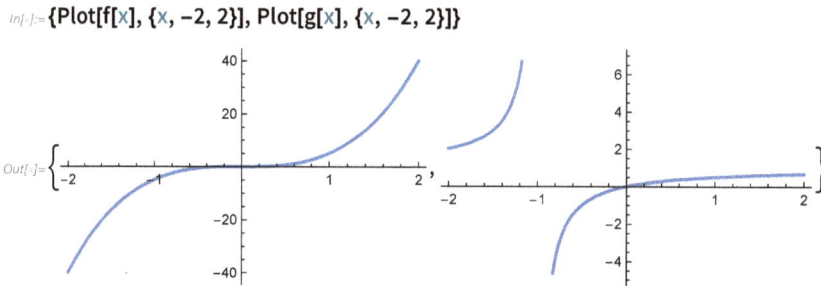

Discontinuous Functions 1

A function that is not continuous at some point is said to be **discontinuous** and have a **discontinuity** at the point.

Find the discontinuities for the following functions:

In[·]:= **f[x_] := (x^2 - 5 x + 4) / (x - 1)**

In[·]:= **g[x_] := Piecewise[{{ -2, x == 1}}, (x^2 - 5 x + 4) / (x - 1)]**

It is clear that the first function has a discontinuity at $x = -1$:

In[ ]:= **f[1] // Quiet**

Out[ ]= Indeterminate

This does not change when you plug in -2 at the discontinuity, since the limit at -1 will still be different from the function value at -1.

This can be seen in the second function:

In[ ]:= **{g[1], Limit[g[x], x → 1]}**

Out[ ]= {−2, −3}

Plot the functions:

In[ ]:= **{Plot[f[x], {x, 0, 2}], Plot[g[x], {x, 0, 2}, Epilog → {Black, PointSize[Large], Point[{1, g[1]}]}]}**

Discontinuous Functions 2

Find the discontinuities for the following functions:

In[ ]:= **f[x_] := Piecewise[{{0, x == 0}}, 1 / x^4]**

In[ ]:= **g[x_] := Floor[x]**

f has limit ∞ at 0, but its value is 0 there, so f has a discontinuity at 0:

In[ ]:= **{Limit[f[x], x → 0], f[0]}**

Out[ ]= {∞, 0}

The floor function **Floor** clearly has discontinuities at the integers.

For example, consider the point $x = 0$. The limit does not exist there:

In[ ]:= **Limit[g[x], x → 0]**

Out[ ]= Indeterminate

The value is 0 there, so it has a discontinuity at 0:

In[·]:= **g[0]**

Out[·]= 0

Here are graphs for the two functions:

In[·]:= **{Plot[f[x], {x, −1, 1}], Plot[g[x], {x, −2, 2}, Axes → {False, True}]}**

Out[·]=

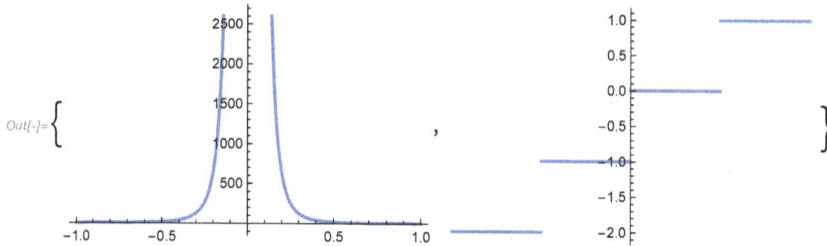

Continuity from Different Directions

In general, a function is considered continuous from the right (resp. left) at a point if its right-hand (left-hand) limit equals the function value at that point.

Look at the floor function again. It is continuous from the right, but not continuous from the left at 0 (and all integers):

In[·]:= **Limit[Floor[x], x → 0, Direction → "FromAbove"] == Floor[0]**

Out[·]= True

In[·]:= **Limit[Floor[x], x → 0, Direction → "FromBelow"] == Floor[0]**

Out[·]= False

Here is a plot:

In[·]:= **Plot[Floor[x], {x, −5, 5}]**

Out[·]=

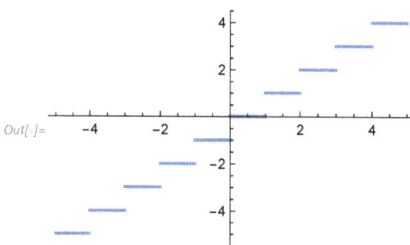

In the range − 5 to 5, you can see that it is continuous from the right at − 5, − 4, − 3, − 2, − 1, 0, 1, 2, 3, 4 and 5.

Root Functions

To find the limit of a root function, you just take the root of the limit to be calculated by the power law. Root functions of polynomials are therefore continuous everywhere in their domain.

Consider the square root function, for example:

In[ ]:= **f[x_] := Sqrt[x]**

It is the square root of the function x, so it is continuous everywhere in its domain. For example, the limit at $x = 9$ equals the square root at 9:

In[ ]:= **{Limit[Sqrt[x], x → 9], Sqrt[9]}**

Out[ ]= **{3, 3}**

It is not continuous at – 1 because it is imaginary there:

In[ ]:= **Sqrt[−1]**

Out[ ]= **_i_**

Here is the graph of $\sqrt{x}$:

In[ ]:= **Plot[Sqrt[x], {x, −1, 10}]**

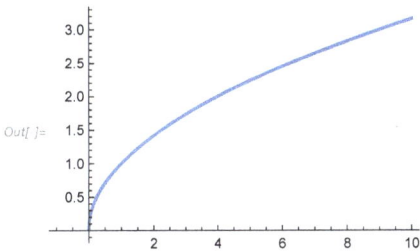

Trigonometric Functions

The sine and cosine functions are both continuous at 0:

In[ ]:= **Limit[Sin[x], x → 0] == Sin[0]**

Out[ ]= **True**

In[ ]:= **Limit[Cos[x], x → 0] == Cos[0]**

Out[ ]= **True**

From the addition formulas for the sine and cosine, it can be deduced that they are continuous everywhere.

You can see this in their graphs:

In[ ]:= **Plot[{Sin[x], Cos[x]}, {x, −2 π, 2 π}, PlotLegends → "Expressions"]**

Out[ ]=

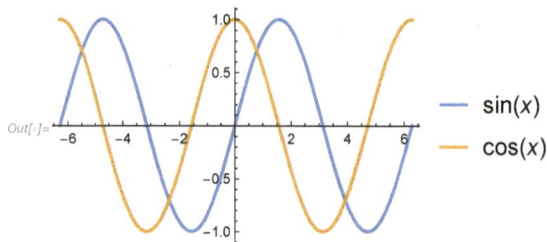

Since tangent $= \frac{\text{sine}}{\text{cosine}}$, it is not continuous wherever the cosine is 0 (since its limit does not exist there).

A plot of the tangent illustrates this:

In[ ]:= **Plot[Tan[x], {x, −2 π, 2 π}]**

Out[ ]=

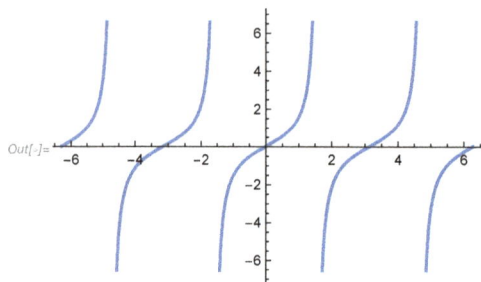

Continuity Laws

If functions f and g are continuous at a point a and c is a constant, then these **continuity laws** follow.

Sum law: $f + g$ is continuous at a.

Difference law: $f − g$ is continuous at a.

Scalar multiplication law: $c * f$ is continuous at a.

Product law: $f * g$ is continuous at a.

Quotient law: $\frac{f}{g}$ is continuous at a if $g[a] \neq 0$.

Calculate the limit for the following function as x approaches π:

In[ ]:= **f[x_] := Cos[x]/(2 + Sin[x])**

You know that the cosine and sine are continuous.

From the continuity laws, you know $Sin[x] + 2$ is also continuous, because it is the sum of two continuous functions.

The original function is a quotient, and the denominator is never 0 because $2 + Sin[x]$ lies between 1 and 3, so f is also continuous.

Therefore plugging in π will gives the limit:

In[ ]:= **f[π]**

Out[ ]= $-\dfrac{1}{2}$

Limit agrees:

In[ ]:= **Limit[f[x], x → π] == −1/2**

Out[ ]= **True**

Compositions

If f is continuous at b and the limit of g at a is b, then the limit of $f[g[x]]$ at a is $f[b]$:

$$\lim_{x \to a} f(g(x)) = f\left(\lim_{x \to a} g(x) \right)$$

In other words, if f is continuous, then finding the limit of a composition is as easy as evaluating f at the limit of its argument. Moreover, if f is continuous at $g[a]$ and g is continuous at a, then their composition $f[g[x]]$ is also continuous at a.

Find the limit of the following function at π:

In[ ]:= **f[x_] := Cos[x^2 − 7 x + 10]**

The function is the composition of the cosine and $x^2 - 7x + 10$. Here are their plots:

In[ ]:= **Plot[{f[x], Cos[x], x^2 − 7 x + 10}, {x, 0, 2 π}, PlotLegends → "Expressions"]**

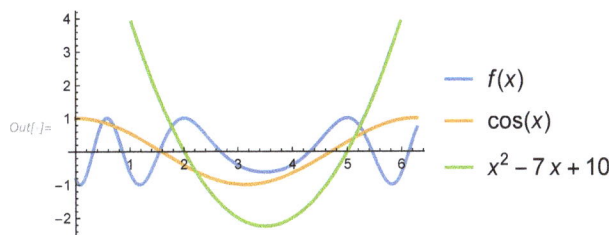

Both are continuous functions, so just plug in π. Limit agrees:

In[ ]:= **{f[π], Limit[f[x], x → π] == f[π]}**

Out[ ]= $\left\{-Cos\left[10 + \pi^2\right], True\right\}$

Intermediate Value Theorem

Continuous functions let you use a very valuable theorem, known as the **intermediate value theorem**.

Let f be continuous on the closed interval $[a, b]$ and let $\mathcal{N}$ be any number between $f[a]$ and $f[b]$, where $f[a] \neq f[b]$. Then there exists a number c in (a, b) such that $f(c) = \mathcal{N}$.

In other words, f takes on every value between $f[a]$ and $f[b]$ at least once.

With this, you can show that there is a root for the following function between 1 and 2:

In[ ]:= **f[x_] := 3 x^3 − 5 x^2 + 2 x − 3**

You need to find a value c between 1, 2 so that $f[c] = 0$.

Since the function is continuous, evaluate it at 1 and 2:

In[ ]:= **f[{1, 2}]**

Out[ ]= **{−3, 5}**

Since $-3 < 0 < 5$, you know a root exists in the interval $[1, 2]$.

Plot the function and find the root with Solve:

In[ ]:= **Plot[f[x], {x, 1, 2}]**

Out[ ]=

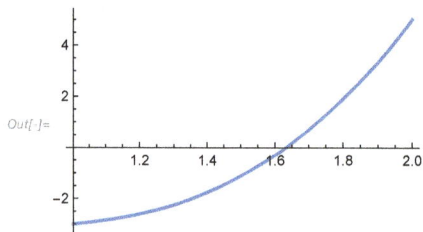

In[ ]:= **Solve[f[x] == 0, x, Reals] // N**

Out[ ]= **{{x → 1.63334}}**

Summary

Continuous functions are very common in the real world.

A continuous function can be thought to have no gaps or breaks in its graph.

Polynomials and root functions are continuous.

Some rational functions and trigonometric functions are not continuous because of discontinuities.

The intermediate value theorem is an interesting consequence of continuity, and helps you find roots of functions.

The next lesson will begin differential calculus and use limits to introduce the concept of a derivative.

Exercises

Exercise 1—Polynomial

Compute the limit for the following function as x approaches 11:

In[·]:= **f[x_] := 4 x^8 + 7 x^4 + 9**

Solution

The function is continuous, so plug in 11:

In[·]:= **f[11]**

Out[·]= 857 538 020

Limit agrees:

In[·]:= **Limit[f[x], x → 11] == f[11]**

Out[·]= True

Here is a graph:

In[·]:= **Plot[f[x], {x, 6, 16}]**

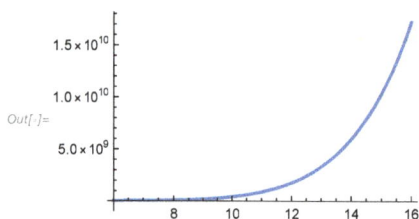

Exercise 2—Rational Function

Compute the following limit as x approaches 5:

In[·]:= **f[x_] := (5 x^4 + 7 x^2 + 10)/(x^3 − 2 x − 10)**

Solution

The function is continuous at 5, so plug in 5:

In[·]:= **f[5]**

Out[·]= $\dfrac{662}{21}$

Limit agrees:

In[ ]:= **Limit[f[x], x → 5] == f[5]**

Out[ ]= **True**

Here is a graph:

In[ ]:= **Plot[f[x], {x, 0, 10}]**

Out[·]=

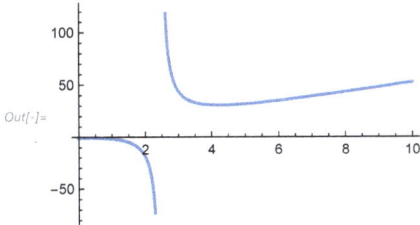

Exercise 3—Finding Discontinuities 1

Find the discontinuities of the following function:

In[ ]:= **f[x_] := $x/(x^2 - 1)$**

Solution

Plot the function:

In[ ]:= **Plot[f[x], {x, −2, 2}]**

Out[ ]=

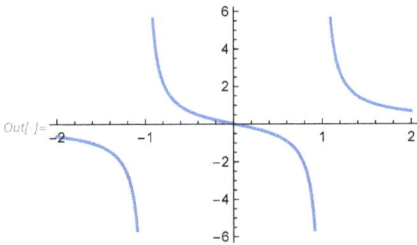

By inspection, the function has discontinuities at −1 and 1.

Confirm with Limit:

In[ ]:= **{Limit[f[x], x → −1] === f[−1], Limit[f[x], x → 1] === f[1]} // Quiet**

Out[ ]= **{False, False}**

Exercise 4—Finding Discontinuities 2

Find the discontinuities of the following function:

In[ ]:= **f[x_] := (x^4 − 10 x^3 + 35 x^2 − 50 x + 24) / (x^2 − 5 x + 6)**

Solution

Factor the denominator to find out where it is 0:

In[ ]:= **Factor[x^2 − 5 x + 6]**

Out[ ]= **(−3 + x) (−2 + x)**

The function has discontinuities at 2 and 3 because it is not defined there:

In[ ]:= **{Limit[f[x], x → 2] === f[2], Limit[f[x], x → 3] === f[3]} // Quiet**

Out[ ]= **{False, False}**

Solve also gives the discontinuities:

In[ ]:= **Solve[x^2 − 5 x + 6 == 0, x]**

Out[ ]= **{{x → 2}, {x → 3}}**

Plot the function:

In[ ]:= **Plot[f[x], {x, 0, 5}]**

Out[ ]=

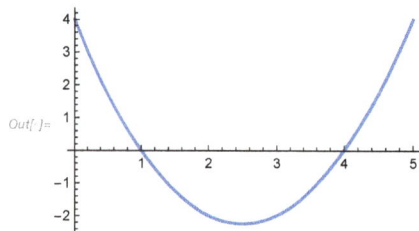

Exercise 5—Trig Functions

Compute the following limit as x approaches 0:

In[ ]:= **f[x_] := 1 / (1 + Cos[x])**

Solution

The cosine is continuous at 0, so the function is continuous at 0.

Plug in 0:

In[·]:= **f[0]**

Out[·]= $\dfrac{1}{2}$

Limit agrees:

In[·]:= **Limit[f[x], x → 0] == 1/2**

Out[·]= **True**

Here is a plot of the function:

In[·]:= **Plot[f[x], {x, −1, 1}]**

Out[ ]=

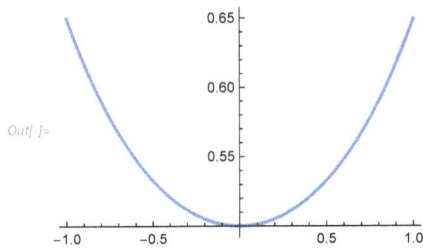

Exercise 6—Composition

Find the limit of the following function as x approaches 2:

In[·]:= **f[*x*_] := Sin[Cos[*x*^4]]**

Solution

The sine, cosine and x^4 are all continuous, so plugging in 2 will give the answer:

In[·]:= **f[2]**

Out[·]= **Sin[Cos[16]]**

Limit agrees:

In[·]:= **Limit[f[x], x → 2] == f[2]**

Out[·]= **True**

Here is a plot of the function:

In[·]:= **Plot[f[x], {x, 1, 3}]**

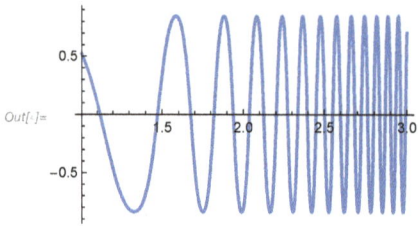

Out[·]=

Exercise 7—Intermediate Value Theorem

Confirm a root exists in the interval $[-1, 0]$ for the following function:

In[·]:= **f[x_] := x^6 + 6 x + 4**

Solution

The function is continuous, so find the values at -1 and 0:

In[·]:= **{f[−1], f[0]}**

Out[·]= **{−1, 4}**

Since $-1 < 0 < 4$, there is a root in the region:

In[·]:= **Solve[f[x] == 0 && −1 ≤ x ≤ 4, x] // N**

Out[·]= **{{x → −0.683688}}**

Here is a plot of the function:

In[·]:= **Plot[f[x], {x, −1, 0}]**

Out[·]=

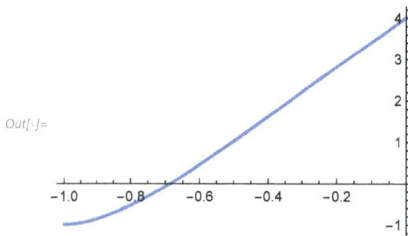

Exercise 8—Gravitational Force

The gravitational force exerted by the planet Earth on a unit mass at a distance r from the center of the planet is given by the following function:

```
In[·]:= force[r_] := Piecewise[{{G M r / R^3, r < R}}, G M / r^2]
   G = G;
   M = 5.9721986`8.*^24;
   R = 3958.7608367135926191043`7.;
```

Is the function a continuous function of r?

Solution

Since the individual pieces are continuous in their domains, check if the function is continuous at R. That is the only place where there can be a discontinuity.

Compute the limit of the function and compare it to its value:

```
In[·]:= force[R] == Limit[force[r], r → R]
```

```
Out[·]= True
```

Therefore the function is continuous.

Plot the function:

```
In[·]:= Plot[force[r], {r, 0, 2 R}]
```

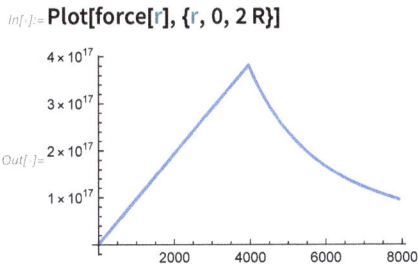

7 | Derivatives and Rates of Change

Overview

Since calculus is concerned primarily with change, a way to quantify such change is needed.

There are functions, which say how one quantity changes with respect to other quantities:

In[ ]:= `f[x_] := 8 x^4 - 8 Sin[x] + 5 / x`

And such functions can be visualized in graphs:

In[ ]:= `Plot[f[x], {x, -1, 1}]`

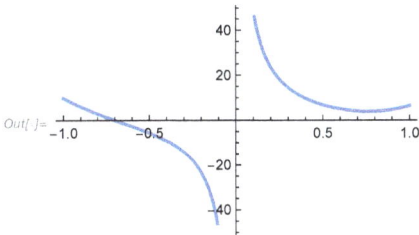

However, more is needed: to find out how the function changes from point to point.

Algebra gives a way to do this for lines through slope. The slope of the line following is 3; go forward one unit, and the line goes up 3 units:

In[ ]:= `Plot[3 x + 4, {x, -2, 2}]`

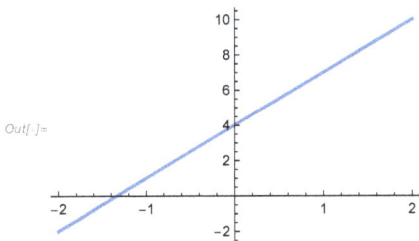

This lesson will show how to do this with more general functions, with the concept known as the **derivative**.

Notion of a Derivative

The derivative is a special kind of limit that is tied to finding the tangent to a curve. How do you find the tangent line to a curve?

To find the equation of a line, you just need its slope and its y-intercept. The y-intercept can easily be found from the slope and point-slope form, so all you really need is the slope.

To find the slope of a line given two points x and a, use the following formula:

$$m = \frac{\text{Rise}}{\text{Run}} = \frac{y - y_0}{x - x_0} = \frac{f(x) - f(a)}{x - a}$$

For general functions, this only gives the slope of the **secant** line that goes through the two points.

For example, the slope of the secant line going through the following function at the points $(1/2, 2)$ and $(3, 1/3)$ is:

```
In[·]:= f[x_] := 1 / x
```

```
In[·]:= (1 / 3 - 2) / (3 - 1 / 2)
```

$$Out[\cdot] = -\frac{2}{3}$$

But the line made using this slope is not tangent to **either point**:

```
In[·]:= Plot[{f[x], 1 / 3 - 2 / 3 (x - 3)}, {x, 0, 4}]
```

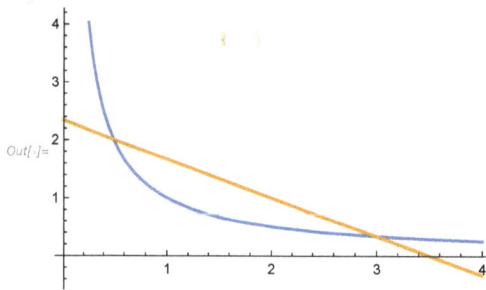

Limits of Secant Lines

To find the tangent line to a curve at a point, take the **limit** of secant lines. One point will stay the same, while the other will get closer and closer to the first point.

In other words, you have the following equation:

$$m = \lim_{x \to a} \frac{f(x) - f(a)}{x - a}$$

This will give the slope of the **tangent** line that only goes through the point $(a, f[a])$.

Here is an example. *b* changes where the tangent line is, and *h* controls the secant line as it gets closer to the tangent:

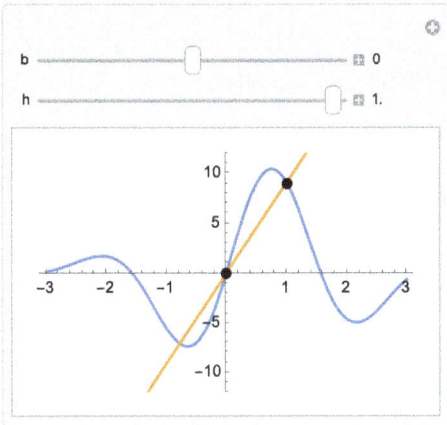

Tangent Line of a Polynomial

Find an equation of the tangent line to the cubic function at the point $(1, 1)$:

In[ ]:= **f[x_] := x^3**

$a = 1$ and $f[x] = x^3$, so the slope is:

In[ ]:= **m = Limit[(f[x] − f[1])/(x − 1), x → 1]**

Out[ ]= 3

With point-slope form, the equation is:

$$y - 1 = 3(x - 1)$$
$$y = 3x - 2$$

Here are the function and the tangent line:

In[ ]:= **Plot[{f[x], m (x − 1) + f[1]}, {x, −2, 2}, Epilog → {PointSize[Large], Point[{1, f[1]}]}]**

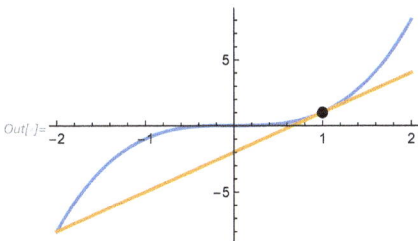

Difference Quotient

The equation for the tangent to a function has another form, using something known as the difference quotient:

In[ ]:= **Limit[DifferenceQuotient[f[x], {x, h}], h → 0] // TraditionalForm**

Out[ ]= $3\,x^2$

The Wolfram Language function DifferenceQuotient helps to automate this computation.

Find the equation of the tangent line to the hyperbola $\frac{1}{x}$ at the point (1, 1):

In[ ]:= **f[x_] := 1 / x**

In[ ]:= **m = Limit[DifferenceQuotient[f[x], {x, h}], h → 0] /. x → 1**

Out[ ]= -1

With point-slope form, the equation is:

$$y - 1 = 1 - x$$
$$y = 2 - x$$

Here are the function and the tangent line:

In[ ]:= **Plot[{f[x], m x + 2}, {x, −2, 2}, Epilog → {PointSize[Large], Point[{1, f[1]}]}]**

Out[ ]=

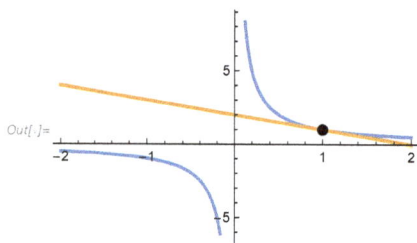

Velocity

Suppose you know the position for an object given a time t and want to find the instantaneous velocity for it at any time. If you denote the position as a function of time, the instantaneous velocity of a function is given by the slope of the tangent line to the position curve.

For example, consider the following position function:

In[ ]:= **s[t_] := 5 t**

The instantaneous velocity is given by the limit of the difference quotient:

In[ ]:= **v[*t_*] := Limit[DifferenceQuotient[s[x], {x, h}], h → 0] /. x → *t***

Here is the plot of the position and velocity:

In[ ]:= **Plot[{s[t], v[t]}, {t, 0, 5}, PlotLegends → {"position", "velocity"}]**

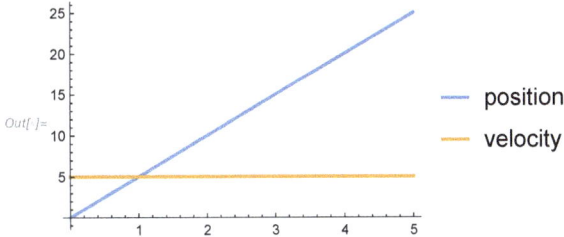

Here it makes sense that the velocity is constant.

Physics Problem

A ball is dropped from a building, 100 meters above the ground. Find the velocity of the ball after two seconds, then find how fast it is going as it hits the ground.

Using the equation of motion for displacement $s[t] = 100 - 4.9\,t^2$, the velocity is easily found at 2:

In[ ]:= **s[*t_*] := 100 − 4.9 *t*^2;**
Limit[DifferenceQuotient[s[t], {t, h}], h → 0] /. t → 2

Out[ ]= −19.6

The velocity is negative because the ball is going down.

Here is a plot with a tangent at $(2, s[2])$:

In[ ]:= **Plot[{s[t], −19.6 (t − 2) + s[2]}, {t, 0, 5},**
Epilog → {PointSize[Large], Point[{2, s[2]}]}, PlotLegends → {"position", "velocity"}]

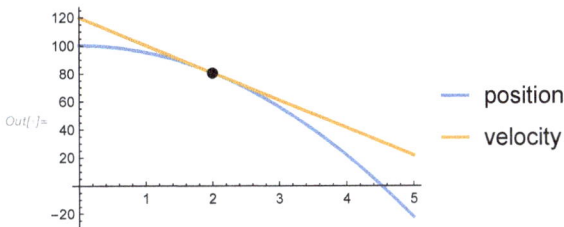

The ball will hit the ground when the displacement is 0, so just solve for the time:

In[ ]:= **sol = Solve[{s[t] == 0, t > 0}, t]**

Out[ ]= {{t → 4.51754}}

The velocity when it hits the ground is -44.3 meters per second:

In[·]:= **Limit[DifferenceQuotient[s[t], {t, h}], h → 0] /. t → sol〚1, 1, 2〛]**

Out[·]:= **−44.2719**

Derivatives at Points

The slope of the tangent line at a point is used so often that it is given a special name: the **derivative at the point**. The Wolfram Language even has its own notation to find the derivatives of functions at points.

For a function $f[x]$, you can also use the shorthand $f'[x]$:

In[·]:= **f[x_] := x^3**

In[·]:= **f'[1]**

Out[·]:= **3**

You can see that the derivative of the function x^3 at the point (1, 1) is 3.

Plot x^3 and the tangent line at (1, 1):

In[·]:= **Plot[{f[x], f'[1] (x − 1) + f[1]}, {x, −2, 2}, Epilog → {PointSize[Large], Point[{1, f[1]}]}]**

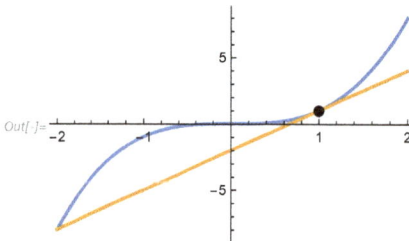

Tangent Line of a Parabola

Find the tangent line of the following function at $x = 2$:

In[·]:= **f[x_] := 2 x^2 + 3 x − 10**

The slope at the tangent line is just the derivative at 2, so finding the tangent line is easy:

In[·]:= **tangent[x_] := f'[2] (x − 2) + f[2]**

In[·]:= **tangent[x] // Simplify**

Out[·]:= **−18 + 11 x**

Plot the function, its derivative and the tangent line:

In[ ]:= **Plot[{f[x], tangent[x]}, {x, 0, 4}, Epilog → {PointSize[Large], Point[{2, f[2]}]}]**

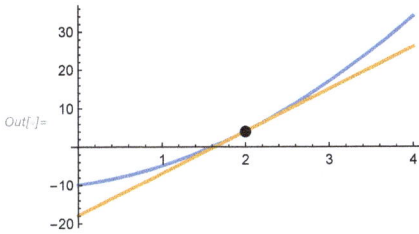

Out[ ]=

Average Rate of Change

Suppose you are given a value y that depends on the value x. Then y is a function of x (i.e. $y = f[x]$). When x changes from x_1 to x_2, y changes from y_1 to y_2. The change is called the **increment** of x (y resp.) and has symbol $\Delta x(\Delta y)$.

Δx equals $x_2 - x_1$, and Δy equals $y_2 - y_1$.

The difference quotient $\Delta y / \Delta x$ is called the **average rate of change of y with respect to x** over the interval $[x_1, x_2]$. This can be interpreted as the slope of the secant line between (x_1, y_1) and (x_2, y_2).

The average rate of change of x^2 with respect to x over the interval $[-0.5, 0.25]$ is $\left(0.25^2 - (-0.5)^2\right) / (0.25 - (-0.5)) = -0.25$:

In[ ]:= **avg = ((0.25)^2 – (–0.5)^2) / (0.25 – (–0.5))**

Out[ ]= **–0.25**

Here is a plot of the secant line that goes through $(-0.5, 0.25)$ and $(0.25, 0.0625)$:

In[ ]:= **Plot[{x^2, (–0.5)^2 + avg (x + 0.5)}, {x, –1, 1}, PlotRange → {–0.1, 0.5}]**

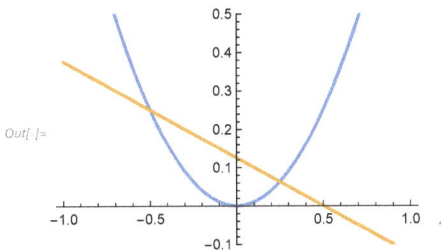

Out[ ]=

Instantaneous Rate of Change

When you take the limit as Δx goes to 0, you get the **instantaneous rate of change of y with respect to x** at $x = x_1$. This can be interpreted as the slope of the tangent line to the curve $y = f[x]$ at $(x_1, f[x_1])$.

Here is a slope of the tangent line to x^2 at $(-0.5, 0.25)$:

In[•]:= **Plot[{x^2, (−0.5)^2 + Limit[(h^2 − (−0.5)^2)/(h − (−0.5)), h → −0.5] (x + 0.5)},**
  {x, −1, 1}, Epilog → {PointSize[Large], Point[{−0.5, 0.25}]}]

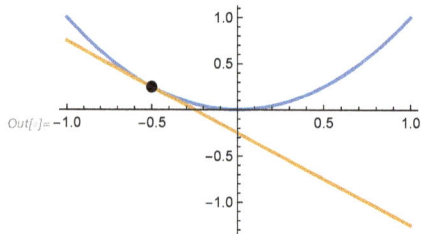

Out[•]=

The instantaneous rate of change of y with respect to x is just the derivative of y at x.

Finding the Derivative Using a Table

You can approximate the derivative using average rates of changes and a table.

The US national debt (in billions) at time t is plotted in the graph:

In[•]:= **ListLinePlot[{{1980, 930.2}, {1985, 1945.9}, {1990, 3233.3}, {1995, 4974.0},**
  {2000, 5674.2}, {2005, 7932.7}, {2010, 14030}, {2015, 18920}}]

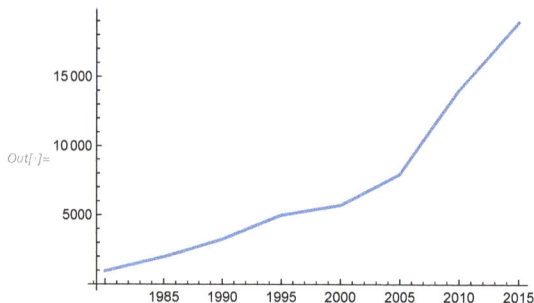

Out[•]=

Estimate the instantaneous rate of change of the debt with respect to time in the year 2010.

First, make a piecewise function that is defined at each of the points in the graph:

In[ ]:= **debtdata[*x*_] = Piecewise[**
   {{930.2, x == 1980}, {1945.9, x == 1985}, {3233.3, x == 1990}, {4974.0, x == 1995},
     {5674.2, x == 2000}, {7932.7, x == 2005},
     {14 030, x == 2010}, {18 920, x == 2015}}, Undefined];

Since you are trying to estimate the derivative, use the function:

In[ ]:= **debtDerivative[*x*_] := (debtdata[*x*] − debtdata[2010])/(*x* − 2010)**

You cannot actually find the derivative of the function at (2010, 14 030) because the table only has eight values.

However, you can interpolate the derivative at (2010, 14 030) by taking the average of the values from years 2005 and 2015:

In[ ]:= **(debtDerivative[2005] + debtDerivative[2015])/2**

Out[ ]= **1098.73**

So the US national debt was increasing at a rate of roughly 1.1 trillion dollars per year in 2010.

Summary

The derivative of a function at a point lets you find the "slope" of the function at that point.

The derivative is found by taking the limit of the slopes of secant lines that get closer and closer to the desired point.

Derivatives are useful in many quantitative subjects, like physics and economics.

The derivative can be approximated from a table of values.

The next lesson will show how to express the derivative itself as a function.

Exercises

Exercise 1—Tangent Line to a Polynomial

Find the tangent curve to the given function at the point (3, 36):

In[·]:= **f[x_] := x^4 − 6 x^2 + 3 x**

Solution

Compute the derivative of the function and plug in 3:

In[·]:= **m = f'[3]**

Out[·]= **75**

With point-slope form, the equation is:

$$y - 36 = 75\,(x - 3)$$
$$y = 75\,x - 189$$

Here is a plot of the function and the tangent line:

In[·]:= **Plot[{f[x], f[3] + m (x − 3)}, {x, 0, 6}, Epilog → {PointSize[Large], Point[{3, f[3]}]}]**

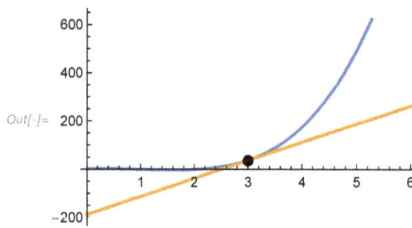

Exercise 2—Tangent Line to an Algebraic Function

Find the tangent line to the given function at the point $\left(2, \frac{1}{2}\right)$:

In[·]:= **f[x_] := 1 / Sqrt[2 x]**

Solution

Compute the derivative of the function and plug in 2:

In[·]:= **m = f'[2]**

Out[·]= $-\dfrac{1}{8}$

With point-slope form, the equation is:

$$y - \frac{1}{2} = \frac{2-x}{8}$$
$$y = -\frac{x}{8} + \frac{3}{4}$$

Here is a plot of the function and the tangent line:

In[ ]:= **Plot[{f[x], f[2] + m (x − 2)}, {x, 0, 6}, Epilog → {PointSize[Large], Point[{2, f[2]}]}]**

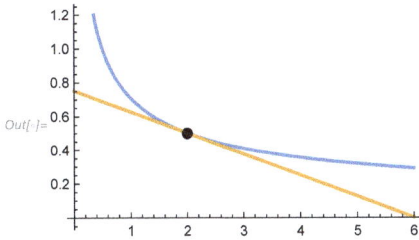

Out[ ]=

Exercise 3—Velocity of a Cannonball

A cannonball is shot vertically into the air with a velocity of 50 meters per second. Its height (in meters) after t seconds can be given by the following function:

In[ ]:= **h[t_] := 50 t − 9.8 t ^ 2**

Find the velocity at three seconds and when it hits the ground again.

Solution

The velocity at three seconds will just be the derivative of the height function at three seconds:

In[ ]:= **h'[3]**

Out[ ]= **−8.8**

To find the velocity when the cannonball hits the ground again, set the height function to 0 and solve for t:

In[ ]:= **sol = Solve[{h[t] == 0, t > 0}, t]**

Out[ ]= **{{t → 5.10204}}**

Then find the derivative of the height at that time:

In[ ]:= **h'[sol[[1, 1, 2]]]**

Out[ ]= **−50.**

So the velocity of the cannonball will be − 8.8 meters per second at three seconds and − 50 meters per second when it hits the ground again.

Here is a plot of the position and velocity of the cannonball:

In[ ]:= **Plot[{h[t], h'[t]}, {t, 0, sol[[1, 1, 2]]}]**

Out[ ]=

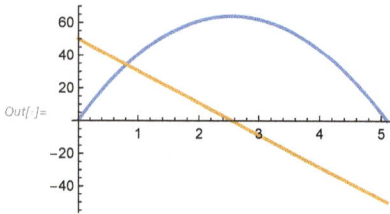

Exercise 4—Slope of an Algebraic Function

Find the slope of the following function at 1:

In[ ]:= **f[x_] := 2 x^3 − 9 x + Sqrt[x]**

Solution

Compute the derivative of the function and plug in 1:

In[ ]:= **f'[1]**

Out[ ]= $-\dfrac{5}{2}$

Here is a plot of the function and its tangent line:

In[ ]:= **Plot[{f[x], f'[1] (x − 1) + f[1]}, {x, 0, 2}, Epilog → {PointSize[Large], Point[{1, f[1]}]}]**

Out[ ]=

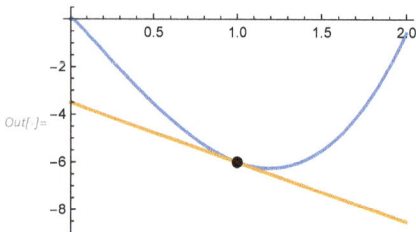

Exercise 5—Finding a Derivative Using a Table

The following table gives the temperature (in °F) in Champaign IL, t hours after 12am recorded every two hours (all obtained from Wolfram | Alpha®).

t	0	2	4	6	8	10	12	14
Temperature	62	59	57	59	65	71	75	78

Estimate the instantaneous rate of change of the temperature at 10am.

Solution

Make a plot for the data:

In[ ]:= **ListLinePlot[Transpose@{{0, 2, 4, 6, 8, 10, 12, 14}, {62, 59, 57, 59, 65, 71, 75, 78}}]**

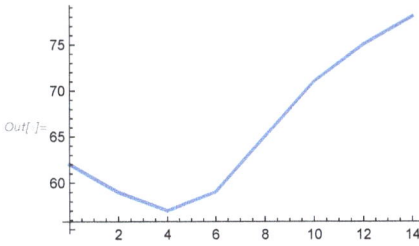

The derivative at 10am can be estimated with the following functions:

In[ ]:= **temperatureData[$t\_$] := Piecewise[{{62, t == 0}, {59, t == 2}, {57, t == 4}, {59, t == 6}, {65, t == 8}, {71, t == 10}, {75, t == 12}, {78, t == 14}}, Undefined];**

In[ ]:= **temperatureDerivative[$t\_$] := (temperatureData[t] – temperatureData[10]) / (t – 10)**

Take the average of the slopes at the values 8 and 12 to estimate the temperature is rising at a rate of $5/2$°F per two hours:

In[ ]:= **(temperatureDerivative[8] + temperatureDerivative[12]) / 2**

Out[ ]= $\dfrac{5}{2}$

8 | The Derivative as a Function

Overview

Previously, you have only been finding the derivative at particular points of a function. The derivative itself can also be considered a **function**. The Wolfram Language uses D to calculate the derivative of a function as a function.

Consider the following function:

$In[\ ]:=$ **f[x_] := 4 x^3 − 9 x**

Here is its derivative:

$In[\ ]:=$ **D[f[x], x]**

$Out[\ ]=$ $-9 + 12 x^2$

Here is a plot of the function and its derivative:

$In[\ ]:=$ **Plot[{Evaluate[f[x]], Evaluate[∂_x f[x]]}, {x, −3, 3},**
PlotLegends → "Expressions"]

$Out[\ ]=$

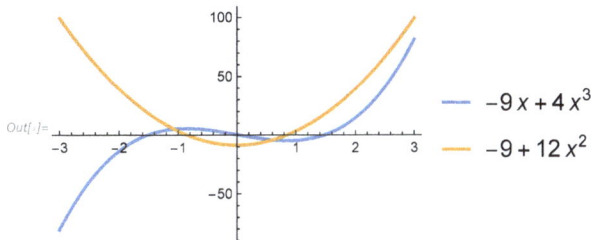

Finding the Derivative of a Function

Find the derivative of the function given:

$In[\ ]:=$ **f[x_] := x^3 + x**

Use D:

$In[\ ]:=$ **D[f[x], x]**

$Out[\ ]=$ $1 + 3 x^2$

$f\,'[x]$ also works:

$In[\ ]:=$ **f'[x]**

$Out[\ ]=$ $1 + 3 x^2$

Plot the function and its derivative:

In[ ]:= **Plot[{f[x], f'[x]}, {x, −3, 3}, PlotLegends → "Expressions"]**

Out[ ]=

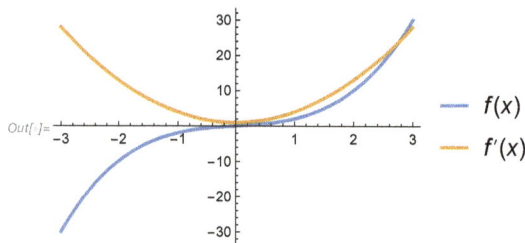

Derivative Notation

There are many ways to express the derivative of a function.

If the function is f, f' is used to indicate the derivative of f:

In[ ]:= **f[x_] := x^2 − 1**

In[ ]:= **f'[x]**

Out[ ]= **2 x**

If y is a function f of x, other notations include y', $\frac{dy}{dx}$, $\frac{df}{dx}$, $\frac{d}{dx}(f[x])$ and $D_x f[x]$.

D and $\frac{d}{dx}$ are known as **differentiation operators** because they indicate differentiation. **Differentiation** is the process of calculating a derivative.

The Wolfram Language uses D and f' to calculate derivatives:

```
In[ ]:= {Plot[{f[x], Evaluate[∂ₓ f[x]]}, {x, −1, 1},
        PlotLegends → "Expressions"],
      Plot[{f[x], Derivative[1][f][x]}, {x, −1, 1},
        PlotLegends → "Expressions"]}
```

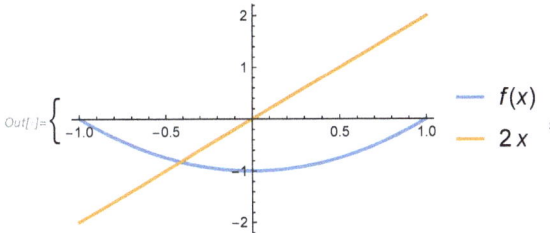

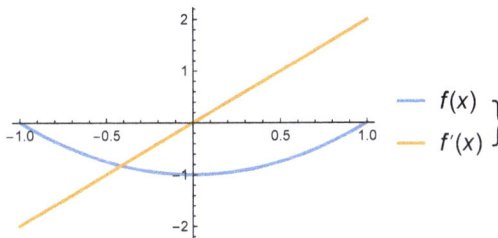

One-Sided Derivatives

Since the derivative is a limit, there is also a notion of left-hand and right-hand derivatives at a point:

```
In[ ]:= g[x_] := 3 RealAbs[x − 5]
```

You can manually calculate the left-hand and right-hand derivative of the function at 5 with Limit and DifferenceQuotient:

```
In[ ]:= Limit[DifferenceQuotient[g[x], {x, h}], h → 0, Direction → 1] /. x → 5
```

```
Out[ ]= −3
```

```
In[ ]:= Limit[DifferenceQuotient[g[x], {x, h}], h → 0, Direction → −1] /. x → 5
```

```
Out[ ]= 3
```

This can also be seen in the plot of g:

In[·]:= **Plot[{g[x], Evaluate[∂ₓ g[x]]}, {x, 4, 6},**
    PlotLegends → "Expressions"]

Out[·]=

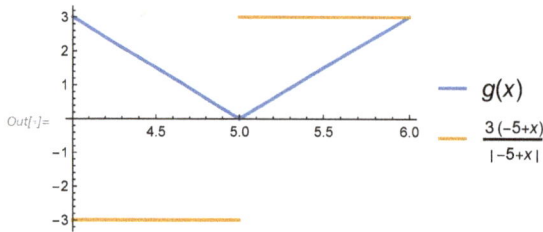

Legend:
— $g(x)$
— $\dfrac{3(-5+x)}{|-5+x|}$

Differentiability

A function f is **differentiable at a** if $f'[a]$ exists (i.e. if its left-hand and right-hand derivatives exist and are equal). It is **differentiable on an open interval** (a, b), (a, ∞) or $(-\infty, b)$ if it is differentiable at every number in the interval.

Check if the following function is differentiable:

In[·]:= **f[x_] := 2 RealAbs[x]**

Find out with D:

In[·]:= **D[f[x], x]**

Out[·]= $\dfrac{2x}{\text{RealAbs}[x]}$

The function is differentiable everywhere except at 0, since its left-hand and right-hand derivatives are not equal there:

In[·]:= **Limit[DifferenceQuotient[f[x], {x, h}], h → 0, Direction → 1] /. x → 0**

Out[·]= **−2**

In[·]:= **Limit[DifferenceQuotient[f[x], {x, h}], h → 0, Direction → −1] /. x → 0**

Out[·]= **2**

Plugging in 0 for the derivative also shows it to be undefined:

In[·]:= **f'[0] // Quiet**

Out[·]= **Indeterminate**

Non-differentiability 1

Where can a function be non-differentiable?

1. It will be non-differentiable at a **corner**:

In[ ]:= **D[3 RealAbs[x], x] /. x → 0 // Quiet**

Out[ ]= Indeterminate

In[ ]:= **Plot[{3 RealAbs[x], Evaluate[∂ₓ (3 RealAbs[x])]}, {x, −1, 1}, PlotLegends → {"function", "derivative"}]**

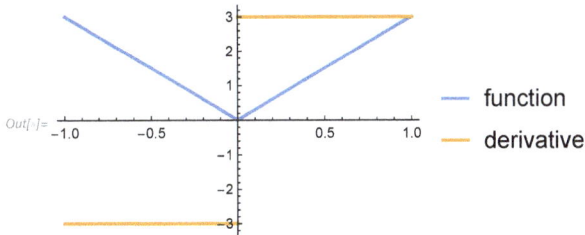

2. It will also be non-differentiable at a **cusp**:

In[ ]:= **D[CubeRoot[x]^2, x] /. x → 0 // Quiet**

Out[ ]= ComplexInfinity

In[ ]:= **Plot[{Surd[x, 3]^2, Evaluate[∂ₓ $\sqrt[3]{x}^2$]}, {x, −2, 2}, PlotLegends → {"function", "derivative"}]**

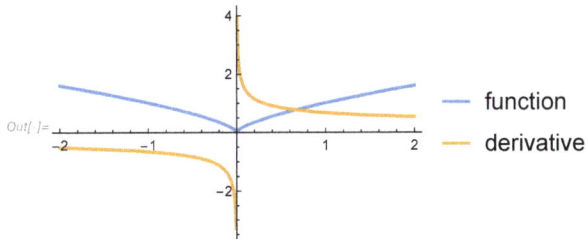

Non-differentiability 2

3. A function is non-differentiable at a **discontinuity**:

In[ ]:= **f[x] := Piecewise[{{x^2 + 1, x < 0}}, −x^2 − 1]**

In[ ]:= **D[f[x], x] /. x → 0 // Quiet**

Out[ ]= Indeterminate

In[·]:= **Plot[{Evaluate[f[x]], Evaluate[∂ₓ f[x]]}, {x, −2, 2},**
    PlotLegends → {"function", "derivative"}]

Out[·]=

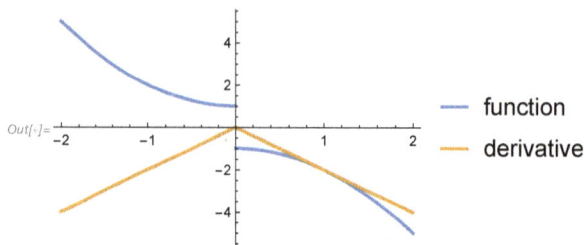

4. A function is non-differentiable at a **vertical tangent**:

In[·]:= **D[CubeRoot[x], x] /. x → 0 // Quiet**

Out[·]= **ComplexInfinity**

In[·]:= **Plot[{Surd[x, 3], Evaluate[∂ₓ $\sqrt[3]{x}$]}, {x, −2, 2},**
    PlotLegends → {"function", "derivative"}]

Out[·]=

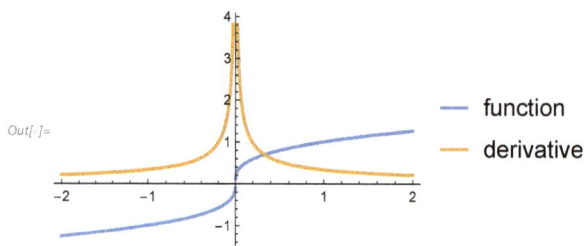

Higher Derivatives 1

Since the derivative of a function is also a function, it also has a derivative. It is denoted by $(f')' = f''$:

In[·]:= **Clear[f]**

In[·]:= **f[x_] := 4 x^3**

Here is the first derivative of the function:

In[·]:= **firstderivative = D[f[x], x]**

Out[·]= $12 x^2$

Here is the second derivative:

In[·]:= **secondderivative = D[firstderivative, x]**

Out[·]= $24 x$

You could also find the second derivative using the original function:

In[ ]:= **D[f[x], {x, 2}]**

Out[ ]:= 24 x

Here are the plots of the function, its first derivative and its second derivative:

In[ ]:= **Plot[{f[x], firstderivative, secondderivative}, {x, −1, 1},**
    PlotLegends → {"f[x]", "first derivative", "second derivative"}]

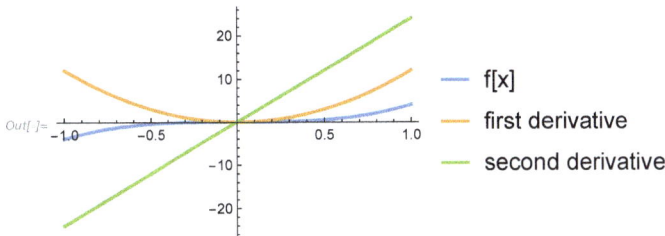

Higher Derivatives 2

The Wolfram Language can calculate the nth derivative of a function for any value of n:

In[ ]:= **f[x_] := x^6**

In[ ]:= **{d0, d1, d2, d3, d4, d5, d6} = Table[D[f[x], {x, i}], {i, 0, 6}]**

Out[ ]:= $\left\{x^6, 6\,x^5, 30\,x^4, 120\,x^3, 360\,x^2, 720\,x, 720\right\}$

Here are plots for the function and its first six derivatives:

In[ ]:= **Plot[{d0, d1, d2, d3, d4, d5, d6}, {x, −3, 3}, ⋯ ◆]**

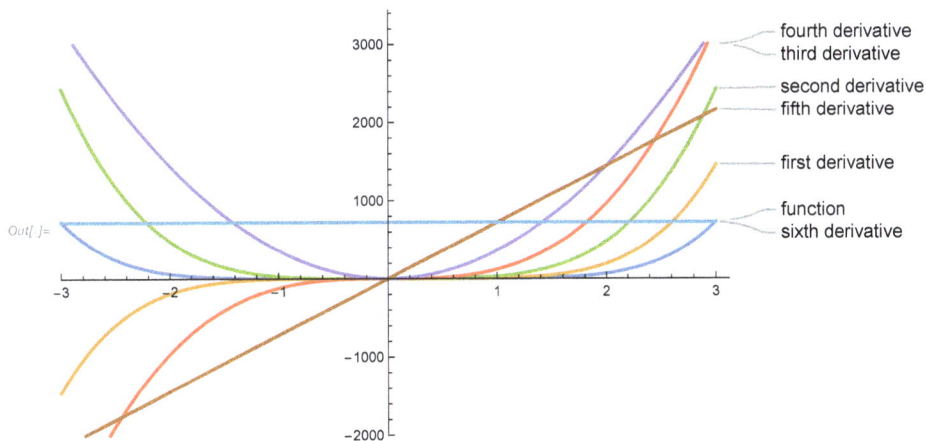

Application from Physics

The nth derivatives for a position function $s[t]$ are given special names:

In[ ]:= **s[_t__] := _t_^5 − 11 _t_ + Sin[3 _t_]/12**

The first six are velocity, acceleration, jerk, snap, crackle and pop:

In[ ]:= **{velocity, acceleration, jerk, snap, crackle, pop} = Table[D[s[t], {t, i}], {i, 6}];**

Here are their plots, along with position:

In[ ]:= **Plot[Evaluate[{s[t], velocity, acceleration, jerk, snap, crackle, pop}], {t, 0, 3}, ⋯ ＋]**

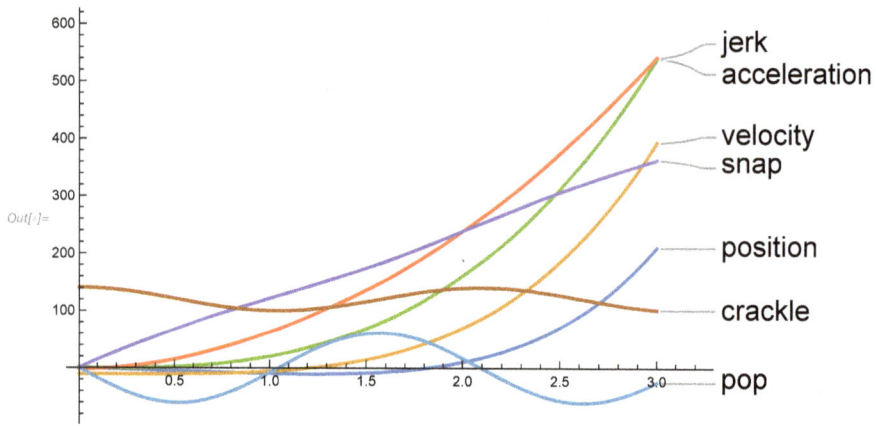

Summary

The derivative can be calculated for functions.

Since the derivative is a limit, you can calculate left and right derivatives.

If the left and right derivatives are not equal, then the derivative does not exist at the given point.

If the derivative does exist, then the function is said to be differentiable at that point.

A function can be non-differentiable in many different ways.

Derivatives of derivatives can be found, and are called higher-order derivatives.

Higher-order derivatives are useful when analyzing the motion of a particle.

The next lesson will cover rules to make differentiation by hand easier.

Exercises

Exercise 1—Derivative of a Function

Find the derivative of the function and plot it on a graph:

In[ ]:= **f[x_] := (1 − 3 x) / (2 + x)**

Solution

Find the derivative:

In[ ]:= **D[f[x], x]**

$$Out[\ ]=\ -\frac{1-3x}{(2+x)^2}\ -\ \frac{3}{2+x}$$

Then plot it on a graph:

In[ ]:= **Plot[{f[x], Evaluate[D[f[x], x]]}, {x, −5, 5}, PlotLegends → {"function", "derivative"}]**

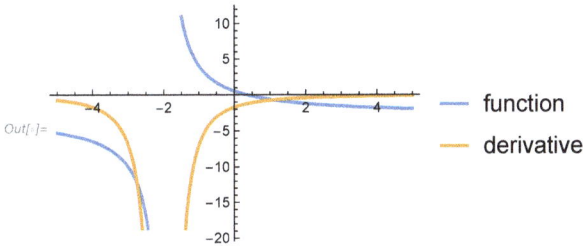

Exercise 2—Differentiability

Where is the following function differentiable? Characterize the places where it is not differentiable:

In[ ]:= **Clear[f]; f[x_] := 4 CubeRoot[x] + 2 RealAbs[x − 1] − 3 CubeRoot[x + 2]^2**

Solution

Plot the function and its derivative:

In[·]:= **Plot[{f[x], Evaluate[D[f[x], x]]}, {x, −5, 5}, PlotLegends → {"function", "derivative"}]**

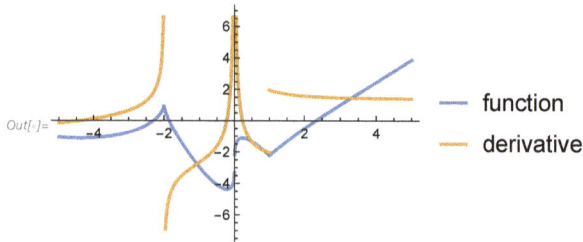

The function is differentiable everywhere except at 0, 1 and −2.

There is a vertical tangent at 0:

In[·]:= **f '[0] // Quiet**

Out[·]= ComplexInfinity

A corner at 1:

In[·]:= **f '[1] // Quiet**

Out[·]= Indeterminate

And a cusp at −2:

In[·]:= **f '[−2] // Quiet**

Out[·]= ComplexInfinity

Exercise 3—One-Sided Derivatives

Find the left- and right-hand derivatives of the following function at 10:

In[·]:= **f[x_] := CubeRoot[x − 10]**

Solution

Use Limit and DifferenceQuotient:

In[·]:= **Limit[DifferenceQuotient[f[x], {x, h}], h → 0, Direction → "FromBelow"] /. x → 10**

Out[·]= ∞

In[·]:= **Limit[DifferenceQuotient[f[x], {x, h}], h → 0, Direction → "FromAbove"] /. x → 10 // Quiet**

Out[·]= ComplexInfinity

You can see it has a vertical tangent at 10:

In[ ]:= **Plot[{f[x], Evaluate[D[f[x], x]]}, {x, 9, 11}, PlotLegends → {"function", "derivative"}]**

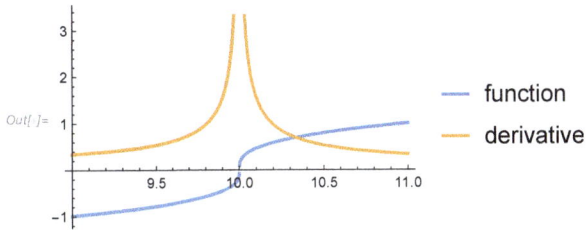

Out[ ]=

Exercise 4—Higher-Order Derivatives

Find and plot the fourth derivative of the following function:

In[ ]:= **f[x_] := 1 / x**

Solution

Use D:

In[ ]:= **D[f[x], {x, 4}]**

Out[ ]= $\dfrac{24}{x^5}$

Here is the plot:

In[ ]:= **Plot[{f[x], Evaluate[D[f[x], {x, 4}]]}, {x, −1, 1}, PlotLegends → "Expressions"]**

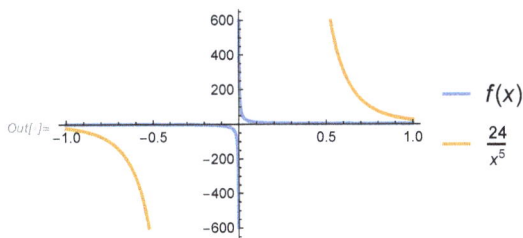

Out[ ]=

Exercise 5—An Application from Physics

Find (and plot) the acceleration, jerk and crackle of the following position function:

In[ ]:= **position[*t*_] := (5 − 4 *t*) / (3 + *t*)**

Solution

The acceleration is the second derivative, the jerk is the third derivative, and the crackle is the fifth derivative:

In[ ]:= **{accel, jerk, crackle} = {D[position[t], {t, 2}], D[position[t], {t, 3}], D[position[t], {t, 5}]}**

$$Out[ ]=\left\{\frac{2\,(5-4\,t)}{(3+t)^3}+\frac{8}{(3+t)^2},\ -\frac{6\,(5-4\,t)}{(3+t)^4}-\frac{24}{(3+t)^3},\ -\frac{120\,(5-4\,t)}{(3+t)^6}-\frac{480}{(3+t)^5}\right\}$$

Here are their plots, with the position function:

In[ ]:= **Plot[{position[t], accel, jerk, crackle}, {t, −5, −1},**
    PlotLegends → {"position", "acceleration", "jerk", "crackle"}]

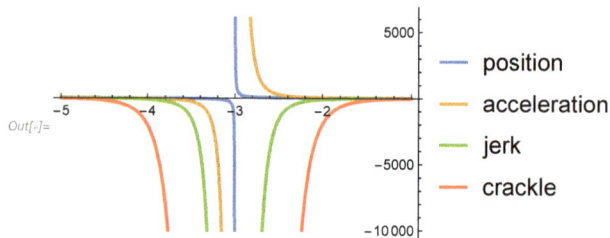

9 | Differentiation Formulas

Overview

Finding derivatives using the limit definition alone would be tedious and tricky for a lot of functions. Luckily, there are formulas for finding the derivatives of many (if not most) functions.

For example, the limit of a constant function is always 0:

```
In[·]:= f[x_] := c
```

```
In[·]:= D[f[x], x]
```

```
Out[·]= 0
```

You can see this in their graphs too:

```
In[·]:= Plot[{5, Evaluate[∂ₓ 5]}, {x, −5, 5}, ··· ◆]
```

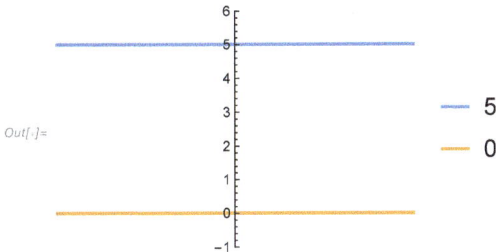

```
Out[·]=
```

This lesson will cover the basic differentiation formulas you should know.

Power Functions 1

If f has the form x^n, then its derivative is $n * x^{n-1}$:

```
In[·]:= f[x_] := x^n
```

```
In[·]:= D[f[x], x]
```

```
Out[·]= n x^{-1+n}
```

For example, for the quadratic function:

```
In[·]:= g[x_] := x^2
```

Its derivative is $2\,x^{2-1} = 2\,x$:

In[·]:= **D[g[x], x]**

Out[·]= **2 x**

Here are their plots:

In[·]:= **Plot[{g[x], Evaluate[∂ₓ g[x]]}, {x, −1, 1},**
    PlotLegends → "Expressions"]

Out[·]=

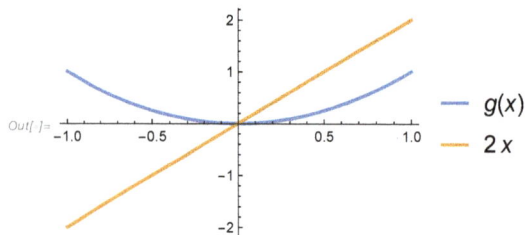

Power Functions 2

n can be **positive or negative**:

In[·]:= **f[x_] := x^−1**

The derivative of x^{-1} is $-1\,x^{-1-1} = -x^{-2}$:

In[·]:= **D[f[x], x]**

$$Out[·]= -\frac{1}{x^2}$$

Here is a plot of the function and its derivative:

In[·]:= **Plot[{f[x], Evaluate[∂ₓ f[x]]}, {x, −1, 1},**
    PlotLegends → "Expressions"]

Out[·]=

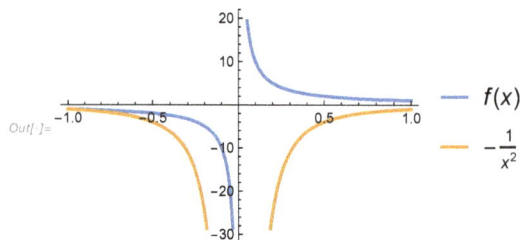

n does not need to be an integer either:

In[ ]:= **g[x_] := x^-1.3**

In[ ]:= **D[g[x], x]**

Out[ ]= $-\dfrac{1.3}{x^{2.3}}$

The Constant Multiple Rule

If c is a **constant** and f is a function:

In[ ]:= **f[x_] := 5 x^3**

Then the derivative of $c * f$ is $c * f\,'$:

In[ ]:= **{D[f[x], x], D[f[x], x] == 5 * 3 x^2}**

Out[ ]= $\{15\,x^2,\ \text{True}\}$

Here is a plot of $5\,x^3$ and its derivative:

In[ ]:= **Plot[{f[x], Evaluate[∂ₓ f[x]]}, {x, -1, 1},**
   PlotLegends → "Expressions"]

Out[ ]=

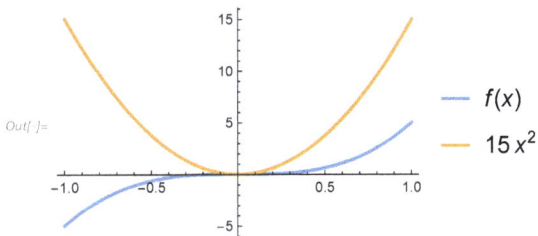

The Sum Rule

The derivative of the sum of two functions f and g...:

In[ ]:= **f[x_] := 2 x**

In[ ]:= **g[x_] := 3 x^2**

In[ ]:= **sum[x_] := f[x] + g[x]**

... is the sum of their derivatives, $f\,' + g\,'$:

In[ ]:= **{f'[x], g'[x]}**

Out[ ]= $\{2,\ 6\,x\}$

In[•]:= **sum '[x]**

Out[•]:= $2 + 6 x$

Here are the plots of the derivatives:

In[•]:= **Plot[{f '[x], g'[x], sum '[x]}, {x, −1, 1}, PlotLegends → "Expressions"]**

Out[•]:=

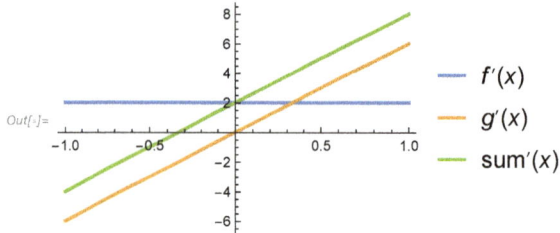

The Difference Rule

The derivative of the difference of two functions f and g...:

In[•]:= **f[x_] := 1 / x**

In[•]:= **g[x_] := 4 x^4**

In[•]:= **difference[x_] := f[x] − g[x]**

... is the difference of the derivatives of the functions, $f' − g'$:

In[•]:= **{f '[x], g'[x]}**

Out[•]:= $\left\{ -\dfrac{1}{x^2},\ 16 x^3 \right\}$

In[•]:= **difference '[x]**

Out[•]:= $-\dfrac{1}{x^2} - 16 x^3$

Here are the plots of the derivatives:

In[•]:= **Plot[{f '[x], g'[x], difference '[x]}, {x, −1, 1}, PlotLegends → "Expressions"]**

Out[•]:=

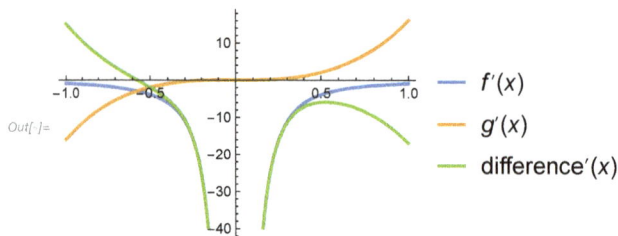

Product Rule

The derivative of the product of functions f and g...:

```
In[ ]:= f[x_] := Sqrt[x]
       g[x_] := x^2
       product[x_] := f[x] * g[x]
```

... is NOT the product of the derivatives, $f' * g'$:

```
In[ ]:= f'[x] * g'[x]
```
$$Out[ ]= \sqrt{x}$$

```
In[ ]:= product'[x]
```
$$Out[ ]= \frac{5 x^{3/2}}{2}$$

It is $f\, g' + g\, f'$:

```
In[ ]:= f[x] g'[x] + g[x] f'[x]
```
$$Out[ ]= \frac{5 x^{3/2}}{2}$$

Here are the plots of the derivatives:

```
In[ ]:= Plot[{f'[x], g'[x], product'[x]}, {x, −1, 1}, PlotLegends → "Expressions"]
```

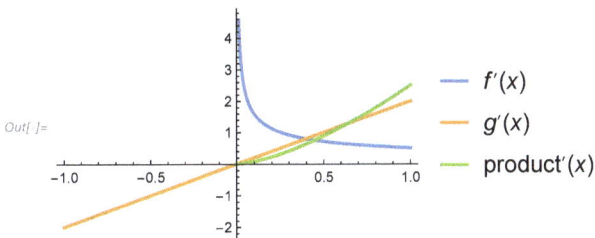

Quotient Rule

The derivative of the quotient of functions f and g...:

```
In[ ]:= f[x_] := 3 x
       g[x_] := 4 x^3
       quotient[x_] := f[x] / g[x]
```

... is NOT the quotient of the derivatives, $\frac{f'}{g'}$:

In[·]:= **f'[x]/g'[x]**

Out[·]= $\dfrac{1}{4\,x^2}$

In[·]:= **quotient'[x]**

Out[·]= $-\dfrac{3}{2\,x^3}$

It is $\frac{g\,f'-f\,g'}{g^2}$:

In[·]:= **(g[x] f'[x] − f[x] g'[x])/g[x]^2**

Out[·]= $-\dfrac{3}{2\,x^3}$

Here are the plots of the derivatives:

In[·]:= **Plot[{f'[x], g'[x], quotient'[x]}, {x, −1, 1}, PlotLegends → "Expressions"]**

Out[·]=

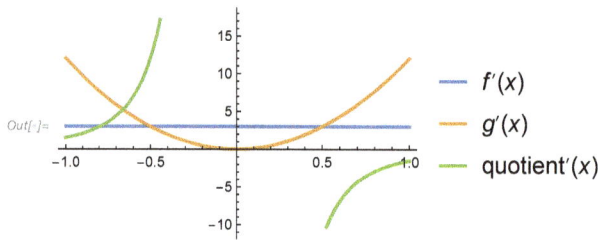

— $f'(x)$
— $g'(x)$
— quotient'(x)

Summary

The derivative is a useful tool, and differentiation rules help find derivatives.

There are rules for sums, differences, products, quotients and more.

The power rule is especially useful for finding derivatives of power functions.

The next lesson will go over the differentiation rules for trigonometric functions.

Exercises

Exercise 1—Constant Multiple and Constants

1) Find the derivative of the following function:

In[ ]:= **f[x_] := 2^100**

2) For the unknown function g, given that $g[5] = 20$ and $g'[5] = 7$, find $(4\,g[5])'$.

Solution

1) Since f is a constant function, its derivative is 0:

In[ ]:= **f'[x]**

Out[ ]= **0**

2) Define the known properties of g:

In[ ]:= **Clear[g]**

In[ ]:= **g[5] = 20;**
  g'[5] = 7;

By the constant multiple rule, $(4\,g[5])' = 4\,g'[5] = 28$:

In[ ]:= **4 g'[5]**

Out[ ]= **28**

Exercise 2—Sum and Difference

Given the unknown functions f and g, with $f'[3] = 4$ and $g'[3] = 2$, find $(f + g)'[3]$ and $(f - g)'[3]$.

Solution

We define:

In[ ]:= **Clear[f, g]**

In[ ]:= **f'[3] = 4; g'[3] = 2**

Out[ ]= **2**

By the sum rule, $(f + g)'[3] = f'[3] + g'[3] = 6$:

In[·]:= **f'[3] + g'[3]**

Out[·]:= **6**

By the difference rule, $(f - g)'[3] = f'[3] - g'[3] = 2$:

In[·]:= **f'[3] - g'[3]**

Out[·]:= **2**

Exercise 3—Product and Quotient

Given the unknown function f with $f[8] = 3$ and $f'[8] = 7$...:

In[·]:= **Clear[f]**

In[·]:= **f[8] = 3;**
    f'[8] = 7;

... and the unknown function g with $g[8] = 8$ and $g'[8] = 5$:

In[·]:= **Clear[g]**

In[·]:= **g[8] = 8;**
    g'[8] = 5;

Find $(f * g)'[8]$ and $\left(\frac{f}{g}\right)'[8]$.

Solution

By the product rule, $(f * g)'[8] = f[8]\, g'[8] + g[8]\, f'[8] = 71$:

In[·]:= **f[8] g'[8] + g[8] f'[8]**

Out[·]:= **71**

By the quotient rule, $\left(\frac{f}{g}\right)'[8] = \frac{g[8]\, f'[8] - f[8]\, g'[8]}{g[8]^2} = \frac{41}{64}$:

In[·]:= **(g[8] f'[8] - f[8] g'[8]) / g[8]^2**

Out[·]:= $\dfrac{41}{64}$

Exercise 4—General Functions

Given the unknown function f:

In[·]:= **Clear[f]**

Find the derivative of $x^3\, f[x]$ and $\dfrac{f(x)}{x^3}$.

Solution

For the first expression, use the product rule:

In[ ]:= **f[x] * D[x^3, x] + x^3 * D[f[x], x]**

Out[ ]= $3 x^2 f[x] + x^3 f'[x]$

Confirm with D:

In[ ]:= **D[x^3 f[x], x]**

Out[ ]= $3 x^2 f[x] + x^3 f'[x]$

For the second expression, use the quotient rule:

In[ ]:= **(x^3 * D[f[x], x] − f[x] * D[x^3, x]) / ((x^3)^2) // Simplify**

Out[ ]= $\dfrac{-3 f[x] + x f'[x]}{x^4}$

Confirm with D:

In[ ]:= **D[f[x] / x^3, x] // Simplify**

Out[ ]= $\dfrac{-3 f[x] + x f'[x]}{x^4}$

Exercise 5—Witch of Maria Agnesi

The following function is known as a **witch of Maria Agnesi**:

In[ ]:= **f[x_] := 2 / (2 + x^2)**

It is named after the first female math professor at a university in the Western world, Maria Gaetana Agnesi.

Find the tangent line to the curve at the point $(-1, 2/3)$ using the quotient rule.

Solution

Use the quotient rule to find the slope at -1:

$\textit{In[\cdot]:=}$ `((2 + x^2) * D[2, x] - 2 * D[2 + x^2, x]) / ((2 + x^2)^2) /. x → -1`

$\textit{Out[\cdot]=}$ $\dfrac{4}{9}$

Confirm with D:

$\textit{In[\cdot]:=}$ `D[f[x], x] /. x → -1`

$\textit{Out[\cdot]=}$ $\dfrac{4}{9}$

Now plot the tangent line and the curve using point-slope form:

$\textit{In[\cdot]:=}$ `Plot[{f[x], f[-1] + f'[-1] (x + 1)}, {x, -2, 2}, Epilog → {PointSize[Large], Point[{-1, f[-1]}]}]`

$\textit{Out[\cdot]=}$
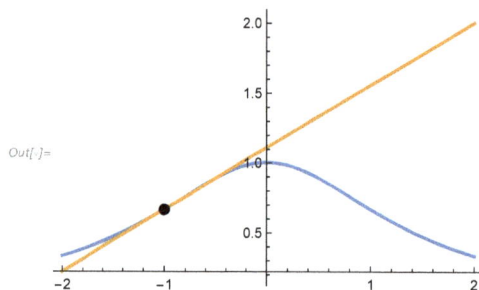

10 | Derivatives of Trigonometric Functions

Overview

The derivatives of trigonometric functions are very useful. By nature, trigonometric functions are oscillatory, so you can expect their derivatives to be oscillatory too.

Consider the plot of this trigonometric function and its derivative, for example:

In[ ]:= `f[x_] := Sin[3 x] + Cos[2 x]`

In[ ]:= `Plot[{f[x], f'[x]}, {x, -5, 5}, PlotLegends → "Expressions"]`

Out[ ]=

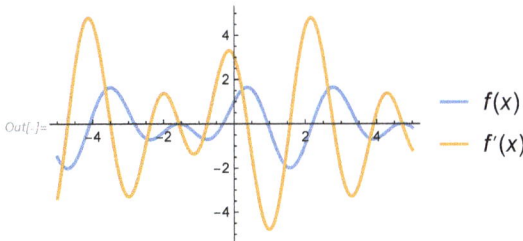

The goal of this lesson is to develop a table of derivatives for the sine, cosine and other trigonometric functions.

Sine

Consider the sine function:

In[ ]:= `f[x_] := Sin[x]`

Using Limit and DifferenceQuotient, you can find its derivative to be the cosine:

In[ ]:= `Limit[DifferenceQuotient[Sin[x], {x, h}], h → 0]`

Out[ ]= `Cos[x]`

Confirm with D:

In[ ]:= `D[f[x], x]`

Out[ ]= `Cos[x]`

Here is a plot of the sine and its derivative the cosine:

In[·]:= **Plot[{f[x], f'[x]}, {x, −5, 5}, PlotLegends → {"Sin[x]", "Cos[x]"}]**

Out[·]=

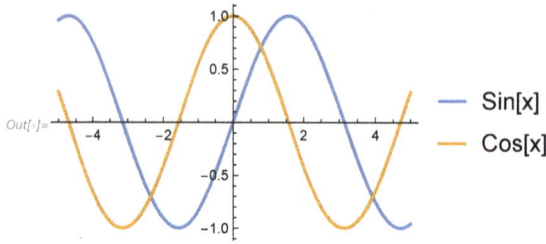

Cosine

Consider the cosine function:

In[·]:= **f[x_] := Cos[x]**

Using Limit and DifferenceQuotient, you can find its derivative to be negative of the sine:

In[·]:= **Limit[DifferenceQuotient[f[x], {x, h}], h → 0]**

Out[·]= **−Sin[x]**

Confirm with D:

In[·]:= **D[f[x], x]**

Out[·]= **−Sin[x]**

Here is a plot of the cosine and its derivative negative of the sine:

In[·]:= **Plot[{f[x], f'[x]}, {x, −5, 5}, PlotLegends → {"Cos[x]", "−Sin[x]"}]**

Out[·]=

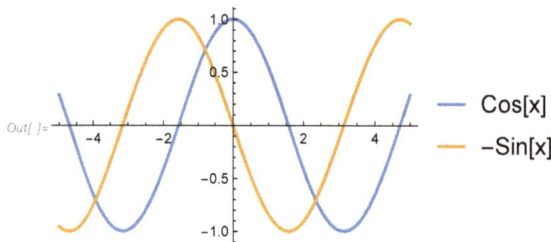

Tangent

Consider the tangent function:

In[ ]:= **f[x_] := Tan[x]**

Using Limit and DifferenceQuotient, you can find its derivative to be $\text{Sec}^2[x]$:

In[ ]:= **Limit[DifferenceQuotient[f[x], {x, h}], h → 0]**

Out[ ]= $\text{Sec}[x]^2$

Confirm with D:

In[ ]:= **D[f[x], x]**

Out[ ]= $\text{Sec}[x]^2$

Here is a plot of tangent and its derivative $\text{Sec}^2[x]$:

In[ ]:= **Plot[{f[x], f'[x]}, {x, −5, 5}, PlotLegends → {"Tan[x]", "Sec^2[x]"}]**

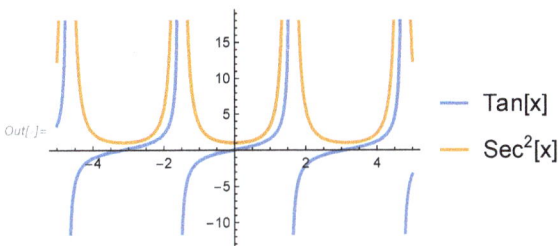

Secant

Consider the secant function:

In[ ]:= **f[x_] := Sec[x]**

Using Limit and DifferenceQuotient, you can find its derivative to be $\text{Sec}[x]\,\text{Tan}[x]$:

In[ ]:= **Limit[DifferenceQuotient[f[x], {x, h}], h → 0]**

Out[ ]= $\text{Sec}[x]\,\text{Tan}[x]$

Confirm with D:

In[ ]:= **D[f[x], x]**

Out[ ]= $\text{Sec}[x]\,\text{Tan}[x]$

Here is a plot of the secant and its derivative Sec[*x*] Tan[*x*]:

In[ ]:= **Plot[{f[x], f'[x]}, {x, −5, 5}, PlotLegends → {"Sec[x]", "Sec[x]Tan[x]"}]**

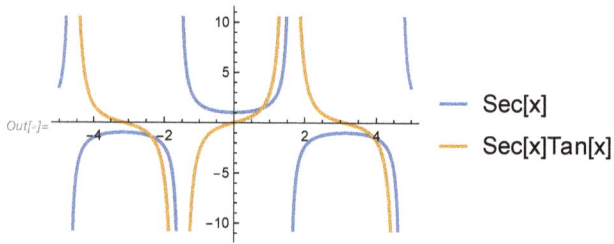

Cosecant

Consider the cosecant function:

In[ ]:= **f[x_] := Csc[x]**

Using Limit and DifferenceQuotient, you can find its derivative to be −Csc[*x*] Cot[*x*]:

In[ ]:= **Limit[DifferenceQuotient[f[x], {x, h}], h → 0]**

Out[ ]= −Cot[x] Csc[x]

Confirm with D:

In[ ]:= **D[f[x], x]**

Out[ ]= −Cot[x] Csc[x]

Here is a plot of the cosecant and its derivative −Csc[*x*] Cot[*x*]:

In[ ]:= **Plot[{f[x], f'[x]}, {x, −5, 5}, PlotLegends → {"Csc[x]", "−Csc[x]Cot[x]"}]**

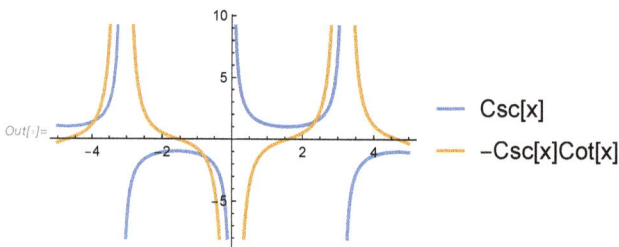

Cotangent

Consider the cotangent function:

In[ ]:= **f[x_] := Cot[x]**

Using Limit and DifferenceQuotient, you can find its derivative to be $-\text{Csc}^2[x]$:

In[ ]:= **Limit[DifferenceQuotient[f[x], {x, h}], h → 0]**

Out[ ]= $-\text{Csc}[x]^2$

Confirm with D:

In[ ]:= **D[f[x], x]**

Out[ ]= $-\text{Csc}[x]^2$

Here is a plot of the cotangent and its derivative $-\text{Csc}^2[x]$:

In[ ]:= **Plot[{f[x], f'[x]}, {x, −5, 5}, PlotLegends → {"Cot[x]", "−Csc^2[x]"}]**

Out[ ]=

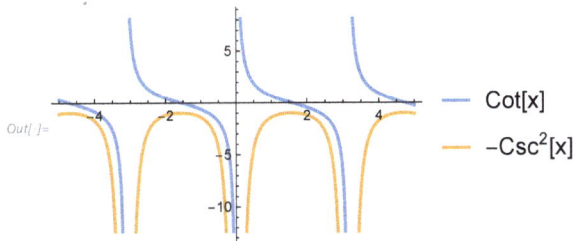

Table of Derivatives

Here is a list of all the basic trigonometric functions and their derivatives:

In[ ]:= **triglist = {Sin[x], Cos[x], Tan[x], Sec[x], Csc[x], Cot[x]};**

In[ ]:= **TableForm[Table[{triglist[[i]], D[triglist[[i]], x]}, {i, 1, 6}],**
   TableHeadings → {None, {"f[x]", "f'[x]"}}]

f[x]	f'[x]
Sin[x]	Cos[x]
Cos[x]	−Sin[x]
Tan[x]	Sec[x]^2
Sec[x]	Sec[x] Tan[x]
Csc[x]	−Cot[x] Csc[x]
Cot[x]	−Csc[x]^2

Out[ ]=

Here are their plots again:

In[·]:= **Table[Plot[{triglist⟦i⟧, Evaluate[∂ₓ triglist⟦i⟧]}, {x, −5,**
    5}, PlotLegends → {Part[triglist, i], D[Part[triglist, i], x]}], {i, 1, 6}]

Out[·]= {

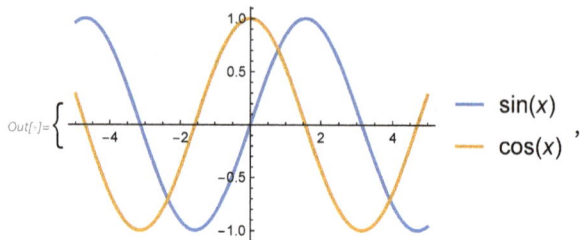

— sin(x)
— cos(x)

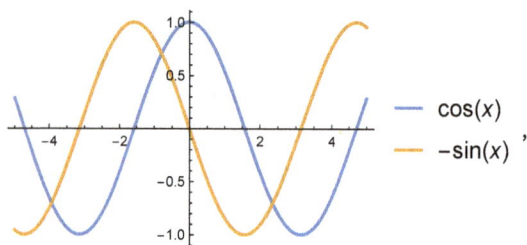

— cos(x)
— −sin(x)

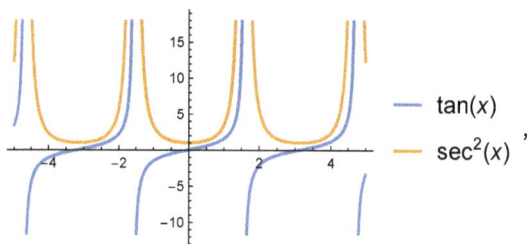

— tan(x)
— sec²(x)

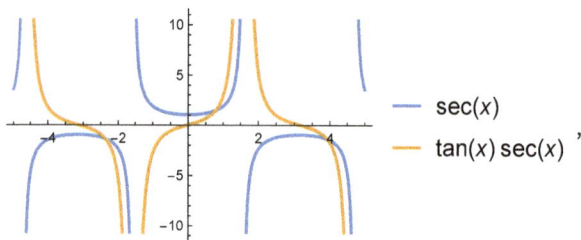

— sec(x)
— tan(x) sec(x)

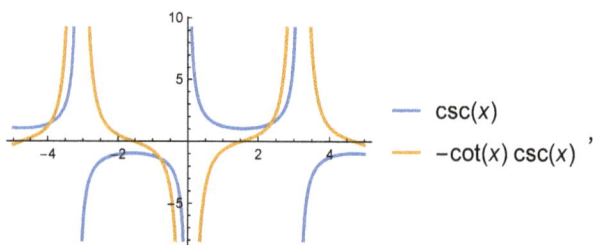

— csc(x)
— −cot(x) csc(x)

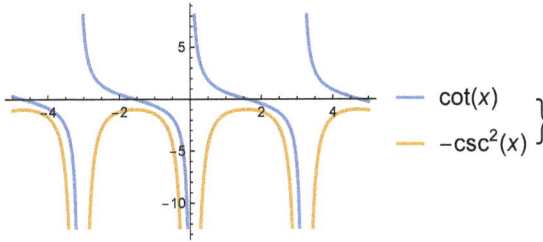

Higher-Order Derivatives

Due to the cyclical nature of the trigonometric functions, taking their higher derivatives can lead to interesting patterns.

For example, the fourth derivative of the sine is the sine, so the pattern repeats:

In[ ]:= **sinderivativelist = Table[D[Sin[x], {x, i}], {i, 0, 4}]**

Out[ ]= {Sin[x], Cos[x], −Sin[x], −Cos[x], Sin[x]}

Therefore, the 1001th derivative of the sine would just be the cosine (1001 divided by
4 gives remainder 1, and the first derivative of the sine is the cosine):

In[ ]:= **D[Sin[x], {x, 1001}]**

Out[ ]= Cos[x]

D can symbolically find the nth derivative of the sine, and the following expression can
be confirmed with the standard trigonometry laws:

In[ ]:= **D[Sin[x], {x, n}]**

Out[ ]= $\text{Sin}\left[\dfrac{n\,\pi}{2} + x\right]$

Here is a plot of the first four derivatives. Notice that the fourth derivative overlaps with the sine function:

```
In[·]:= Plot[sinderivativelist, {x, –5, 5},
         PlotLabels → (Style[♯, Large] & /@ {"Sine", "Cosine", "–Sine", "–Cosine", "Sine"})]
```

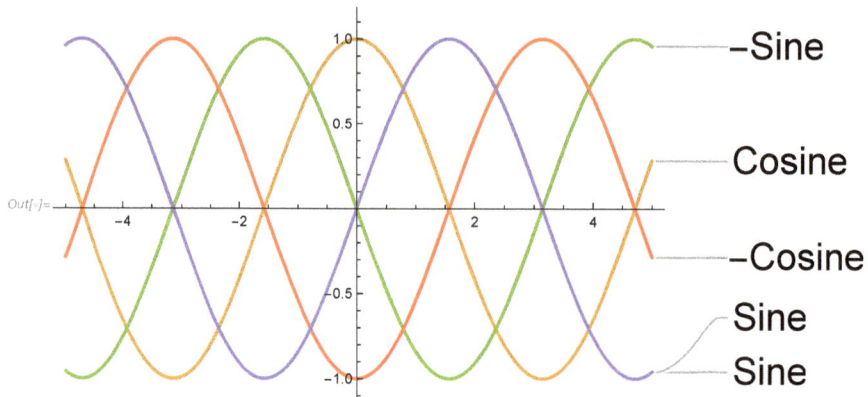

Summary

The periodic trigonometric functions have periodic derivatives.

The derivative of sine is cosine.

The derivative of cosine is negative sine.

The derivative of tangent is secant squared.

The derivative of secant is secant times tangent.

The derivative of cosecant is negative cosecant times cotangent.

The derivative of cotangent is negative cosecant squared.

The next lesson will cover the chain rule, which is an easy way to find the derivatives of compositions of functions.

Exercises

Exercise 1—Calculate a Derivative

Find the derivative of the following function:

In[ ]:= **f[x_] := x Sin[x] − Sec[x]**

Solution

Use the trig rules, the difference rule and the product rule:

In[ ]:= **D[x, x] Sin[x] + D[Sin[x], x] x − D[Sec[x], x]**

Out[ ]= x Cos[x] + Sin[x] − Sec[x] Tan[x]

D confirms:

In[ ]:= **D[f[x], x]**

Out[ ]= x Cos[x] + Sin[x] − Sec[x] Tan[x]

Here is a plot of the function and its derivative:

In[ ]:= **Plot[{f[x], f'[x]}, {x, −5, 5}, PlotLegends → {"function", "derivative"}]**

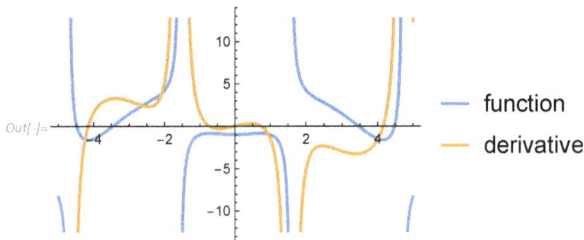

Exercise 2—Find the Tangent Line to a Curve

Find the tangent line to the following function at the point $(\pi/3, \pi/2)$:

In[ ]:= **f[x_] := 3 x Cos[x]**

Solution

Find the derivative at $\frac{\pi}{3}$:

In[ ]:= **f'[π/3]**

Out[ ]= $\dfrac{3}{2} - \dfrac{\sqrt{3}\,\pi}{2}$

Use point-slope form to find the equation of the tangent line:

In[ ]:= **tangent[x_] := f[π/3] + f'[π/3] (x − π/3)**

Plot the function and the tangent line:

In[ ]:= **Plot[{f[x], tangent[x]}, {x, −π, π}, Epilog → {PointSize[Large], Point[{π/3, f[π/3]}]}]**

Out[ ]=

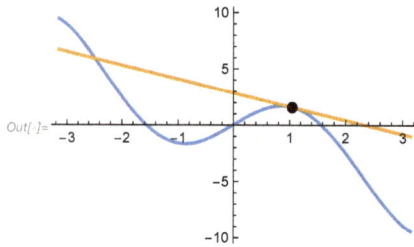

Exercise 3—Horizontal Tangents

Find the places where the following function has a horizontal tangent in the range −5 to 5:

In[ ]:= **f[x_] := Sin[x]/(2 + Cos[x])**

Solution

The function will have horizontal tangents wherever its slope is 0:

In[ ]:= **sol = Solve[f'[x] == 0 && −5 ≤ x ≤ 5, x]**

Out[ ]= $\left\{ \left\{ x \to -\dfrac{4\pi}{3} \right\}, \left\{ x \to -\dfrac{2\pi}{3} \right\}, \left\{ x \to \dfrac{2\pi}{3} \right\}, \left\{ x \to \dfrac{4\pi}{3} \right\} \right\}$

Here is a plot of the function:

In[ ]:= **Plot[f[x], {x, −5, 5}, Epilog → {PointSize[Large], Point[{x, f[x]} /. sol]}]**

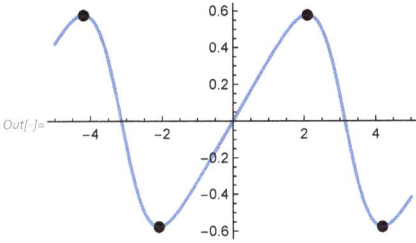

Out[ ]=

Exercise 4—Spring

A spring is hung on a hook and a mass is hung on the lower end. When the mass is pulled downward and then released, its position can be modeled by the following function:

In[ ]:= **spring[t_] := 3 Cos[t] + 2 Sin[t]**

Find when its speed is greatest and when it is furthest from the equilibrium point in the range $0 \leq t \leq 10$.

Solution

The speed will be greatest wherever the derivative of velocity, acceleration, is 0:

In[ ]:= **sol = Solve[{spring''[t] == 0, 0 ≤ t ≤ 10}, t] // N**

Out[ ]= **{{t → 5.30039}, {t → 2.1588}, {t → 8.44198}}**

This is also where the position of the spring is 0:

In[ ]:= **Solve[spring[t] == 0 && 0 ≤ t ≤ 10] // N**

Out[ ]= **{{t → 5.30039}, {t → 2.1588}, {t → 8.44198}}**

It is furthest from the equilibrium point wherever the velocity is 0:

In[ ]:= **sol1 = Solve[{spring'[t] == 0, 0 ≤ t ≤ 10}, t] // N**

Out[ ]= **{{t → 3.7296}, {t → 0.588003}, {t → 6.87119}}**

Here is a plot of the position, velocity and acceleration, with the places of greatest speed and 0 speed found in the preceding:

In[ ]:= **Plot[{spring[t], spring '[t], spring ''[t]}, {t, 0, 10}, ⸬ +]**

Out[ ]=

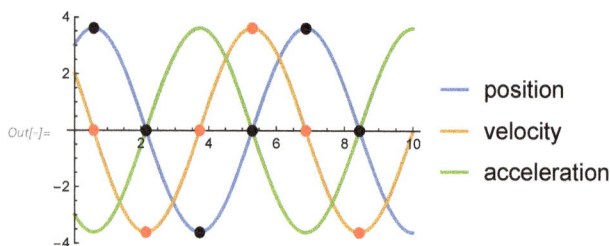

— position
— velocity
— acceleration

Exercise 5—Friction

A rope makes an angle θ with a horizontal plane and drags an object with weight W. The magnitude of the force is given by:

In[ ]:= **force[θ_] := μ w / (μ Sin[θ] + Cos[θ])**

Here μ is known as the **coefficient of friction**.

Suppose the weight is 100 lbs and the coefficient of friction is 0.4:

In[ ]:= **w = 100; μ = 0.4;**

Find when the rate of change of the force equals 0 in the range $0 \le \theta \le 2\pi$. Then plot the corresponding points against the graph of the force function.

Solution

To find when the rate of change is 0, use Solve:

In[ ]:= **sol = Solve[force'[θ] == 0 && 0 $\le$ θ < 2 π, θ] // Quiet**

Out[ ]= **{{$\theta \to$ 0.380506}, {$\theta \to$ 3.5221}}**

Then plot the corresponding points against the graph of the function with Plot:

In[ ]:= **Plot[force[θ], {θ, 0, 2 π}, Epilog $\to$ {PointSize[Large], Point[{θ, force[θ]} /. sol]}]**

Out[ ]=

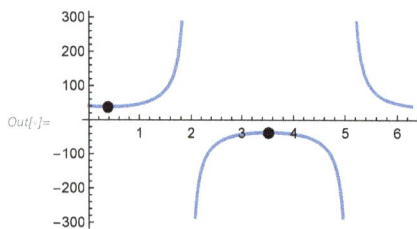

11 | The Chain Rule

Overview

Recall that composing the functions f and g to form $f \circ g$ will take values from g and plug them into f:

In[ ]:= **f[x_] := x^2**

In[ ]:= **g[x_] := Sin[x]**

In[ ]:= **f[g[x]]**

Out[ ]= $\text{Sin}[x]^2$

The rules discussed so far cannot be used to find the derivatives of compositions:

In[ ]:= **D[f[g[x]], x]**

Out[ ]= 2 Cos[x] Sin[x]

This issue can be resolved with the **chain rule**. This lesson introduces the chain rule and shows how to use it in various examples.

Chain Rule

The chain rule says that if f is differentiable at $g[x]$, and g is differentiable at x, then $f[g[x]]$ is differentiable at x and is equal to $f'[g[x]] * g'[x]$:

In[ ]:= **Clear[f, g]**

In[ ]:= **D[f[g[x]], x]**

Out[ ]= $f'[g[x]] \, g'[x]$

So the derivative of $\text{Sin}^2[x]$ is:

In[ ]:= **(D[x^2, x] /. x → Sin[x]) * D[Sin[x], x]**

Out[ ]= 2 Cos[x] Sin[x]

Here is a plot of $\mathrm{Sin}^2[x]$ and its derivative:

$In[\cdot]:=$ **Plot[{Sin[x]^2, Evaluate[∂_xSin[x]^2]}, {x, $-\pi$, π}, $\cdots$ $+$]**

$Out[\cdot]=$

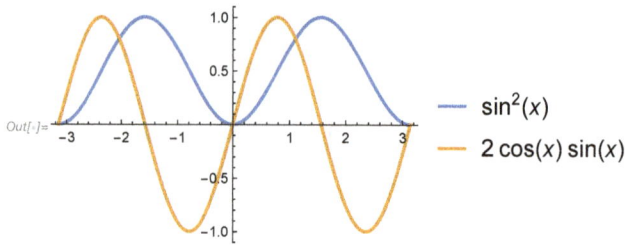

— $\sin^2(x)$

— $2\cos(x)\sin(x)$

Algebraic Function

Find the derivative of the following function:

$In[\cdot]:=$ **f[x_] := Sqrt[x^2 − 1]**

It is a composition of the functions $g[x] = \sqrt{x}$ and $h[x] = x^2 - 1$:

$In[\cdot]:=$ **g[x_] := Sqrt[x]**
h[x_] := x^2 − 1
g[h[x]]

$Out[\cdot]=$ $\sqrt{-1 + x^2}$

By the chain rule, its derivative is:

$In[\cdot]:=$ **Sqrt'[x^2 − 1] * 2 x**

$Out[\cdot]=$ $\dfrac{x}{\sqrt{-1 + x^2}}$

D agrees:

$In[\cdot]:=$ **D[f[x], x]**

$Out[\cdot]=$ $\dfrac{x}{\sqrt{-1 + x^2}}$

Here is a plot of the function and its derivative:

In[ ]:= **Plot[{f[x], f'[x]}, {x, −1, 4}, PlotLegends → "Expressions"]**

Out[ ]=

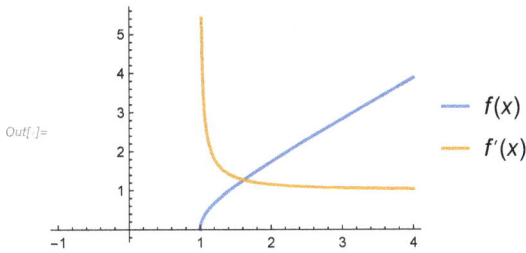

— $f(x)$
— $f'(x)$

Trig Function

Find the derivative of the following function:

In[ ]:= **f[x_] := Cos[x^2]**

It is a composition of the functions $g[x] = \text{Cos}[x]$ and $h[x] = x^2$:

In[ ]:= **g[x_] := Cos[x]**
h[x_] := x^2
g[h[x]]

Out[ ]= $\text{Cos}[x^2]$

Using the chain rule, its derivative is:

In[ ]:= **−Sin[x^2] * 2 x**

Out[ ]= $-2 \times \text{Sin}[x^2]$

D agrees:

In[ ]:= **D[f[x], x]**

Out[ ]= $-2 \times \text{Sin}[x^2]$

Here is a plot of the function and its derivative:

In[ ]:= **Plot[{f[x], f'[x]}, {x, −1, 4}, PlotLegends → "Expressions", PlotRange → All]**

Out[ ]=

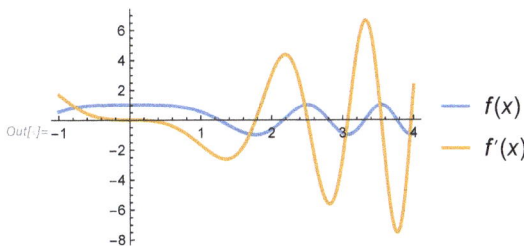

— $f(x)$
— $f'(x)$

Power Rule and Chain Rule

If you let the outer function be x^n:

In[ ]:= **f[x_] := x^n**

Then you can find the derivative of any power of a function g:

In[ ]:= **Clear[g]**

In[ ]:= **D[f[g[x]], x]**

Out[ ]= $n \, g[x]^{-1+n} \, g'[x]$

For example, the derivative of $(x^4 + 1)^{100}$ is easily found:

In[ ]:= **D[(x^4 + 1)^100, x]**

Out[ ]= $400 \, x^3 \left(1 + x^4\right)^{99}$

It also works for fractions:

In[ ]:= **D[(Sin[x])^(5/4), x]**

Out[ ]= $\dfrac{5}{4} \, Cos[x] \, Sin[x]^{1/4}$

As well as negative numbers:

In[ ]:= **D[(Cos[x] – x^2)^–55, x]**

Out[ ]= $-\dfrac{55 \, (-2 \, x - Sin[x])}{\left(-x^2 + Cos[x]\right)^{56}}$

Product Rule and Chain Rule

Find the derivative of the following function:

In[ ]:= **f[x_] := (3 x – 2)^3 (9 x^4 – x + 3)^5**

You *could* expand the function and find the derivative of each term separately:

In[ ]:= **Expand[f[x]]**

Out[ ]= $-1944 + 11\,988 \, x - 29\,862 \, x^2 + 38\,871 \, x^3 - 58\,035 \, x^4 + 182\,798 \, x^5 - 394\,506 \, x^6 +$
$453\,114 \, x^7 - 457\,227 \, x^8 + 1\,058\,670 \, x^9 - 2\,043\,630 \, x^{10} + 2\,041\,605 \, x^{11} - 1\,538\,190 \, x^{12} +$
$2\,952\,450 \, x^{13} - 5\,197\,770 \, x^{14} + 4\,395\,870 \, x^{15} - 2\,361\,960 \, x^{16} + 4\,002\,210 \, x^{17} -$
$6\,495\,390 \, x^{18} + 4\,428\,675 \, x^{19} - 1\,358\,127 \, x^{20} + 2\,125\,764 \, x^{21} - 3\,188\,646 \, x^{22} + 1\,594\,323 \, x^{23}$

In[ ]:= **long = D[%, x];**

Or you could find the derivative of each factor separately and put them together with the product rule:

In[ ]:= **(3 x − 2)^3 ∗ D[(9 x^4 − x + 3)^5, x] + (9 x^4 − x + 3)^5 ∗ D[(3 x − 2)^3, x]**

Out[ ]= $5 (-2 + 3 x)^3 (-1 + 36 x^3) (3 - x + 9 x^4)^4 + 9 (-2 + 3 x)^2 (3 - x + 9 x^4)^5$

In[ ]:= **Simplify[long == %]**

Out[ ]= **True**

D prefers the latter (and you should too!):

In[ ]:= **D[f[x], x]**

Out[ ]= $5 (-2 + 3 x)^3 (-1 + 36 x^3) (3 - x + 9 x^4)^4 + 9 (-2 + 3 x)^2 (3 - x + 9 x^4)^5$

Multiple Compositions

When using the chain rule, you start with the outer function and work your way inward as much as possible:

In[ ]:= **Clear[f, g]**

In[ ]:= **D[f[g[x]], x]**

Out[ ]= $f'[g[x]] g'[x]$

This extends to even more compositions:

In[ ]:= **Clear[h]**

In[ ]:= **D[f[g[h[x]]], x]**

Out[ ]= $f'[g[h[x]]] g'[h[x]] h'[x]$

Consider the function $\text{Sin}[\text{Cos}[x^2]]$, for example:

In[ ]:= **f[x_] := Sin[Cos[x^2]]**

Its derivative is $\text{Cos}[\text{Cos}[x^2]] * -\text{Sin}[x^2] * 2\ x$:

In[ ]:= **Cos[Cos[x^2]] ∗ −Sin[x^2] ∗ 2 x**

Out[ ]= $-2 x \text{Cos}[\text{Cos}[x^2]] \text{Sin}[x^2]$

Confirm with D:

In[ ]:= **D[f[x], x]**

Out[ ]= $-2 x \text{Cos}[\text{Cos}[x^2]] \text{Sin}[x^2]$

A Composition of a Composition

Find the derivative of the following function:

In[·]:= **f[x_] := CubeRoot[Cos[x^2]]**

It is the composition of the functions $\sqrt[3]{x}$, cosine and x^2, so its derivative is:

In[·]:= **(D[CubeRoot[x], x] /. x → Cos[x^2]) * (D[Cos[x], x] /. x → x^2) * D[x^2, x]**

Out[·]= $-\dfrac{2 x \, Sin[x^2]}{3 \, \sqrt[3]{Cos[x^2]}^{\,2}}$

Confirm with D:

In[·]:= **D[f[x], x]**

Out[·]= $-\dfrac{2 x \, Sin[x^2]}{3 \, \sqrt[3]{Cos[x^2]}^{\,2}}$

Here is a plot of the function and its derivative:

In[·]:= **Plot[{f[x], f'[x]}, {x, −5, 5}, PlotLegends → "Expressions"]**

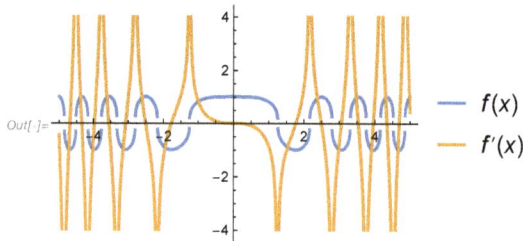

Simple Harmonic Motion

A particle undergoes simple harmonic motion if its position can be modeled by the following equation:

In[·]:= **s[t_] := A Cos[ω t + δ]**

where A is the amplitude, ω is the frequency and δ is the phase shift.

The velocity will be its first derivative, and the acceleration will be its second derivative:

In[ ]:= **v[$t\_$] := s'[t]**
    a[$t\_$] := s''[t]

In[ ]:= **{v[t], a[t]}**

Out[ ]= $\left\{-A\,\omega\,\mathrm{Sin}[\delta + t\,\omega], -A\,\omega^2\,\mathrm{Cos}[\delta + t\,\omega]\right\}$

It satisfies the following equation (called a **differential equation**):

In[ ]:= **HoldForm[s''[t] + ω^2 s[t] == 0]**

Out[ ]= $s''[t] + \omega^2\,s[t] == 0$

Using HoldForm keeps the equation from evaluating. When you release it using ReleaseHold, you can see that the equation is true:

In[ ]:= **ReleaseHold[%]**

Out[ ]= **True**

Summary

The chain rule lets you find the derivatives of many more functions than before.

The chain rule can be used more than once when a function is the composition of more than two functions.

The next lesson will cover implicit differentiation.

Exercises

Exercise 1—Composition

Find the derivative of the following function:

$In[\cdot]:=$ **f[x_] := (2 x^3 – x^5)^75**

Solution

Using the chain rule, you can find the answer:

$In[\cdot]:=$ **answer = 75 (2 x^3 – x^5)^74 * D[(2 x^3 – x^5), x]**

$Out[\cdot]=$ $75 \left(6 x^2 - 5 x^4\right) \left(2 x^3 - x^5\right)^{74}$

D confirms:

$In[\cdot]:=$ **answer == D[f[x], x]**

$Out[\cdot]=$ **True**

Exercise 2—Bullet-Nose Curve

The following function is known as a **bullet-nose curve**:

$In[\cdot]:=$ **f[x_] := RealAbs[x] / Sqrt[1 – x^2]**

Find the tangent line to the bullet-nose curve at $\left(\frac{\sqrt{3}}{2}, \sqrt{3}\right)$.

Solution

By the quotient rule and the chain rule, the derivative of the function at $\frac{\sqrt{3}}{2}$ is 8:

$In[\cdot]:=$ **slope =**
   ((Sqrt[1 – x^2] * D[RealAbs[x], x] – RealAbs[x] * (D[Sqrt[x], x] /. (x → 1 – x^2)) * D[1 – x^2,
      x]) / (Sqrt[1 – x^2]^2)) /. x → Sqrt[3] / 2

$Out[\cdot]=$ **8**

Confirm with D:

$In[\cdot]:=$ **slope == D[f[x], x] /. x → Sqrt[3] / 2**

$Out[\cdot]=$ **True**

With point-slope form, the tangent line is:

In[ ]:= **tangent[x_] := Sqrt[3] + slope (x − Sqrt[3]/2)**

Here is a plot of the function and its tangent line at $\left(\frac{\sqrt{3}}{2}, \sqrt{3}\right)$:

In[ ]:= **Plot[{f[x], tangent[x]}, {x, −2, 2}, PlotLegends → "Expressions",**
  Epilog → {PointSize[Large], Point[{Sqrt[3]/2, f[Sqrt[3]/2]}]}]

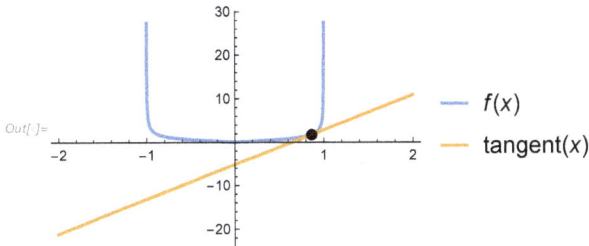

Exercise 3—Multiple Compositions

Find the derivative of the following function:

In[ ]:= **f[x_] := Cos[Cos[Cos[x]]]**

Solution

Using the chain rule, you can find the answer:

In[ ]:= **answer = −Sin[Cos[Cos[x]]] * −Sin[Cos[x]] * −Sin[x]**

Out[ ]= −Sin[x] Sin[Cos[x]] Sin[Cos[Cos[x]]]

Confirm with D:

In[ ]:= **answer == D[f[x], x]**

Out[ ]= True

Here is a plot of the function and its derivative:

In[ ]:= **Plot[{f[x], f'[x]}, {x, −π, π}, PlotLegends → "Expressions"]**

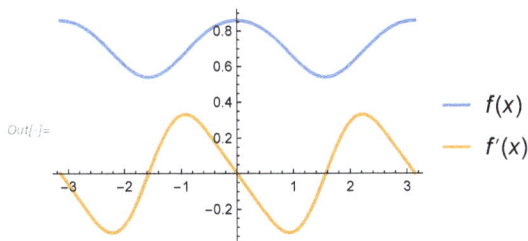

Exercise 4—Absolute Value Function

Given that the function $|x|$ can be expressed as $\sqrt{x^2}$, find its derivative.

Solution

Write the function:

$In[\cdot]:=$ **f[x_] := Sqrt[x^2]**

By using the chain rule, the answer is:

$In[\cdot]:=$ **answer = 1/2 (x^2)^(−1/2) * 2 x**

$Out[\cdot]=\dfrac{x}{\sqrt{x^2}}$

Confirm with D:

$In[\cdot]:=$ **answer == D[f[x], x]**

$Out[\cdot]=$ **True**

Here are the function and its derivative in a graph:

$In[\cdot]:=$ **Plot[{f[x], f'[x]}, {x, −2, 2}, PlotLegends → "Expressions"]**

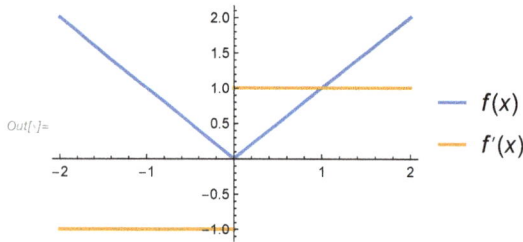

Exercise 5—Simple Harmonic Motion

A particle undergoes simple harmonic motion:

$In[\cdot]:=$ **s[t_] := A Cos[ω t + δ]**

Find its velocity and acceleration when it reaches its maximum A.

Solution

First, solve for t when the function equals A:

In[ ]:= **Solve[s[t] == A, t]**

$$Out[ ]= \left\{ \left\{ t \rightarrow \boxed{\frac{-\delta + 2\pi c_1}{\omega} \quad \text{if} \quad c_1 \in \mathbb{Z}} \right\} \right\}$$

Use $C[1] = 1$ for t. The velocity will be its first derivative, and the acceleration will be its second derivative:

In[ ]:= **v[t_] := s'[t]**
a[t_] := s''[t]

Plug in $t = \frac{(-\delta + 2\pi)}{\omega}$:

In[ ]:= $\{v[\dfrac{-\delta + 2\pi}{\omega}], a[\dfrac{-\delta + 2\pi}{\omega}]\}$

$Out[ ]= \{0, -A\omega^2\}$

You can see that it has velocity 0 and acceleration $-A\omega^2$ at its peak A.

12 | Implicit Differentiation

Overview

So far, all the functions dealt with have an explicit form, such as:

In[ ]:= $f[x\_] := x\wedge2 - x + 1$

Some functions have a different form and are defined **implicitly**.

Consider the following equation, where y is a function of x:

$$x^2 + y^2 = 4$$

If you solve for y, you get two functions f and g:

In[ ]:= $\text{sol} = \text{Solve}[x\wedge2 + y\wedge2 == 4, y]$

Out[ ]= $\left\{\left\{y \to -\sqrt{4-x^2}\right\}, \left\{y \to \sqrt{4-x^2}\right\}\right\}$

In[ ]:= $\text{Plot}[\{-\sqrt{4-x^2}, \sqrt{4-x^2}\}, \{x, -3, 3\}, \cdots \,]$

Out[ ]=

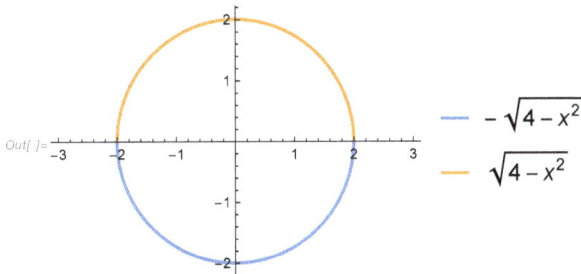

The goal of this lesson is to learn how to calculate derivatives implicitly in the Wolfram Language.

Implicit Functions

You can find the derivatives of each of the previous functions separately:

In[ ]:= $\{D[-\sqrt{4-x^2}, x], D[\sqrt{4-x^2}, x]\}$

Out[ ]= $\left\{\dfrac{x}{\sqrt{4-x^2}}, -\dfrac{x}{\sqrt{4-x^2}}\right\}$

Note that each expression generally has the form $-x/y$, where y is the function:

$In[\cdot]:=$ $\{-x/-\sqrt{4-x^2}, -x/\sqrt{4-x^2}\}$

$Out[\cdot]=$ $\left\{\dfrac{x}{\sqrt{4-x^2}}, -\dfrac{x}{\sqrt{4-x^2}}\right\}$

The tangent line to the curve at $\left(\sqrt{2}, \sqrt{2}\right)$ has slope -1. Here is a plot of the tangent line and the expression:

$In[\cdot]:=$ **Show[Plot[$\sqrt{2}$ +−1 (x − $\sqrt{2}$), {x, −3, 3}, PlotStyle → Red, $\boxed{\cdots +}$],**

ContourPlot[x^2 + y^2 == 4, {x, −2, 2}, {y, −2, 2}]]

$Out[\cdot]=$

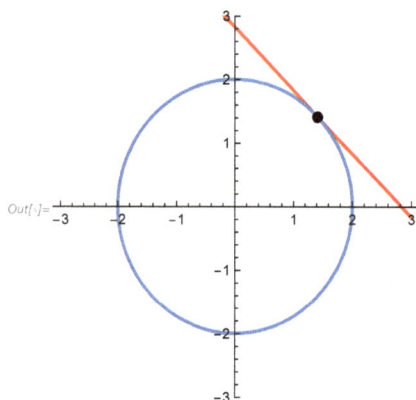

Implicit Differentiation

To differentiate an expression implicitly with respect to the x, you take derivatives as before. However, any time you encounter a variable that is not x (like y), you treat it as a function $y(x)$ of x, whose derivative is dy/dx. Then you solve for dy/dx to find the derivative.

For the equation $x^2 + y^2 = 4$, taking the derivative gives $2x + 2y\,dy/dx = 0$. You can see that the derivative of y^2 is $2\left(y^{2-1}\right) * dy/dx$. You are essentially using the chain rule on y.

Solving for dy/dx, you can see that $dy/dx = -x/y$, which was noted earlier.

Now you will see how to do it in the Wolfram Language.

Circle 1

There are two ways to enact implicit differentiation in the Wolfram Language. The first way expresses y as a function of x:

In[ ]:= **eqn = x^2 + y[x]^2 == 4;**

In[ ]:= **Solve[D[eqn, x], y'[x]]**

Out[ ]= $\left\{\left\{y'[x] \rightarrow -\dfrac{x}{y[x]}\right\}\right\}$

You can see that the derivative of any point $(x, \ y)$ on the curve is simply $-x/y$, as before. So at the point $\left(-1, \ \sqrt{3}\right)$, the equation of the tangent line to the curve is $1\big/\sqrt{3}$:

In[ ]:= **tangent[x_] := Sqrt[3] + 1/Sqrt[3] (x + 1)**

In[ ]:= **Show[Plot[tangent[x], {x, -3, 3}, PlotStyle → Red, ⋯ ⚬],**
  ContourPlot[x^2 + y^2 == 4, {x, -2, 2}, {y, -2, 2}]]

Out[ ]=

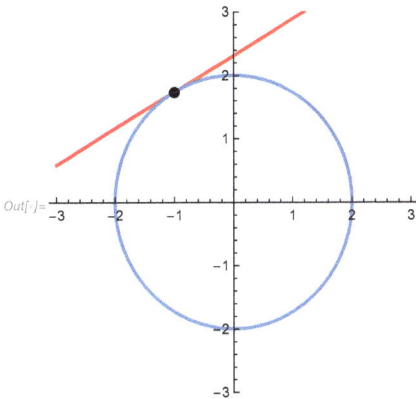

Circle 2

The second way to do implicit differentiation in the Wolfram Language uses the option NonConstants in D.

Find the derivative again, this time using NonConstants:

In[ ]:= **eqn = x^2 + y^2 == 4;**

In[ ]:= **D[eqn, x, NonConstants → {y}]**

Out[ ]= **2 x + 2 y D[y, x, NonConstants → {y}] == 0**

Now solve the previous function for D[y, x, NonConstants→{y}]:

In[ ]:= **Solve[%, D[y, x, NonConstants → {y}]]**

Out[ ]= $\left\{\left\{\text{D[y, x, NonConstants → {y}]} \to -\dfrac{x}{y}\right\}\right\}$

You again see that the derivative of any point $(x,\ y)$ on the curve is simply $-x/y$. Usually, the first way will be used when doing implicit differentiation, but the second way will be used in one of the exercises.

Folium of Descartes

Find the tangent line to the curve for the following implicitly defined function at $(2,\ 2)$:

$$x^3 + y^3 = 4\,x\,y$$

Compute the derivative with implicit differentiation and plug in your values:

In[ ]:= **eqn = x^3 + y[x]^3 == 4 x y[x];**

In[ ]:= **Solve[D[eqn, x], y '[x]] /. {x → 2, y[x] → 2}**

Out[ ]= **{{y'[2] → −1}}**

Here are the curve and the tangent line in a plot:

In[ ]:= **tangent[x_] := 2 − (x − 2)**

In[ ]:= **Show[Plot[tangent[x], {x, −3, 3}, PlotStyle → Red, ⋯ +],**
    ContourPlot[x^3 + y^3 == 4 x y, {x, −3, 3}, {y, −3, 3}]]

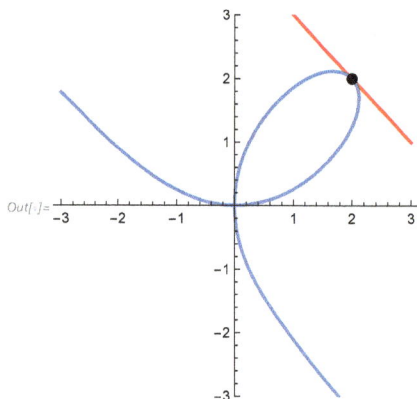

Ellipse

Find the tangent line to the curve for the following implicitly defined function at $\left(1, \frac{-\sqrt{3}}{2}\right)$:

$$x^2 + 4y^2 = 4$$

Compute the derivative with implicit differentiation and plug in your values:

In[ ]:= **eqn = x^2 + 4 y[x]^2 == 4;**

In[ ]:= **Solve[D[eqn, x], y'[x]] /. {x → 1, y[x] → $-\dfrac{\sqrt{3}}{2}$}**

Out[ ]= $\left\{\left\{y'[1] \to \dfrac{1}{2\sqrt{3}}\right\}\right\}$

Here are the curve (an ellipse) and the tangent line in a plot:

In[ ]:= **tangent[x_] := -Sqrt[3]/2 + 1/(2 Sqrt[3]) (x − 1)**

In[ ]:= **Show[Plot[tangent[x], {x, −3, 3}, PlotStyle → Red, ⋯ +],**
ContourPlot[x^2 + 4 y^2 == 4, {x, −3, 3}, {y, −3, 3}]]

Out[ ]=

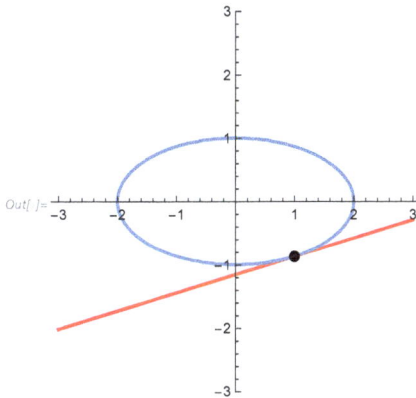

Hyperbola

Find the tangent line to the curve for the following implicitly defined function at $\left(\frac{-3}{2}, \frac{-\sqrt{5}}{2}\right)$:

$$x^2 - y^2 = 1$$

Compute the derivative with implicit differentiation:

$In[\cdot]:=$ **eqn = x^2 − y[x]^2 == 1;**

$In[\cdot]:=$ **Solve[D[eqn, x], y'[x]] /. {x → −3/2, y[x] → −Sqrt[5]/2}**

$Out[\cdot]:=$ $\left\{\left\{y'\left[-\dfrac{3}{2}\right] \to \dfrac{3}{\sqrt{5}}\right\}\right\}$

Here are the curve (a hyperbola) and the tangent line in a plot:

$In[\cdot]:=$ **tangent[x_] := −Sqrt[5]/2 + 3/Sqrt[5] (x + 3/2)**

$In[\cdot]:=$ **Show[Plot[tangent[x], {x, −2, 2}, PlotStyle → Red, ⋯ ⊕],**
        ContourPlot[x^2 − y^2 == 1, {x, −5, 5}, {y, −5, 5}]]

$Out[\cdot]:=$

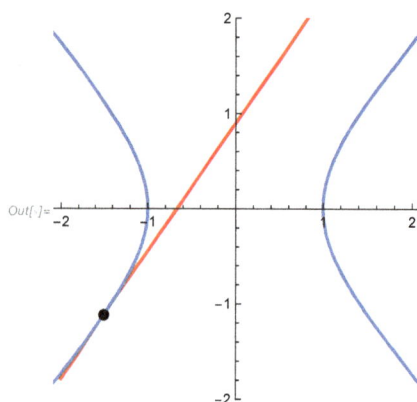

Cardioid

Find the tangent line to the curve for the following implicitly defined function at $\left(0, -\frac{1}{3}\right)$:

$$x^2 + y^2 = (3x^2 + 3y^2 - x)^2$$

Compute the derivative with implicit differentiation:

$In[\cdot]:=$ **eqn = x^2 + y[x]^2 == (3 x^2 + 3 y[x]^2 − x)^2;**

$In[\cdot]:=$ **Solve[D[eqn, x], y'[x]] /. {x → 0, y[x] → 1/3}**

$Out[\cdot]:=$ $\{\{y'[0] \to 1\}\}$

Here are the curve (a cardioid) and the tangent line in a plot:

```
In[ ]:= tangent[x_] := 1/3 + x
```

```
In[ ]:= Show[Plot[tangent[x], {x, -1, 1}, PlotStyle → Red, ⋯ + ],
        ContourPlot[x^2 + y^2 == (3 x^2 + 3 y^2 - x)^2, {x, -1, 1}, {y, -1, 1}]]
```

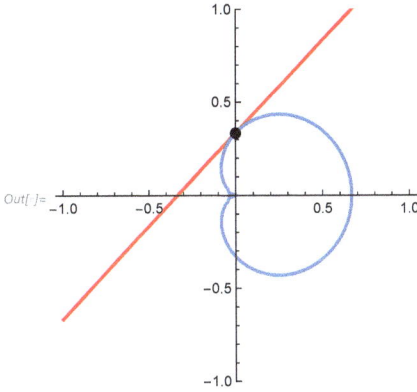

Devil's Curve

Find the tangent line to the curve for the following implicitly defined function at $\left(\frac{1}{2}, \sqrt{1 - \frac{\sqrt{5}}{4}}\right)$:

$$y^2 \left(y^2 - 2\right) = x^2 \left(x^2 - 3\right)$$

Compute the derivative with implicit differentiation:

```
In[ ]:= eqn = y[x]^2 (y[x]^2 - 2) == x^2 (x^2 - 3);
```

```
In[ ]:= Simplify[Solve[D[eqn, x], y'[x]] /. {x → 1/2, y[x] → Sqrt[1 - Sqrt[5]/4]}]
```

$$Out[ ]= \left\{\left\{y'\left[\frac{1}{2}\right] \to \sqrt{\frac{5}{4 - \sqrt{5}}}\right\}\right\}$$

Here are the curve (called a devil's curve) and the tangent line in a plot:

In[·]:= **tangent[x_] := Sqrt[1 − Sqrt[5] / 4] + Sqrt[5 / (4 − Sqrt[5])] (x − 1/2)**

In[·]:= **Show[Plot[tangent[x], {x, −5, 5}, PlotStyle → Red, ⋯ ⊕],**

    ContourPlot[y^2 (y^2 − 2) == x^2 (x^2 − 3), {x, −5, 5}, {y, −5, 5}]]

Out[·]=

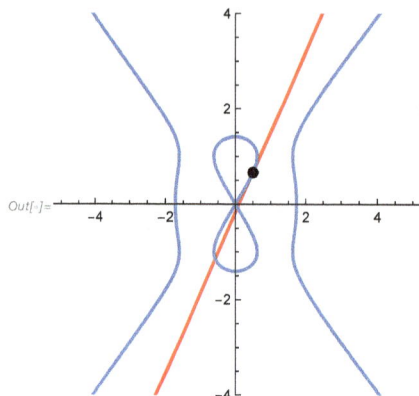

Kampyle of Eudoxus

Find the tangent line to the curve for the following implicitly defined function at $\left(\frac{-3}{2}, \frac{3\sqrt{5}}{4}\right)$:

$$y^2 = x^4 - x^2$$

Compute the derivative with implicit differentiation:

In[·]:= **eqn = y[x]^2 == x^4 − x^2;**

In[·]:= **Solve[D[eqn, x], y'[x]] /. {x → −3/2, y[x] → 3 Sqrt[5]/4}**

Out[·]= $\left\{\left\{y'\left[-\frac{3}{2}\right] \rightarrow -\frac{7}{\sqrt{5}}\right\}\right\}$

Here are the curve (called a kampyle of Eudoxus) and the tangent line in a plot:

In[ ]:= **tangent[x_] := 3 Sqrt[5] / 4 − 7 / Sqrt[5] (x + 3 / 2)**

In[ ]:= **Show[Plot[tangent[x], {x, −3, 3}, PlotStyle → Red, ⋯ ☆],**
  ContourPlot[y^2 == x^4 − x^2, {x, −3, 3}, {y, −3, 3}]]

Out[ ]=

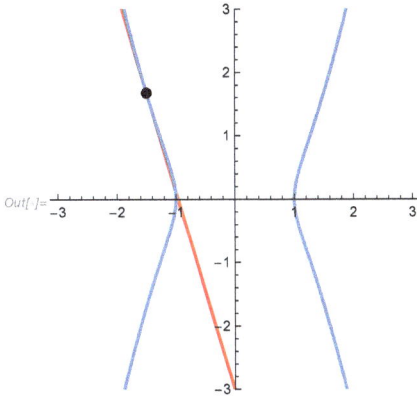

Multiple Derivatives

Finding higher-order derivatives is simple. Consider the following curve:

$$x^3 + y^3 = 8$$

Compute the first derivative and second derivative with implicit differentiation:

In[ ]:= **eqn = x^3 + y[x]^3 == 8;**

In[ ]:= **sol1 = Solve[D[eqn, x], y'[x]]**

Out[ ]= $\left\{\left\{y'[x] \rightarrow -\dfrac{x^2}{y[x]^2}\right\}\right\}$

In[ ]:= **sol2 = Solve[D[eqn, {x, 2}], y''[x]]**

Out[ ]= $\left\{\left\{y''[x] \rightarrow -\dfrac{2\left(x + y[x]\, y'[x]^2\right)}{y[x]^2}\right\}\right\}$

Plug in the first derivative wherever it is needed in the second derivative:

In[ ]:= **Simplify[sol2 /. sol1[[1]]]**

Out[ ]= $\left\{\left\{y''[x] \rightarrow -\dfrac{2x\left(x^3 + y[x]^3\right)}{y[x]^5}\right\}\right\}$

Summary

There are two ways to compute implicit derivatives in the Wolfram Language.

You can either:

1. Give an equation and express one of the variables as a function of the others, then differentiate the equation and solve for the derivative that you want.

2. Or you can differentiate the equation without expressing one of the variables as a function of the others by setting the variable you want to find the derivative for to not be a constant with the option NonConstants. Then solve for the NonConstants expression.

Both have their advantages and disadvantages.

Many interesting curves can be expressed implicitly and can be graphed with ContourPlot.

The next lesson will cover rates of change in the various sciences.

Exercises

Exercise 1—Implicit Derivative

Find the derivative of the following function:

$$\cos(x + y) = y^2 \sin(x)$$

Solution

Compute the derivative with implicit differentiation:

$In[\cdot]:=$ **eqn = Cos[x + y[x]] == y[x]^2 Sin[x];**

$In[\cdot]:=$ **Solve[D[eqn, x], y'[x]]**

$$Out[\cdot]=\left\{\left\{y'[x] \rightarrow \frac{-\text{Sin}[x + y[x]] - \text{Cos}[x]\, y[x]^2}{\text{Sin}[x + y[x]] + 2\,\text{Sin}[x]\, y[x]}\right\}\right\}$$

Here are the curve and the tangent line that crosses at the point $\left(\frac{\pi}{2}, 0\right)$ in a plot:

$In[\cdot]:=$ **tangent[$x\_$] := −(x − π/2)**

$In[\cdot]:=$ **Show[Plot[tangent[x], {x, −3, 3}, PlotStyle → Red, ⋯ ⊹],**
 ContourPlot[Cos[x + y] == y^2 Sin[x], {x, −3, 3}, {y, −3, 3}]]

$Out[\cdot]=$

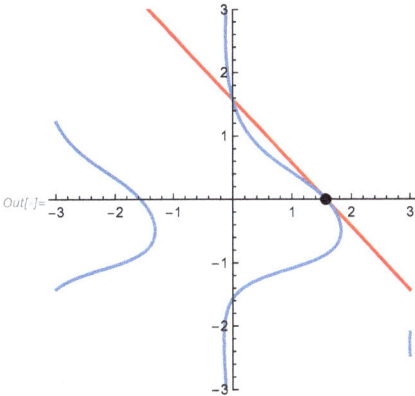

Exercise 2—Tangent to an Ellipse

Find the tangent line to the following ellipse at $\left(0, -\sqrt{2}\right)$:

$In[\cdot]:=$ **ellipse = x^2 − x y[x] + y[x]^2 == 2;**

Solution

Calculate the implicit derivative and plug in $\left(0, -\sqrt{2}\right)$:

In[·]:= **Solve[D[ellipse, x], y'[x]] /. {x → 0, y[x] → −Sqrt[2]}**

Out[·]= $\left\{\left\{y'[0] \to \dfrac{1}{2}\right\}\right\}$

Here are the curve and the tangent line in a plot:

In[·]:= **tangent[x_] := −Sqrt[2] + x / 2**

In[·]:= **Show[Plot[tangent[x], {x, −5, 5}, PlotStyle → Red, ⋯ ＋],**
     ContourPlot[x^2 − x y + y^2 == 2, {x, −5, 5}, {y, −5, 5}]]

Out[·]=

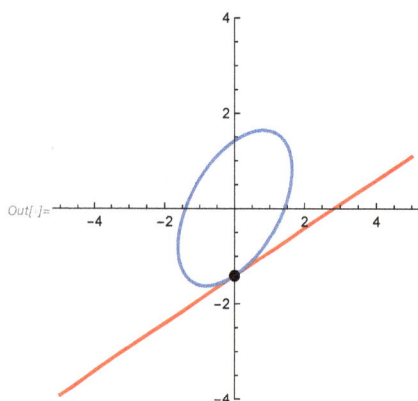

Exercise 3—Tschirnhausen Cubic

Find the tangent line to the following curve at $\left(\frac{3}{2}, \frac{1}{4}\right)$:

In[·]:= **cubic = 6 y[x]^2 == x (x − 2)^2;**

Solution

Calculate the implicit derivative and plug in $\left(\frac{3}{2}, \frac{1}{4}\right)$:

In[·]:= **Solve[D[cubic, x], y'[x]] /. {x → 3 / 2, y[x] → 1 / 4}**

Out[·]= $\left\{\left\{y'\left[\dfrac{3}{2}\right] \to -\dfrac{5}{12}\right\}\right\}$

Here are the curve and the tangent line in a plot:

In[ ]:= **tangent[*x*_] := 1/4 – 5/12 (*x* – 3/2)**

In[ ]:= **Show[Plot[tangent[x], {x, –5, 5}, PlotStyle → Red, ··· +],**
    ContourPlot[6 y^2 == x (x – 2)^2, {x, –5, 5}, {y, –5, 5}]]

Out[ ]=

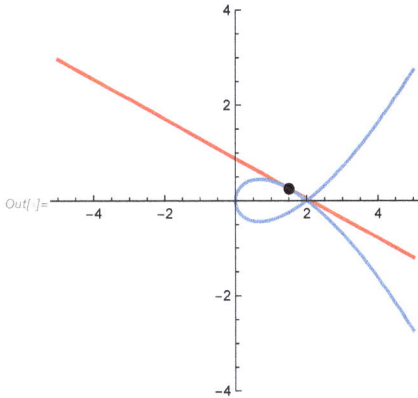

Exercise 4—Multiple Derivatives

Find the third derivative of the following function:

In[ ]:= **eqn = 1/x + 2/y[x] == 1;**

Solution

Calculate the first derivative with D:

In[ ]:= **sol1 = Solve[D[eqn, x], y'[x]]**

Out[ ]= $\left\{\left\{ y'[x] \to -\dfrac{y[x]^2}{2 x^2} \right\}\right\}$

Plug that in when calculating the second derivative with D:

In[ ]:= **sol2 = Simplify[Solve[D[eqn, {x, 2}], y''[x]] /. sol1[[1]]]**

Out[ ]= $\left\{\left\{ y''[x] \to \dfrac{y[x]^2 (2 x + y[x])}{2 x^4} \right\}\right\}$

Plug the first and second derivative in when calculating the third derivative with D:

In[ ]:= **Simplify[Solve[D[eqn, {x, 3}], y'''[x]] /. {sol1[[1, 1]], sol2[[1, 1]]}]**

Out[ ]= $\left\{\left\{ y^{(3)}[x] \to -\dfrac{3 y[x]^2 (2 x + y[x])^2}{4 x^6} \right\}\right\}$

Exercise 5—Van der Waals

The **van der Waals equation** for n moles of a gas is given by the following equation:

In[·]:= **vanderWaals = (P + n^2 a/V^2) (V − n b) == n R T**

Out[·]= $\left(P + \dfrac{a\, n^2}{V^2}\right)(-b\, n + V) == n\, R\, T$

where P is pressure, V is volume, T is temperature, R is the universal gas constant and a and b are positive constants that depend on the gas.

Find the derivative of P with respect to V.

Solution

Here the second way to do implicit differentiation with NonConstants is used:

In[·]:= **D[vanderWaals, V, NonConstants → P]**

Out[·]= $P + \dfrac{a\, n^2}{V^2} + (-b\, n + V)\left(-\dfrac{2\, a\, n^2}{V^3} + D[P, V, \text{NonConstants} \to \{P\}]\right) == 0$

Solve for D[P, V, NonConstants → {P}]:

In[·]:= **Solve[%, D[P, V, NonConstants → {P}]]**

Out[·]= $\left\{\left\{D[P, V, \text{NonConstants} \to \{P\}] \to \dfrac{2\, a\, b\, n^3 - a\, n^2\, V + P\, V^3}{(b\, n - V)\, V^3}\right\}\right\}$

13 | Rates of Change in the Sciences

Overview

Consider the following function, which gives a particle's position s in meters after t seconds:

In[ ]:= $s[t\_] := t\wedge3 + 2\,t - 2$

Its derivative gives the velocity of the particle in meters per second at any given time in its domain:

In[ ]:= $v[t\_] := s'[t]$

As you can see in the following graph, at time $t = 10$ seconds the particle's position is 1018 m and its velocity is 302 m/s:

In[ ]:= $s[10]$

Out[ ]= 1018

In[ ]:= $v[10]$

Out[ ]= 302

In[ ]:= Plot[{s[t], v[t]}, {t, 0, 20}, PlotLegends → "Expressions",
  Epilog → {PointSize[Large], Point[{10, s[10]}], Point[{10, v[10]}]}]

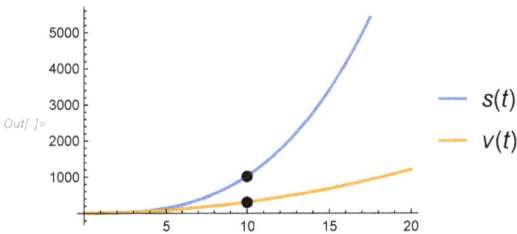

The goal of this lesson is to explore the many applications of differential calculus in the sciences.

Mechanics 1

Consider the following position function s:

In[ ]:= $s[t\_] := t\wedge3 - 4\,t\wedge2 + 4\,t$

Its velocity and acceleration are given by its first and second derivatives, respectively:

$In[\cdot]:=$ **v[$t\_$] := s'[t]**

$In[\cdot]:=$ **a[$t\_$] := s''[t]**

It is at rest when its velocity is 0:

$In[\cdot]:=$ **Solve[v[t] == 0, t]**

$Out[\cdot]=\left\{\left\{t \to \dfrac{2}{3}\right\}, \{t \to 2\}\right\}$

Here is a plot of the position, velocity and acceleration:

$In[\cdot]:=$ **Plot[{s[t], v[t], a[t]}, {t, 0, 3}, PlotLegends → {"position", "velocity", "acceleration"}]**

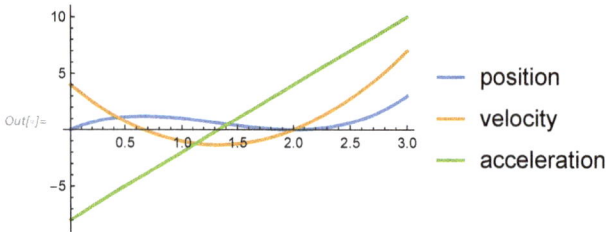

Mechanics 2

The particle is moving forward when the velocity is positive:

$In[\ ]:=$ **Reduce[v[t] > 0, t]**

$Out[\cdot]= t < \dfrac{2}{3} \, || \, t > 2$

The particle speeds up when its acceleration and velocity are the same sign:

$In[\ ]:=$ **Reduce[a[t] * v[t] > 0, t]**

$Out[\cdot]= \dfrac{2}{3} < t < \dfrac{4}{3} \, || \, t > 2$

The **total distance** covered in the time period $t = 0$ to $t = 3$ is found by calculating the distance covered when the particle is moving forward:

$In[\cdot]:=$ **d1 = RealAbs[s[2/3] − s[0]] + RealAbs[s[3] − s[2]]**

$Out[\ ]= \dfrac{113}{27}$

and **adding** the distance covered when the particle is moving backward:

In[ ]:= **d1 + RealAbs[s[2] − s[2 / 3]]**

Out[ ]= $\dfrac{145}{27}$

Linear Density

The **linear density** of a rod is given by its mass divided by its length:

In[ ]:= **lineardensity = mass / length;**

If the mass m (in kg) of a rod is given as a function of its length l (in meters):

In[ ]:= **m[l_] := CubeRoot[l]**

its linear density (in kg/m) at a given length l is given by the derivative of m with respect to l:

In[ ]:= **D[m[l], l]**

Out[ ]= $\dfrac{1}{3 \sqrt[3]{l}^2}$

At length 10 m, the rod has linear density 0.07 kg/m:

In[ ]:= **m '[10] // N**

Out[ ]= **0.0718145**

Here are the mass and the linear density on a graph:

In[ ]:= **Plot[{m[l], m'[l]}, {l, 0, 20}, PlotLegends → {"mass", "linear density"}]**

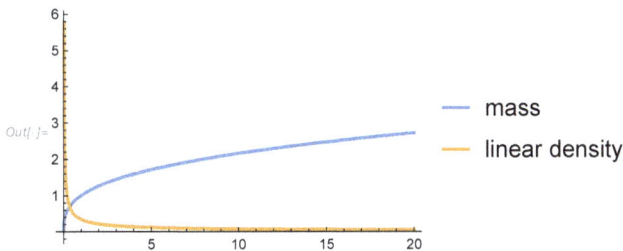

Current

The **current** going through a surface is given by the amount of charge that goes through for a given time period:

In[ ]:= **current = charge / time;**

If the charge q (in coulombs) is a function of time t (in seconds):

In[ ]:= **q[t_] := Sin[t] Cos[5 t]**

then the current (in amperes) at any instant is given by its derivative with respect to time t:

In[ ]:= **D[q[t], t]**

Out[ ]= **Cos[t] Cos[5 t] − 5 Sin[t] Sin[5 t]**

At 5.0 seconds, the current is −0.35 amperes:

In[ ]:= **q '[5.0]**

Out[ ]= **−0.35341**

Here are the charge and current on a graph:

In[ ]:= **Plot[{q[t], q '[t]}, {t, 0, 10}, PlotLegends → {"charge", "current"}]**

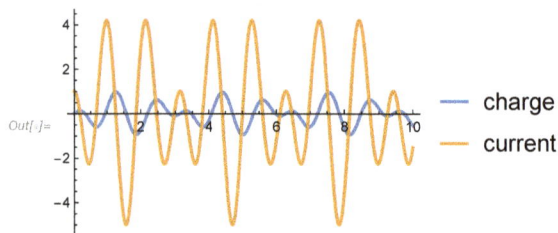

Chemistry—Rate of Reaction

In a chemical reaction, one or more reactants undergo a process to become one or more products.

For example, reactants A and B are transformed into products C and D in the following reaction:

In[ ]:= **reaction = A + B → C + D**

Out[ ]= **A + B → C + D**

The concentration of a substance A (denoted [A] and measured in moles per liter) is a function of time t:

In[ ]:= **concA[t_] := −t**
concB[t_] := −t
concC[t_] := t
concD[t_] := t

The **instantaneous rate of reaction** for a substance is given by the derivative of its concentration with respect to time t.

For the products in the above reaction, the reaction rates are 1 mol/liter sec, and the reactants have the reaction rates −1 mol/liter sec:

In[ ]:= **D[#, t] & /@ {concA[t], concB[t], concC[t], concD[t]}**

Out[ ]= **{−1, −1, 1, 1}**

In general, for a reaction of the form:

In[ ]:= **generalreaction = a A + b B → c C + d D**

Out[ ]= a A + b B → c C + d D

There is the following property:

In[ ]:= **−concA'[t] / a == −concB'[t] / b == concC'[t] / c == concD'[t] / d**

Out[ ]= $\dfrac{1}{a} == \dfrac{1}{b} == \dfrac{1}{c} == \dfrac{1}{d}$

Thermodynamics

If a substance has constant temperature and is compressed in a closed system, then its volume only depends on its pressure:

In[ ]:= **pressure = k / volume;**

Pressure and volume have an inverse relationship; if pressure goes up, then volume goes down, and vice versa. In other words, $\frac{dP}{dV}$ and $\frac{dV}{dP}$ are both negative.

Isothermal compressibility is defined by the following function:

In[ ]:= **β = −volume'[pressure] / volume;**

It determines how fast the volume of a substance decreases per unit volume as the pressure increases at constant temperature.

For example, at 25°C the volume of a sample of air is given by the following equation, where P is the pressure in kilopascals (kPa):

In[ ]:= **volume[P_] := 5.3 / P**

At 100 kPa, the isothermal compressability is 0.01 per kPa:

In[ ]:= **−volume'[100] / volume[100]**

Out[ ]= 0.01

Here is a plot of the volume and isothermal compressibility:

```
In[·]:= Plot[{volume[P], –(Derivative[1][volume][P]/volume[P])},
        {P, 0, 200}, PlotLegends → {"volume", "compressability"}]
```

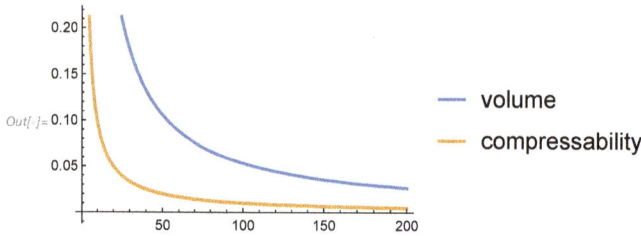

Biology—Growth Rate

The number of individuals in a given population is given by the following function at time t:

```
In[·]:= population[t_] := 2 t^3 + t;
```

The **growth rate** is given by the derivative of the population with respect to time t:

```
In[·]:= population '[t]
```

$$Out[·]= 1 + 6 t^2$$

Here are the function and its derivative in separate graphs:

```
In[·]:= {Plot[population[t], {t, 0, 60}, PlotLegends → {"population"}],
        Plot[population '[t], {t, 0, 60}, PlotLegends → {"growth rate"}]}
```

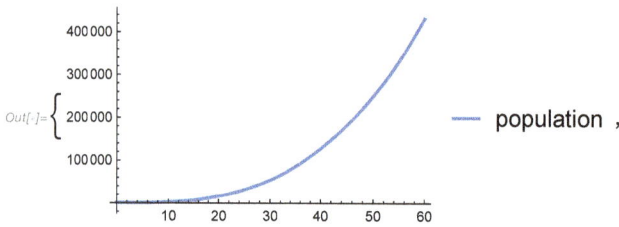

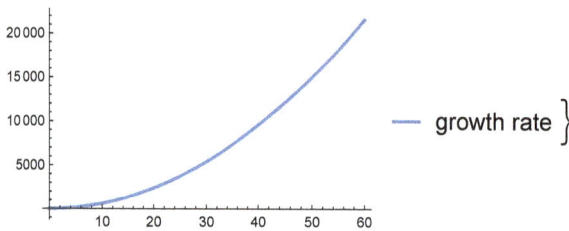

For a population of bacteria with initial population P_0 that doubles every minute, its population is given by the following function:

In[ ]:= **bacteria[t] := P$_0$ 2^t**

This is an example of an **exponential function**.

Biology—Laminar Flow

The French physician Jean-Louis-Marie Poiseuille discovered the law of **laminar flow**:

In[ ]:= **laminar[*r*_] := P (R^2 – *r*^2) / (4 η l)**

It relates the velocity of blood in a blood vessel to the distance r from the center of a vessel given the max radius R, length of the vessel l, pressure P and viscosity η.

The velocity gradient is the derivative of the velocity with respect to the radius r:

In[ ]:= **gradient[*r*_] := laminar '[*r*]**

In[ ]:= **gradient[r]**

Out[ ]= $-\dfrac{P\,r}{2\,l\,\eta}$

Here are plots of the function and its derivative when $\eta = 0.027$, $R = 0.01$ cm, $l = 4$ cm, and $P = \frac{4000 \text{ dynes}}{\text{cm}^2}$:

In[·]:= **{Plot[laminar[r] /. {$\eta \to$ 0.027`, R $\to$ 0.01`, l $\to$ 4, P $\to$ 4000}, {r, 0, 0.01`}, ⋯ +],**
Plot[gradient[r] /. {$\eta \to$ 0.027`, l $\to$ 4, P $\to$ 4000}, {r, 0, 0.01`}, ⋯ +]}

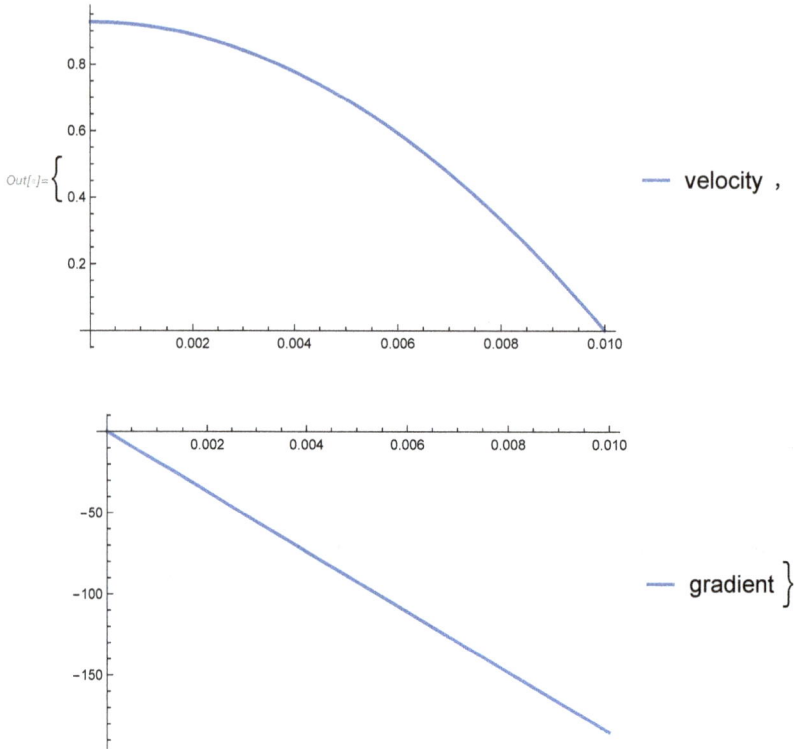

Economics

The following **cost function** calculates a particular company's total cost of producing x units:

In[·]:= **cost[x_] := x^2 + 3 x**

The **marginal cost** is found by taking the derivative:

In[·]:= **cost'[x]**

Out[·]= $3 + 2x$

The marginal cost is approximately the amount of money it costs to produce an extra unit.

In this case, the cost of producing the 1001st unit is \$2003:

In[ ]:= **cost'[1000]**

Out[ ]= 2003

Here are plots of the cost function and its marginal cost:

In[ ]:= **{Plot[cost[x], {x, 0, 2000}, PlotLegends → {"cost"}, ImageSize → Medium],**
     Plot[cost'[x], {x, 0, 2000}, PlotLegends → {"marginal cost"}, ImageSize → Medium]}

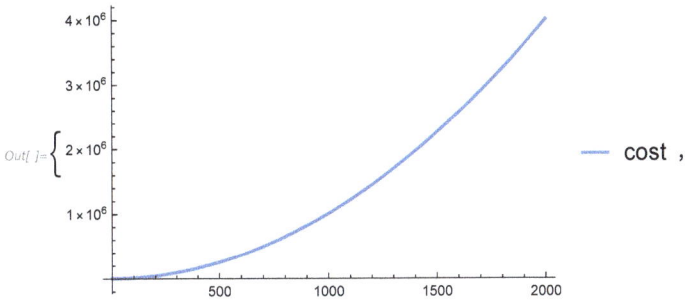

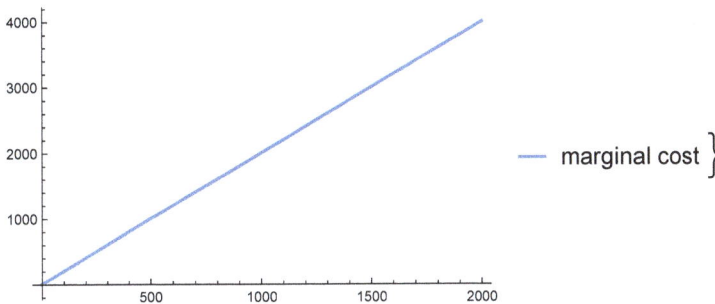

Out[ ]=

Summary

Derivatives are used universally in the sciences.

Physicists use derivatives in many equations relating to motion.

Chemists use derivatives when finding reaction rates.

Biologists use derivatives when calculating growth rates.

Economists use derivatives when calculating quantities such as marginal cost, marginal revenue, marginal demand, etc.

They are many more uses the lesson has not even begun to cover!

Wherever a quantity is changing, derivatives play a key role.

The next lesson will cover related rates.

Exercises

Exercise 1—Frequency

The **frequency** of vibrations of a string is given by:

In[·]:= **freq = 1/(2 L) Sqrt[T / ρ];**

where L is the length of the string, T is its tension, and ρ is its linear density.

Find the derivative of frequency with respect to the length, tension and linear density.

Solution

Derivative with respect to L:

In[·]:= **D[freq, L]**

Out[·]= $-\dfrac{\sqrt{\dfrac{T}{\rho}}}{2\,L^2}$

Derivative with respect to T:

In[·]:= **D[freq, T]**

Out[·]= $\dfrac{1}{4\,L\,\sqrt{\dfrac{T}{\rho}}\,\rho}$

Derivative with respect to ρ:

In[·]:= **D[freq, ρ]**

Out[·]= $-\dfrac{T}{4\,L\,\sqrt{\dfrac{T}{\rho}}\,\rho^2}$

Exercise 2—Boyle's Law

Boyle's law is named in honor of Robert Boyle.

It states that the product of pressure and volume remains constant when a sample of gas is compressed at constant temperature:

In[ ]:= **c = pressure * vol;**

Find the rate of change of volume with respect to pressure. Then find the isothermal compressibility.

Solution

Set *c* to a constant and solve for volume:

In[ ]:= **Solve[c == h, vol]**

Out[ ]= $\left\{\left\{ \text{vol} \rightarrow \dfrac{h}{\text{pressure}} \right\}\right\}$

Take the derivative of volume with respect to pressure:

In[ ]:= **D[$\dfrac{h}{\text{pressure}}$, pressure]**

Out[ ]= $-\dfrac{h}{\text{pressure}^2}$

The isothermal compressibility is $\frac{-1}{V}\frac{dV}{dP}$:

In[ ]:= **-%/($\dfrac{h}{\text{pressure}}$)**

Out[ ]= $\dfrac{1}{\text{pressure}}$

Exercise 3—Rate of Reaction

If a product C is formed from one molecule of A and one molecule of B, and the initial concentrations of A and B are both a, then:

In[ ]:= `concC[t_] := a^2 k t/(a k t + 1)`

where k is a constant.

Find the rate of reaction and graph the rate and concentration on separate graphs from 0 to 10,000 for $a = k = 1$:

Solution

Calculate the derivative of the concentration to find the rate of reaction:

In[ ]:= `concC'[t]`

$$Out[ ]= -\frac{a^3 k^2 t}{(1+a k t)^2} + \frac{a^2 k}{1+a k t}$$

Here is a plot of the concentration and the rate of reaction:

In[ ]:= `{Plot[concC[t] /. {a → 1, k → 1}, {t, 0, 1000}, ⋯ ▸ ],`
`Plot[concC'[t] /. {a → 1, k → 1}, {t, 0, 1000}, ⋯ ▸ ]}`

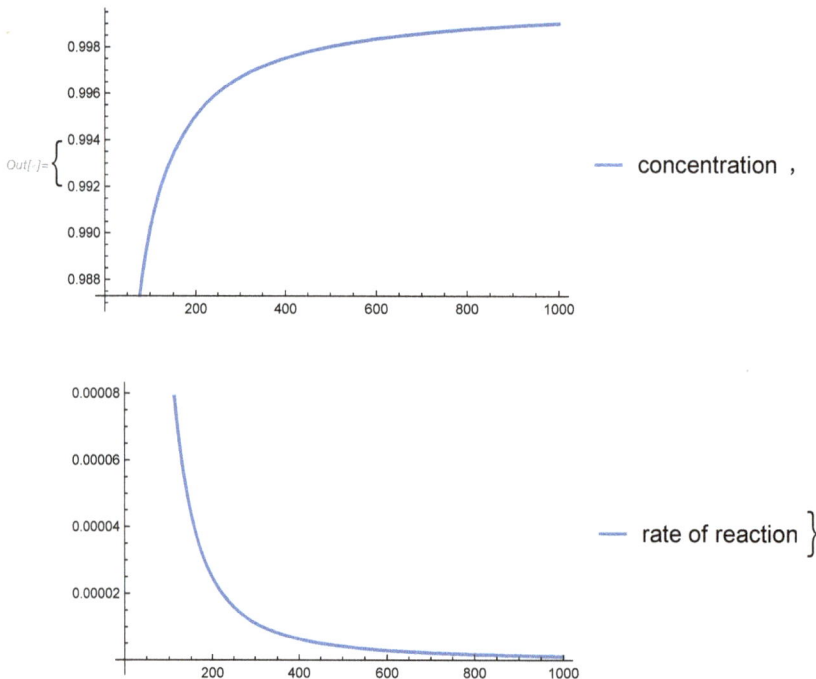

You can see the concentration reaches 1 as t goes to infinity, and the rate of reaction goes to 0.

Exercise 4—Cost

A company has the given cost function to produce x amount of units:

In[ ]:= **cost[x_] := 339 + 12 x − 0.09 x^2 + 0.0005 x^3**

Find the marginal cost, find approximately how much it costs to produce the $10{,}001^{st}$ unit, and graph the cost and the marginal cost.

Solution

The marginal cost is the derivative of the cost function:

In[ ]:= **cost'[x]**

Out[ ]= $12 − 0.18 x + 0.0015 x^2$

The cost to produce the $10{,}001^{st}$ unit is approximately the marginal cost of the $10{,}000^{th}$ unit:

In[ ]:= **cost'[10 000]**

Out[ ]= $148\,212.$

Here is the graph of the cost and marginal cost:

In[ ]:= **{Plot[cost[x], {x, 0, 20 000}, PlotLegends → {"cost"}, ImageSize → Medium],**
    Plot[cost'[x], {x, 0, 20 000}, PlotLegends → {"marginal cost"}, ImageSize → Medium]}

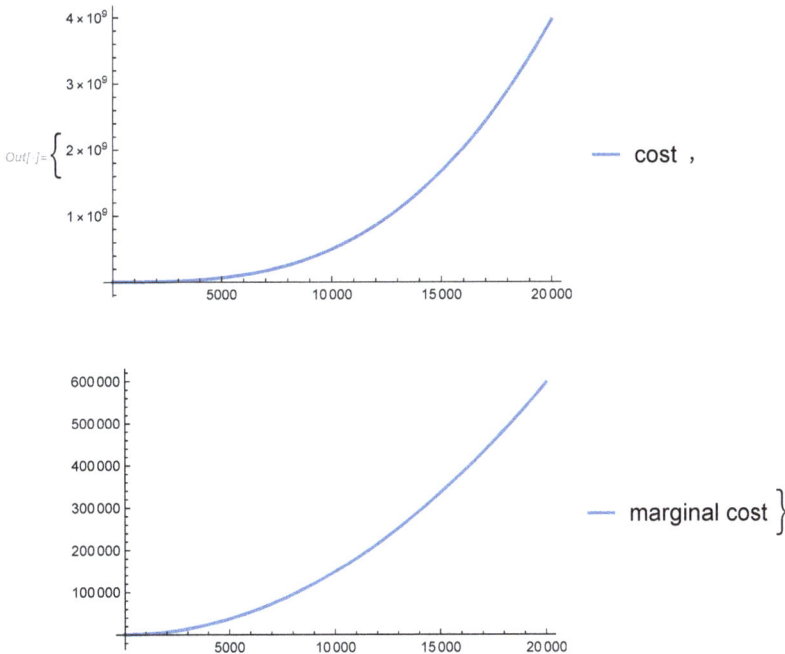

Exercise 5—Fish Farm Population

A fish farm has the following function for the rate of change of the fish population:

$In[\cdot]:=$ **fishchange $= r_0 (1 - P/C) P - B P$**

$Out[\cdot]:= -B P + P \left(1 - \dfrac{P}{C}\right) r_0$

where r_0 is the birth rate, P is the population, C is the maximum population size, and B is the percentage of the population harvested.

Find the **stable population level** when the birth rate is 10% and the harvesting rate is 5%. What happens when the harvesting rate is 10%?

Solution

The population will be stable when the rate of change is 0. Solve for P with the given parameters:

$In[\cdot]:=$ **sol $=$ Solve[(fishchange /. $\{r_0 \to 1/10, B \to 5/100\}$) $== 0$, P]**

$Out[\cdot]:= \left\{ \{P \to 0\}, \left\{P \to \dfrac{C}{2}\right\} \right\}$

You can see that the population will be stable if you set it to half the maximum population size.

When the harvesting rate is 10%:

$In[\cdot]:=$ **Simplify[fishchange /. $\{r_0 \to 1/10, B \to 1/10\}$]**

$Out[\cdot]:= -\dfrac{P^2}{10\,C}$

The rate of change is always negative, so the population will always be decreasing and only be stable when the population is 0.

14 | Related Rates

Overview

In most problems, more than one thing will be changing in relation to something else (for example, time). Consider a ladder falling along a side of a wall.

Its height at the top of the ladder and distance from the base of the ladder to the base of the wall both change as it is falling. The rate at which the height is dropping and the rate at which the distance from the base of the wall is increasing are **related**.

The goal of this lesson is to develop a strategy to solve related rates problems and go over some examples.

Falling Ladder 1

A ladder 5 feet long resting against a vertical wall begins to slide down. When the ladder is sliding away from the wall at a rate of 0.5 ft / s and the bottom of the ladder is 4 feet away from the wall, how fast is the ladder sliding down?

The first step is to assign variables to the changing quantities in the problem. In this case, y will be the height and x will be the distance the bottom of the ladder is from the base of the wall:

In[·]:= **Clear[x, y]**

Next, draw a picture:

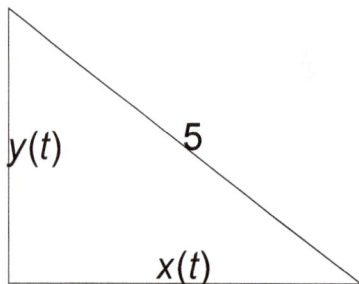

Since x, y and the length of the ladder form a triangle, they are related by the Pythagorean theorem:

In[·]:= **eqn = x[t]^2 + y[t]^2 == 5^2**

Out[·]= $x[t]^2 + y[t]^2 == 25$

Falling Ladder 2

$x[t]$ and $y[t]$ are functions of time, so differentiate the equation with respect to t time:

In[·]:= **sol = D[eqn, t]**

Out[·]= $2 x[t] x'[t] + 2 y[t] y'[t] == 0$

You are solving for $y'[t]$ and have $x[t]$ and $x'[t]$:

In[·]:= **sol /. {x[t] → 4, x'[t] → 0.5}**

Out[·]= $4. + 2 y[t] y'[t] == 0$

From the Pythagorean theorem, you can solve for $y[t]$:

In[ ]:= **sol1 = Solve[(eqn /. x[t] → 4) && y[t] > 0, y[t]]**

Out[ ]= **{{y[t] → 3}}**

You can solve for $y'[t]$ now:

In[ ]:= **Solve[sol /. {x[t] → 4, x'[t] → 0.5, sol1[[1, 1]]}, y'[t]]**

Out[ ]= **{{y'[t] → −0.666667}}**

When the base of the ladder is 4 feet from the base of the wall, the top of the ladder is sliding down the wall at a rate of 0.67 ft / sec. The negative sign in the solution for $y'[t]$ indicates the top of the ladder is going down.

General Strategy

This is the general strategy used in the previous problem:

1. Try to understand the problem.

2. Assign symbols to all variables in the problem that are functions of time.

3. Draw a picture.

4. Find an equation that relates all the variables in the problem. Usually something from geometry will help.

5. Differentiate the equation with respect to time (using the chain rule).

6. Isolate the unknown you are solving for and find any remaining unknown variables in the problem using given information.

7. Substitute the given information and any derived information you found into the new differentiated equation and solve for the unknown variable.

Filling a Balloon 1

A spherical balloon is filled with air at a rate of 50 cm^3 / sec. Find the rate at which the radius is increasing when the diameter is 25 cm.

The volume and radius are functions of time:

In[ ]:= **Clear[v, r]**

Draw a picture:

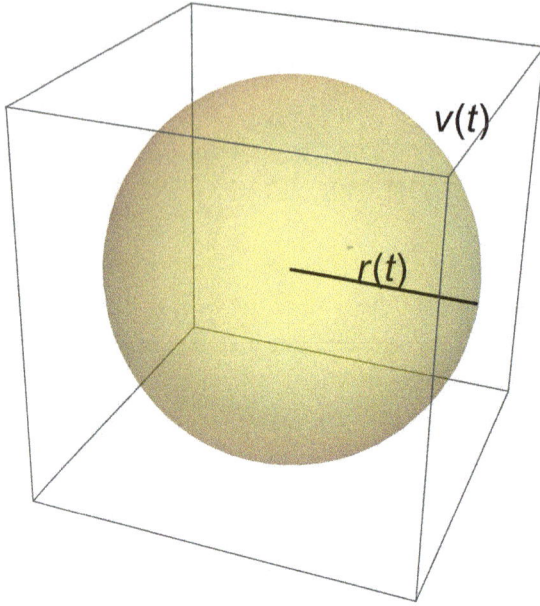

From geometry, you know the equation for the volume of a sphere:

$In[\cdot]:=$ **eqn = v[t] == 4 π r[t]^3 / 3**

$Out[\cdot]=$ v[t] == $\frac{4}{3}$ π r[t]^3

Filling a Balloon 2

Differentiate the equation with respect to time:

$In[\cdot]:=$ **sol = D[eqn, t]**

$Out[\cdot]=$ v'[t] == 4 π r[t]^2 r'[t]

Since you know $v'[t]$ and are solving for $r'[t]$, you first need to find $r[t]$. The diameter is twice the radius, so you know $r[t]$ is 12.5 cm.

Now substitute all the given and derived values and solve for $r'[t]$:

$In[\cdot]:=$ **Solve[sol /. {v'[t] → 50, r[t] → 25/2}, r'[t]]**

$Out[\cdot]=$ $\left\{\left\{r'[t] \to \dfrac{2}{25\,\pi}\right\}\right\}$

When the diameter is 25 cm, the radius is increasing at $2 / 25\,\pi$ (roughly 0.025) cm / sec.

Watchdog

A cat is walking down the street at 3 feet per second. A sitting dog sees the cat and stays focused on it. If the dog is 6 feet from the street and the cat is 8 feet away from the point on the street closest to the dog, how fast is the dog's head turning?

The rate at which the dog's head is turning and the distance the cat is away from the point on the street closest to the dog are both functions of time:

In[ ]:= **Clear[x, θ]**

Draw a picture:

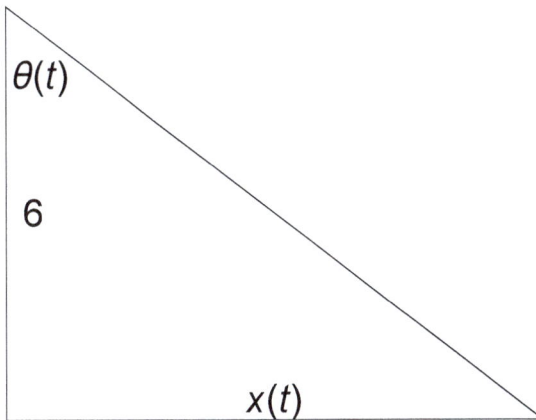

The angle is related to the distance the cat walks by the tangent function:

In[ ]:= **eqn = Tan[θ[t]] == x[t]/6**

Out[ ]= $\text{Tan}[\theta[t]] == \dfrac{x[t]}{6}$

Watchdog 2

Differentiate the equation with respect to time:

In[ ]:= **sol = D[eqn, t]**

Out[ ]= $\text{Sec}[\theta[t]]^2\, \theta'[t] == \dfrac{x'[t]}{6}$

You know $x'[t]$ (which is 3) and are solving for $\theta'[t]$, so you need to find $\text{Sec}[\theta[t]]$.

The secant is hypotenuse/adjacent; you have adjacent (which is 6), and you have the length of the other side (which is $x[t] = 8$), so you find the hypotenuse by the Pythagorean theorem:

In[·]:= **Solve[6^2 + 8^2 == c^2 && c > 0, c]**

Out[·]= **{{c → 10}}**

The hypotenuse is 10, and $\mathrm{Sec}[\theta[t]]$ is $10/6 = 5/3$.

Now solve for $\theta'[t]$:

In[·]:= **Solve[sol /. {Sec[θ[t]] → 5/3, x'[t] → 3}, θ'[t]]**

Out[·]= $\left\{\left\{\theta'[t] \to \dfrac{9}{50}\right\}\right\}$

When the cat is 8 feet from the point on the street closest to the dog, the dog's head is rotating at $27/250$ radians/sec (or 6.188 degrees/sec).

Oil Tank 1

Oil is being poured into an inverted cone at a rate of $3\ \mathrm{m}^3/\mathrm{sec}$. If the cone's height is 10 m and its radius is 5 m, find the rate at which the oil level is rising when the oil is 5 m deep.

The volume and height are functions of time:

In[·]:= **Clear[v, h]**

Draw a picture:

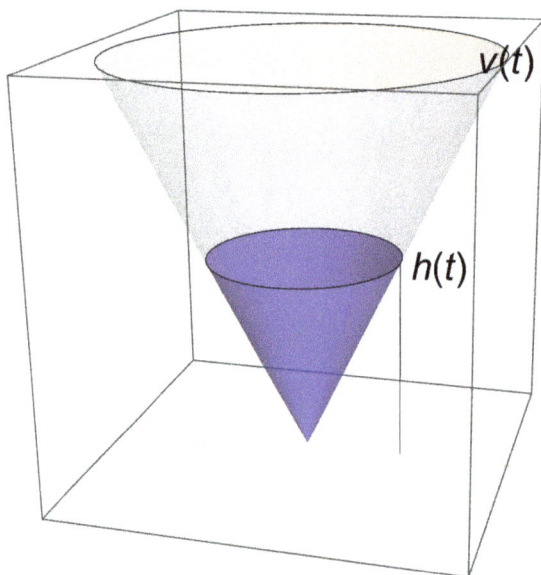

From geometry, you know the volume of a cone:

In[ ]:= **eqn = v[t] == π r[t]^2 h[t] / 3**

Out[ ]:= $v[t] == \dfrac{1}{3} \pi\, h[t]\, r[t]^2$

Oil Tank 2

At this point, you could differentiate the equation with respect to *t*, but it would be helpful to get rid of *r*[*t*] and express it in terms of *h*[*t*]. You can do so with similar triangles.

The radius is half the height in the oil tank, so by similarity any cone within the tank also has radius equal to half the height. Alter the volume equation:

In[ ]:= **eqn = % /. r[t] → h[t] / 2**

Out[ ]:= $v[t] == \dfrac{1}{12} \pi\, h[t]^3$

Now differentiate with respect to *t*:

In[ ]:= **sol = D[eqn, t]**

Out[ ]:= $v'[t] == \dfrac{1}{4} \pi\, h[t]^2\, h'[t]$

You want to find *h*'[*t*]. Now plug in the given values to find *h*'[*t*]:

In[ ]:= **Solve[sol /. {v '[t] → 3, h[t] → 5}, h '[t]]**

Out[ ]:= $\left\{\left\{h'[t] \to \dfrac{12}{25\,\pi}\right\}\right\}$

When the oil tank is 5 m deep, the oil level is rising at $12/25\,\pi$ (roughly 0.15) m/sec.

Summary

In this lesson, the objective was to develop a way to solve problems that involve related rates.

A general strategy for solving such problems was presented.

Then problems involving related rates were solved.

Each problem took advantage of the information given in the problem and any geometry inherent to the problem.

The next lesson will cover linear approximations.

Exercises

Exercise 1—Expanding Cube

A cube is expanding over time with side length increasing at 0.5 m/sec. Find the rate at which the volume is increasing when the side length is 5 m.

Solution

The volume and side length are functions of time:

In[·]:= **Clear[v, s]**

Draw a picture:

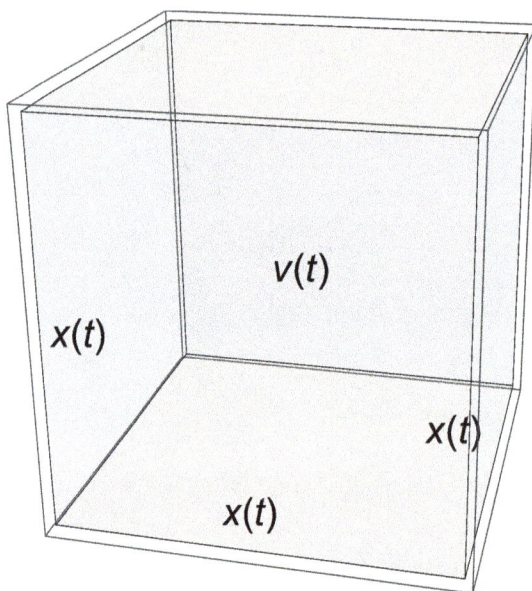

From geometry, you know the volume of a cube:

In[·]:= **eqn = v[t] == x[t]^3**

Out[·]= $v[t] == x[t]^3$

Differentiate the equation with respect to time:

In[·]:= **sol = D[eqn, t]**

Out[·]= $v'[t] == 3 x[t]^2 x'[t]$

You know $x[t]$ and $x'[t]$, and you want $v'[t]$, so calculate it:

In[ ]:= **sol /. {x[t] → 5, x'[t] → 0.5}**

Out[ ]:= v′[t] == 37.5

When the side length is 5 m, the volume is increasing at 37.5 m / sec.

Exercise 2—Cars

Two cars are headed toward an intersection. The first car is traveling east at 40 mph and the second car is traveling south at 30 mph. How fast are the cars approaching each other when the first car is 0.4 miles from the intersection and the second car is 0.2 miles from the intersection?

Solution

The distances the cars are from the intersection and the distance between the cars are all functions of time:

In[ ]:= **Clear[a, b, c]**

Draw a picture:

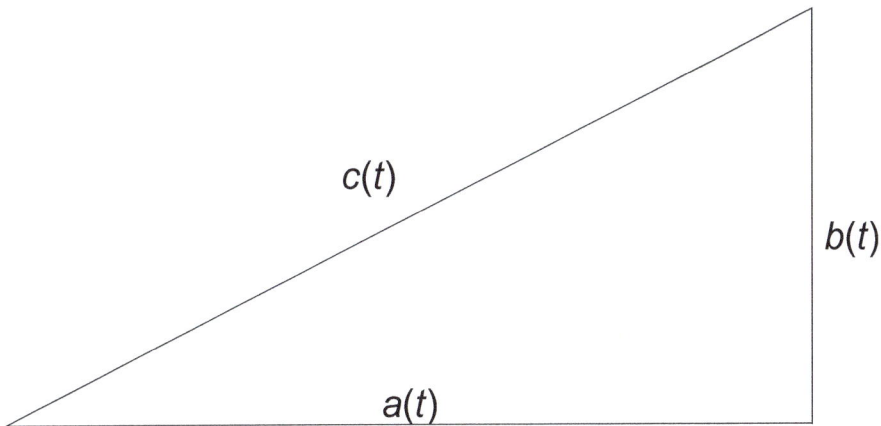

From the Pythagorean theorem, you have an equation relating all three distances:

In[ ]:= **eqn = a[t]^2 + b[t]^2 == c[t]^2**

Out[ ]:= a[t]^2 + b[t]^2 == c[t]^2

Differentiate with respect to t:

In[ ]:= **sol = D[eqn, t]**

Out[ ]:= 2 a[t] a′[t] + 2 b[t] b′[t] == 2 c[t] c′[t]

You know $a[t]$, $a'[t]$, $b[t]$ and $b'[t]$ and want to find $c'[t]$. You need $c[t]$, so solve for it in the original equation:

In[ ]:= **sol1 = Solve[(eqn /. {a[t] → 0.4, b[t] → 0.2}) && c[t] > 0, c[t]]**

Out[ ]= **{{c[t] → 0.447214}}**

Now solve for $c'[t]$ (note that $a'[t]$ and $b'[t]$ are both negative, since they are going toward the intersection):

In[ ]:= **Solve[sol /. {a[t] → 0.4, b[t] → 0.2, a'[t] → −40, b'[t] → −30, sol1[[1]][[1]]}, c'[t]]**

Out[ ]= **{{c'[t] → −49.1935}}**

When the first car is 0.4 miles and the second car is 0.2 miles from the intersection, they are approaching each other at 49.2 mph.

Exercise 3—Tracking Telescope

A tracking telescope on the ground tracks a plane that is 6 km above it. The angle decreases at a rate of 1 radian per minute when the angle of elevation is $\pi/6$. How fast is the plane traveling at that time?

Solution

The distance from the point 6 km above the telescope to the plane and the angle of elevation are functions of time:

In[ ]:= **Clear[d, θ]**

Draw a picture:

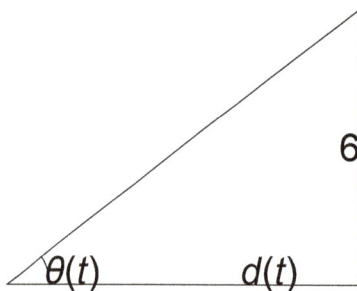

The angle and the distance traveled by the airplane are related by the tangent function:

In[ ]:= **eqn = Tan[θ[t]] == 6 / d[t]**

Out[ ]= $\text{Tan}[\theta[t]] == \dfrac{6}{d[t]}$

Differentiate the equation with respect to t:

In[ ]:= **sol = D[eqn, t]**

Out[ ]:= $\mathsf{Sec[\theta[t]]^2\ \theta'[t] == -\dfrac{6\ d'[t]}{d[t]^2}}$

You know $\theta'[t] = -1$, $\theta[t] = \pi/6$ and $d[t] = 6/\mathrm{Tan}[\theta[t]]$, so just solve for $d'[t]$:

In[ ]:= **Solve[sol /. d[t] → 6 / Tan[θ[t]] /. {θ[t] → π/6, θ'[t] → −1}, d'[t]]**

Out[ ]:= **{{d'[t] → 24}}**

At angle $\pi/6$, the plane is flying away at 24 km / min. (i.e. 1440 km / hr).

Exercise 4—Sand Cone

Coarse sand is thrown onto a pile. The pile takes the shape of a circular cone with base diameter and height always equal. If the sand is thrown on at a rate of 40 m^3 / min, how fast is the height of the pile rising when it is 20 m high?

Solution

The volume of the cone and the height of the cone are functions of time:

In[ ]:= **Clear[v, h]**

Draw a picture:

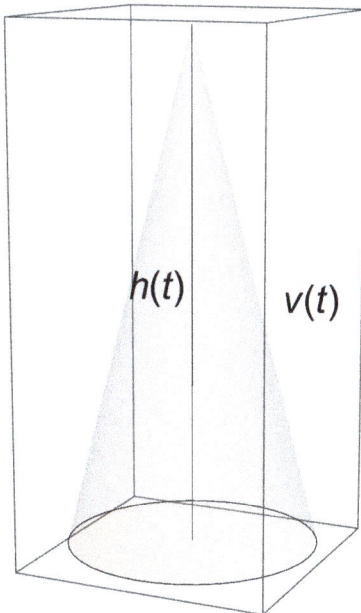

From geometry, you know the volume of a cone:

In[·]:= **eqn = v[t] == (π / 3) h[t] r[t]^2**

Out[·]= $v[t] == \dfrac{1}{3} \pi\, h[t]\, r[t]^2$

Given that the diameter and height are always equal, the radius is half the height. Alter the volume equation:

In[·]:= **eqn = % /. r[t] → h[t] / 2**

Out[·]= $v[t] == \dfrac{1}{12} \pi\, h[t]^3$

Differentiate the equation with respect to *t*:

In[·]:= **sol = D[eqn, t]**

Out[·]= $v'[t] == \dfrac{1}{4} \pi\, h[t]^2\, h'[t]$

You know $v'[t]$ and $h[t]$, so solve for $h'[t]$:

In[·]:= **Solve[sol /. {v'[t] → 40, h[t] → 20}, h'[t]]**

Out[·]= $\left\{\left\{h'[t] \to \dfrac{2}{5\,\pi}\right\}\right\}$

When the height is 20 m, it is increasing at $2/5\,\pi$ (roughly 0.13) m/sec.

15 | Linear Approximations

Overview

The derivative of a function at a point is calculated by "zooming" in on it with limits. As you get closer and closer to the desired point, a differentiable function looks more and more like its tangent line there.

Consider the plot of this function and its tangent line as you zoom in on the point $(1, 1)$:

```
In[ ]:= f[x_] := x^2
```

```
In[ ]:= Plot[{f[x], 1 + 2 (x - 1)}, {x, .5, 1.5},  ⋯  + ]
```

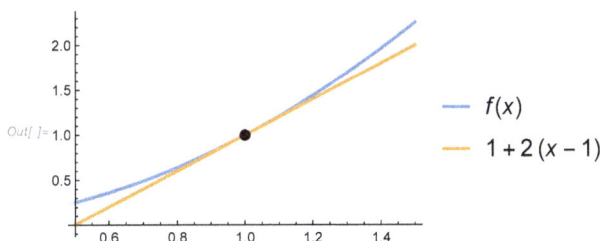

You can use this property to approximate a function near a point using its tangent line!

The goal of this lesson is to go over linear approximations, differentials and their various applications.

Linear Approximation

The equation you have been using for tangent lines comes from point-slope form:

$$y = (x - a) f'(a) + f(a)$$

Replacing y with $f[x]$, you can approximate y with the tangent:

$$f(x) = (x - a) f'(a) + f(a)$$

This approximation is called the **linear approximation** of f at a because a tangent line is used to approximate f:

```
In[ ]:= Clear[f]
```

```
In[ ]:= D[f[x], x]
```

```
Out[ ]= f'[x]
```

Its graph is called the **linearization** of f at a:

In[·]:= **Plot[{x^2, 1 + 2 (x – 1)}, {x, .5, 1.5}, ⋯ ⊕]**

Out[·]=

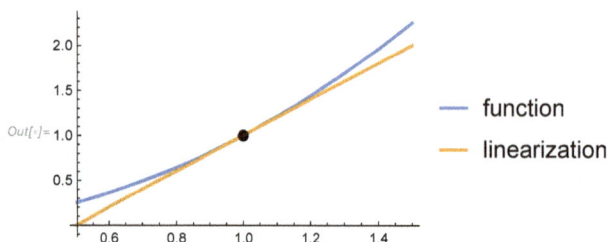

── function
── linearization

Finding Linearizations

Find the linearization of the following function at $a = 1$ and use it to approximate 8.98 and 9.03:

In[·]:= **f[x_] := Sqrt[x + 8]**

Calculate the derivative at $x = 1$:

In[·]:= **f '[1]**

Out[·]= $\dfrac{1}{6}$

The function is 3 when $a = 1$, so this is the linear approximation:

In[·]:= **tangent = Simplify[3 + 1/6 (x – 1)]**

Out[·]= $\dfrac{17 + x}{6}$

Approximate 8.98 and 9.03:

In[·]:= **{tangent /. x → 0.98, tangent /. x → 1.03}**

Out[·]= **{2.99667, 3.005}**

Plot the function, its linearization and the approximations on a graph:

In[·]:= **Plot[{f[x], tangent}, {x, 0, 2}, PlotLegends → {"function", "linearization"}, ⋯ → ⋯ ⊕]**

Out[·]=

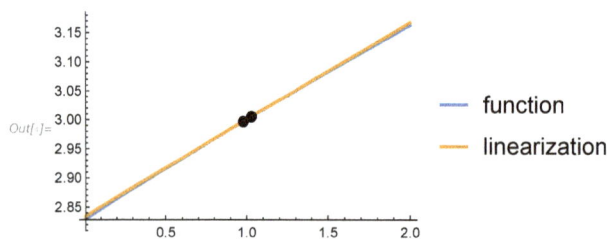

── function
── linearization

You can see that the approximations are very close to the original function.

Accuracy

For what values is the previous approximation accurate to within 0.5 units? In other words, where do the function and its approximation differ by less than 0.5?

This can be expressed using the absolute value function:

$$\left| \frac{1}{6} (-x - 17) + \sqrt{x + 8} \right| < 0.5$$

It is easy to solve using Reduce and RealAbs:

In[ ]:= **Reduce[RealAbs[Sqrt[x + 8] – tangent] < 1/2, x]**

Out[ ]= $2\left(2 - 3\sqrt{3}\right) < x < 2\left(2 + 3\sqrt{3}\right)$

Numerically, x should be between roughly -6.4 and 14.4:

In[ ]:= **% // N**

Out[ ]= $-6.3923 < x < 14.3923$

If you increase the accuracy to be less than 0.1, the interval shrinks:

In[ ]:= **Reduce[RealAbs[Sqrt[x + 8] – tangent] < 1/10, x]**

Out[ ]= $\frac{2}{5}\left(4 - 3\sqrt{15}\right) < x < \frac{2}{5}\left(4 + 3\sqrt{15}\right)$

In[ ]:= **% // N**

Out[ ]= $-3.04758 < x < 6.24758$

Physics Applications

Physicists use linear approximations quite often.

To derive a formula for the period of a pendulum, physics textbooks use the following equation:

In[ ]:= **eqn = a_T == –g Sin[θ]**

Out[ ]= a_T == –g Sin[θ]

where a_T is the tangential acceleration, g is the acceleration due to gravity, and θ is the angle the pendulum makes with the vertical.

At 0, the sine has slope 1 and value 0, so near 0, the sine is approximately:

In[·]:= **Sin[θ] == θ;**

Therefore for small values of θ, physicists use the approximation:

In[·]:= **eqn /. Sin[θ] → θ**

Out[·]= $a_T == -g\,\theta$

Here is a plot of the equation and its approximation near 0 for g = 9.8:

In[·]:= **Plot[{−9.8 Sin[θ], −9.8 θ}, {θ, −1, 1},**

 PlotLegends → {"tangential acceleration", "approximation"}]

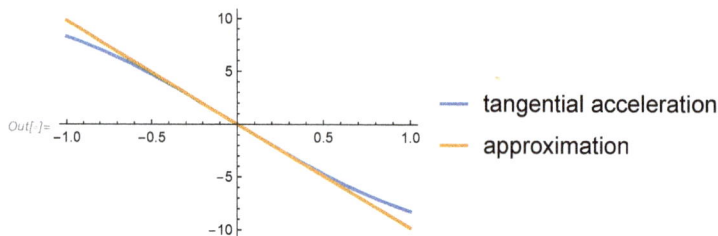

Differentials

Sometimes, the notation of **differentials** is used when doing approximations. Recall that the derivative of a function $y = f[x]$ with respect to x is sometimes denoted $\frac{dy}{dx}$.

If you let dx be an independent variable and let dy be a dependent variable, then you get the equation:

 dy == dx f'[x]

dx and dy are both known as **differentials**. If you are given dx and a value x in the domain of f, then you can find dy.

The following graph helps to illustrate:

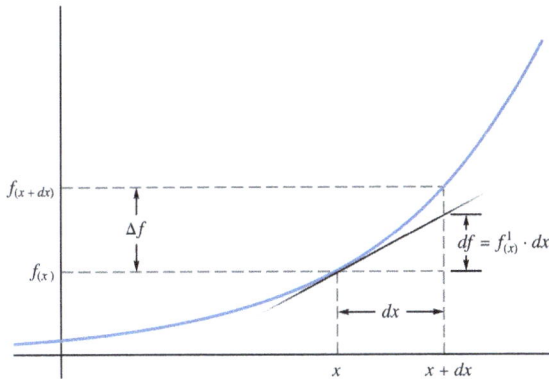

Given a point x, if you translate by $\Delta x = dx$ units, then df represents the change in f if you use the tangent line approximation. Δf represents the actual change in f.

Differentials Example

Consider the following function:

In[ ]:= **f[x_] := x^3 + 2 x^2 − x + 1**

Compare the values of df and Δf as x changes from 3 to 3.05.

Find Δf:

In[ ]:= **deltaf = f[3.05] − f[3]**

Out[ ]= 1.92762

Find df:

In[ ]:= **df = f'[3] (0.05)**

Out[ ]= 1.9

In general, the equation for df is:

In[ ]:= **df = f'[x] dx**

Out[ ]= $dx \left(-1 + 4x + 3x^2\right)$

Here are the function and its tangent line in a graph:

In[·]:= **Plot[{f[x], f[3] + f'[3] (x − 3)}, {x, 0, 6}]**

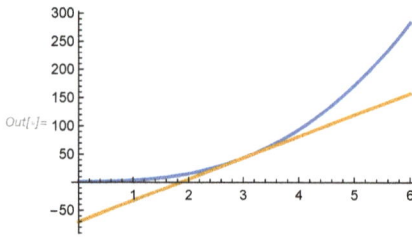

Out[·]=

Importance of Differentials

Differentials are important because the functions may make it impossible to find Δf exactly. Therefore, using the approximation $df \sim \Delta f$ can be very helpful.

For a function f, its linear approximation can be given as:

　　f[a + dx] == dy + f[a]

For example, the very first function:

In[·]:= **f[x_] := Sqrt[x + 8];**

has differential dy:

In[·]:= **dy = f'[x] dx**

$$Out[·]= \frac{dx}{2\sqrt{8+x}}$$

So 8.98 can be approximated with $x = 1$ and $dx = -0.02$:

In[·]:= **f[1] + dy /. {x → 1, dx → −0.02}**

Out[·]= **2.99667**

Compare to the actual answer:

In[·]:= **f[0.98]**

Out[·]= **2.99666**

Error Analysis

Differentials are also useful when estimating errors that occur because of approximate measurements.

For example, the radius of a sphere is measured to be 11 cm with a possible error in measurement of at most 0.1 cm. You want to find the maximum error in using this value to calculate the volume of the sphere.

You know the volume for a general sphere:

In[ ]:= **eqn = vol[r] == (4/3) π r^3**

Out[ ]= vol[r] == $\dfrac{4\pi r^3}{3}$

You also know its derivative with respect to *r*:

In[ ]:= **D[eqn, r]**

Out[ ]= vol'[r] == 4 π r²

This can be expressed in terms of differentials:

In[ ]:= **sol = dvol == 4 π r^2 dr**

Out[ ]= dvol == 4 dr π r²

You know *dr* = 0.1 and *r* = 11, so you can estimate Δ*vol* by calculating *d vol*:

In[ ]:= **sol /. {r → 11, dr → 0.1}**

Out[ ]= dvol == 152.053

Relative Error

The previous error of roughly 152 cm³ seems rather large given such a small radius error (= 0.1 cm)! To better understand such numbers, the **relative error** is often used.

It is calculated by dividing the total error by the total volume:

In[ ]:= **4 π r^2 dr /(4/3 π r^3) /. {r → 11, dr → 0.1}**

Out[ ]= 0.0272727

So the relative error is 0.027. This looks much better than before!

If you simplify the preceding expression, you can find the relative error for volume in terms of the relative error for radius:

In[ ]:= **4 π r^2 dr /(4/3 π r^3)**

Out[ ]= $\dfrac{3\,dr}{r}$

So the relative error for volume (0.027) is three times the relative error for radius (0.009).

When the relative error is expressed as a percentage, you get the **percentage error**. So the percentage error for volume is 2.7% and the percentage error for radius is 0.9%.

Summary

Linear approximation is very important in math and science.

Linearization involves finding the tangent line to a curve at a particular point.

For values close to the point, the approximation is very close to the actual function.

Linear approximations are particularly useful in physics.

Differentials are useful in finding approximations too.

They are especially useful in error analysis.

The next lesson continues investigating the applications of differential calculus and shows how to find maxima and minima.

Exercises

Exercise 1—Linearization

Find the linearization of the following function at $x = 1$:

In[ ]:= **g[x_] := x Sin[x] − x**

Solution

Calculate the derivative at $x = 1$:

In[ ]:= **g'[1]**

Out[ ]= −1 + Cos[1] + Sin[1]

Then calculate the linear approximation of f at 1:

In[ ]:= **tangent = Simplify[g[1] + g'[1] (x − 1)]**

Out[ ]= −Cos[1] + x (−1 + Cos[1] + Sin[1])

Here is a plot of the function and its linearization:

In[ ]:= **Plot[{g[x], tangent}, {x, −1, 3}]**

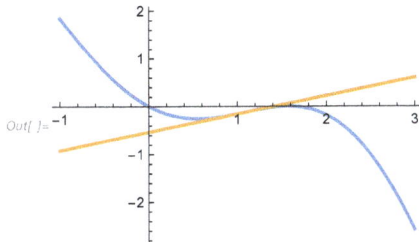

Exercise 2—Estimation and Accuracy

Use linear approximation to estimate the value of $\sqrt[3]{64.1}$ using the function $\sqrt[3]{x + 64}$. Find for what values of x it is accurate to within 0.01.

Solution

Define the function:

In[ ]:= **h[x_] := CubeRoot[x + 64]**

Find the linear approximation of f at $x = 0$:

```
In[·]:= tangent = h[0] + h'[0] x
```

$$Out[·]= 4 + \frac{x}{48}$$

$\sqrt[3]{64.1}$ is found to be approximately:

```
In[·]:= SetPrecision[tangent /. x → 0.1, 10]
```

```
Out[·]= 4.002083333
```

The actual answer is:

```
In[·]:= SetPrecision[CubeRoot[64.1], 10]
```

```
Out[·]= 4.002082249
```

The answers agree up to the hundred thousands place!

Find the range of accuracy for the linearization:

```
In[·]:= Reduce[RealAbs[h[x] – tangent] < 0.01, x] // Quiet
```

```
Out[·]= −576.64 < x < −575.36 || −9.20171 < x < 10.0016
```

So the approximation is good when x is between -576.6 and -575.3 or when x is between -9.2 and 10.0.

Exercise 3—Differentials

For the following function, find df and evaluate df given that $x = 7/2$ and $dx = 0.2$. Then approximate $f[3.7]$.

```
In[·]:= f[x_] := Sin[Cos[π x]]
```

Solution

You know that $df = f'[x]\, dx$:

```
In[·]:= df = f'[x] dx
```

```
Out[·]= −dx π Cos[Cos[π x]] Sin[π x]
```

Now plug in the values:

```
In[·]:= df /. {x → 7/2, dx → 0.2}
```

```
Out[·]= 0.628319
```

$3.7 = 7/2 + 0.2$, so $f[3.7]$ is approximately:

In[ ]:= **f[7 / 2] + df /. {x → 7 / 2, dx → 0.2}**

Out[ ]= 0.628319

The actual value is:

In[ ]:= **f[3.7]**

Out[ ]= 0.554519

Here is a plot with the function, its tangent line and the estimated and actual values of $f[3.7]$:

In[ ]:= **Plot[{f[x], f[7 / 2] + f'[7 / 2] (x − 7 / 2)}, {x, 7 / 2, 7 / 2 + 0.3}, ⋯]**

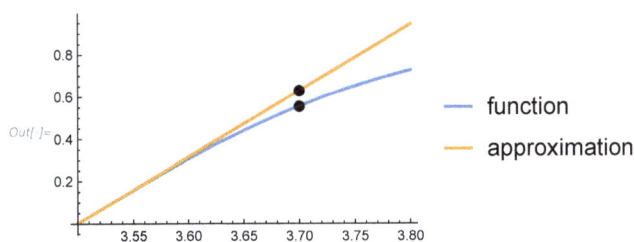

Exercise 4—Error Analysis

The radius of a circular disk is given as 50 cm with a maximum error in measurement of 0.1 cm. Use differentials to find the total error, relative error and percentage error in the calculated area of the disk.

Solution

The area of a circular disk is given by the following formula:

In[ ]:= **eqn = area[r] == π r^2;**

Its derivative is:

In[ ]:= **D[eqn, r]**

Out[ ]= area′[r] == 2 π r

In differential form, this says:

In[ ]:= **sol = *d*area == 2 π r *d*r**

Out[ ]= *d*area == 2 π r *d*r

Plug in $r = 50$, $dr = 0.1$ to find the total error:

In[ ]:= **sol /. {r → 50, dr → 0.1}**

Out[ ]= **d area == 31.4159**

The relative error is the total error divided by the area:

In[ ]:= **2 π r dr / (π r^2) /. {r → 50, dr → 0.1}**

Out[ ]= **0.004**

which as a percentage is 0.4%.

Exercise 5—Binomial Approximation

The binomial theorem states that the expression $(x + y)^n$ can be expressed as:

$$(x + y)^n = \sum_{i=0}^{n} \frac{n! \, x^i \, y^{n-i}}{i! \, (n - i)!}$$

where $x! = x * (x - 1) * (x - 2) * \ldots * 2 * 1$.

Find the linear approximation for $(1 + x)^n$ near 1.

Solution

Linearize the function:

In[ ]:= **binomial[x_] := (1 + x)^n**

… near $x = 0$:

In[ ]:= **D[binomial[x], x] /. x → 0**

Out[ ]= **n**

The resulting linear approximation for $(1 + x)^n$ near 1 is $1 + n\,x$:

In[ ]:= **approx = binomial[0] + binomial'[0] x**

Out[ ]= **1 + n x**

For example, 1.01^{10} is approximately:

In[ ]:= **1 + 10 * 0.001**

Out[ ]= **1.01**

The actual answer is:

In[ ]:= **(1.001)^10**

Out[ ]= **1.01005**

16 | Maxima and Minima

Overview

Many mathematical problems involve getting the best solution for a task. The best solution is called the optimal solution, and the process is called **optimization**.

Consider the plot of this function:

In[ ]:= **f[x_] := −x^2 + 3**

In[ ]:= **Plot[f[x], {x, −5, 5}]**

Out[ ]=

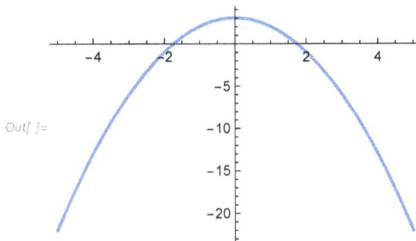

Its highest value is at the point (0, 3).

The goal of this lesson is to develop a process for finding maximum or minimum solutions for problems, using derivatives.

The Importance of Optimization

Optimization is very important in various fields.

Businesses want to maximize their profits and minimize their costs.

Plants grow in such a way as to obtain the maximum amount of sunlight.

GPS systems are designed to get you to your destination in the shortest amount of time (or distance).

In general, optimization problems involve finding the **maximum** or **minimum** value for a given quantity. Differential calculus is highly suited to solving many classes of optimization problems.

Absolute Maxima and Minima

Let c be a number in the domain D of a function f. Then $f[c]$ is the **absolute maximum** value of f on D if $f[c] \geq f[x]$ for all x in D.

$f[c]$ is the **absolute minimum** value of f on D if $f[c] \leq f[x]$ for all x in D.

Consider the sine function in the interval $[0, 2\pi]$:

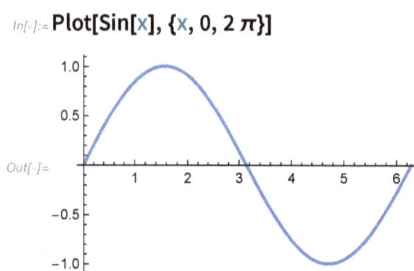

In[·]:= **Plot[Sin[x], {x, 0, 2 𝜋}]**

Out[·]=

In this interval, the sine has an absolute maximum of 1 at $x = \pi/2$ and an absolute minimum of -1 at $x = 3\pi/2$.

The absolute maximum and minimum are sometimes called the **global** maximum and minimum. Both the absolute maximum and absolute minimum are called **extreme values** of f.

Local Maxima and Minima

Consider the following function:

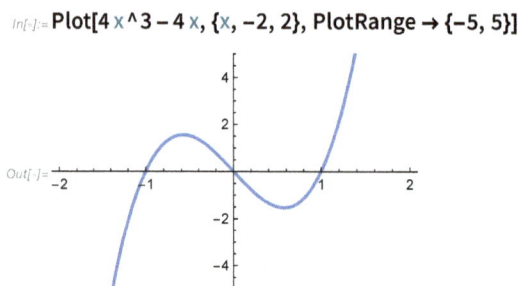

In[·]:= **Plot[4 x^3 – 4 x, {x, –2, 2}, PlotRange → {–5, 5}]**

Out[·]=

There are points in the graph that are not absolute maxima or minima, but they are still the largest or smallest values of the points near them. These points are called **local maxima or minima**.

Let c be a number in the domain D of a function f.

Then $f[c]$ is a **local maximum** value of f on D if $f[c] \geq f[x]$ for x near c.

$f[c]$ is the **local minimum** value of f on D if $f[c] \leq f[x]$ for x near c.

The preceding function has a local maximum at $\left(\frac{-1}{\sqrt{3}}, \frac{8}{3\sqrt{3}} \right)$ and a local minimum at $\left(\frac{1}{\sqrt{3}}, \frac{-8}{3\sqrt{3}} \right)$. An absolute maximum is by definition a local maximum, and an absolute minimum is by definition a local minimum.

Special Cases

Some functions have absolute maxima and minima everywhere in their domain.

Consider the plot of the constant function $f[x] = 5$:

In[ ]:= **Plot[5, {x, −5, 5}]**

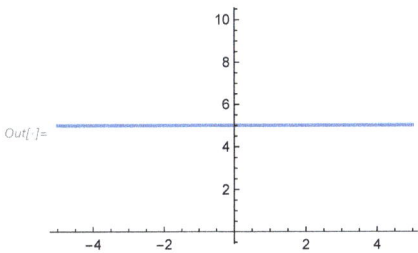

Out[ ]=

$f[c] \geq f[x] = 5$ and $f[c] \leq f[x] = 5$ for every real number c, so every real number is an absolute maximum and minimum!

Some functions do not have any absolute maxima or minima (or even local maxima or minima) in their domain.

Consider the plot of the function $f[x] = x^3$:

In[ ]:= **Plot[x^3, {x, −10, 10}]**

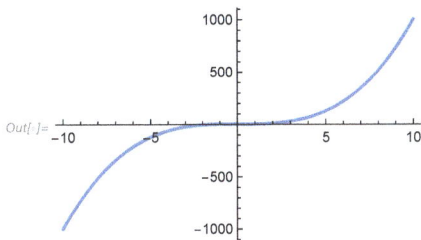

Out[ ]=

From the definitions, you can see that it has no absolute (or local) maxima or minima.

Extreme Value Theorem

Given the previous examples, how can you know precisely when a function has any extreme values?

The **extreme value theorem** gives conditions for when a function is guaranteed to have extreme values. The extreme value theorem states that if f is **continuous** on a **closed interval** $[a, b]$, then f attains an absolute maximum value $f[c]$ and an absolute minimum value $f[d]$ at some numbers c and d in $[a, b]$.

Here are some examples:

In[ ]:= **{Plot[Sin[x], {x, −5, 5}, ⋯ → ⋯ ⋄],**

Plot[1 / x, {x, −5, 5}, ⋯ → ⋯ ⋄], Plot[x, {x, −5, 5}, ⋯ → ⋯ ⋄]}

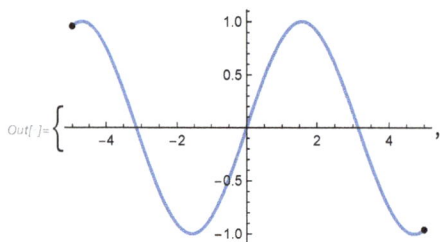

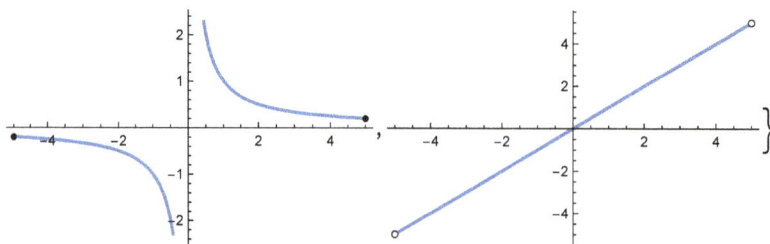

Out[ ]=

The black dots indicate that a side of the interval is closed, and the white dots indicate that a side of the interval is not closed.

The first example shows a continuous function on a closed interval, so the function attains an absolute maximum and minimum on the interval.

The second example shows a discontinuous function on a closed interval, so there is no guarantee the function attains an absolute maximum or minimum (you can see from the graph that it does not).

The last example shows a continuous function on an interval that is not closed, so there is no guarantee the function attains an absolute maximum and minimum (in this case, you can verify from the definitions that it does not).

Fermat's Theorem

The extreme value theorem guarantees when a function has an absolute maximum or minimum, but it does not tell how to find them.

Fermat's theorem gives a way of finding a local maximum or minimum for a differentiable function. It is named in honor of Pierre de Fermat.

The theorem utilizes the observation that local maxima or minima for differentiable functions have horizontal tangents (i.e. the slope is 0). Fermat's theorem states that if f has a local maximum or minimum at c, and if $f'[c]$ exists, then $f'[c] = 0$.

Consider the previous function $4 x^3 - 4 x$:

In[ ]:= **Plot[4 x^3 – 4 x, {x, –2, 2}, PlotRange → {–5, 5}]**

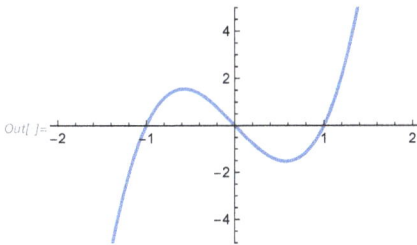

Out[ ]=

It has local maxima and minima at $\left(\frac{-1}{\sqrt{3}}, \frac{8}{3\sqrt{3}} \right)$ and $\left(\frac{1}{\sqrt{3}}, \frac{-8}{3\sqrt{3}} \right)$, respectively.

Confirm the derivatives at those points are zero:

In[ ]:= **{D[4 x^3 – 4 x, x] /. x → –1 / Sqrt[3], D[4 x^3 – 4 x, x] /. x → 1 / Sqrt[3]}**

Out[ ]= **{0, 0}**

Caveats to Fermat's Theorem

Note that the converse of Fermat's theorem may not be true.

Consider the function x^3:

In[ ]:= **Plot[x^3, {x, −2, 2}]**

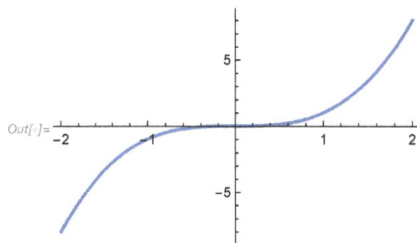

Even though the function has a horizontal tangent at $x = 0$, $(0, 0)$ is neither a local maximum or minimum of the function.

The absolute value function has a local (and absolute) minimum at $(0, 0)$, even though the derivative does not exist there:

In[ ]:= **Plot[RealAbs[x], {x, −1, 1}]**

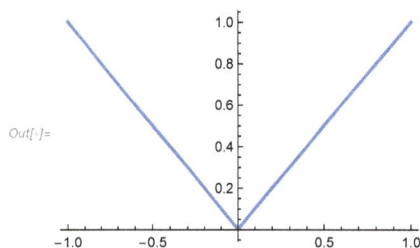

Critical Numbers

Despite the drawbacks of Fermat's theorem, it does give an idea of where to look for local maxima and minima.

A **critical number** of a function f is a number c in the domain of f such that either $f\,'[c] = 0$ or $f\,'[c]$ does not exist.

For example, the function $\sqrt[3]{x}$ has a critical point at 0 because the derivative is not defined there:

In[ ]:= **Plot[CubeRoot[x], {x, −1, 1}]**

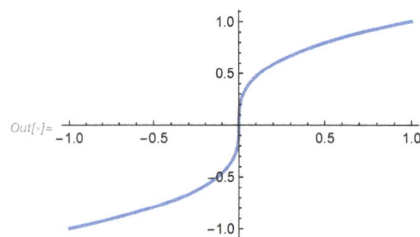

In general, if f has a local maximum or minimum at c, then c is a critical number of f.

Closed Interval Method

Fermat's theorem and the extreme value theorem lead to a general method for finding absolute maxima and minima for closed intervals.

To find the absolute maximum and minimum values of a continuous function f on a closed interval $[a, b]$:

1. Find the values of f at the critical numbers of f in (a, b).

2. Find the values of f at the endpoints of the interval.

3. The largest of the values from 1 and 2 will be the absolute maximum, and the smallest of the values from 1 and 2 will be the absolute minimum.

Consider the function $x^3 - x^2 + x$ on the interval $[0, 2]$:

In[ ]:= **f[x_] := x^3 – x^2 – x**

It has a critical number at $x = 1$:

In[ ]:= **sol = Solve[D[f[x], x] == 0 && 0 ≤ x ≤ 2, x]**

Out[ ]= **{{x → 1}}**

Calculate the function values at 0, 1 and 2:

In[ ]:= **f /@ {0, 1, 2}**

Out[ ]= **{0, –1, 2}**

$f[1] = -1$, $f[0] = 0$, and $f[2] = 2$, so it has an absolute maximum at $(2, 2)$ and an absolute minimum at $(-1, 1)$:

You can confirm with Maximize and Minimize:

In[ ]:= **{Maximize[{f[x], 0 ≤ x ≤ 2}, x], Minimize[{f[x], 0 ≤ x ≤ 2}, x]}**

Out[ ]= **{{2, {x → 2}}, {–1, {x → 1}}}**

Business Example

A business makes a product with the following cost function:

In[ ]:= **cost[x_] := 7134.23 – 1.396 x + 0.000019647 x² + 8.165084*^–9 x³**

Find the number of units that minimizes the cost.

First, find the critical numbers of the function:

In[·]:= **crit = Solve[D[cost[x], x] == 0 && 0 ≤ x ≤ 10 000, x]**

Out[·]= **{{x → 6789.63}}**

Then plug the critical numbers and endpoints into the function:

In[·]:= **cost /@ {0, 10 000, crit[[1, 1, 2]]}**

Out[·]= **{7134.23, 3304.01, 1117.25}**

Producing about 6790 units will minimize the cost.

Here is the function and its minimum in a graph:

In[·]:= **Plot[cost[x], {x, 0, 10 000}, ⋯ ✦]**

Out[·]=

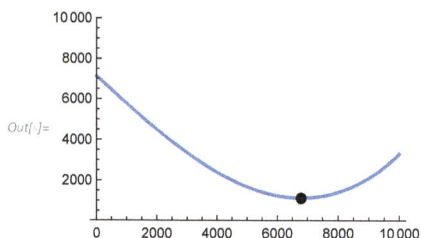

Summary

Optimization is a crucial component of everyday life.

Businesses want to make the most money with the least amount of cost.

Drivers want to get to their destinations in the least amount of time.

Fermat's theorem helps find the critical numbers of a function.

The closed interval method utilizes the critical numbers of a function to find its absolute maxima and minima on closed intervals.

The next lesson will cover the mean value theorem, a very powerful tool in calculus.

Exercises

Exercise 1—Max and Min of a Trig Function

Use the closed interval method to find the absolute maxima and minima in the range [1, 6] for the following function:

In[ ]:= **f[*x*_] := *x* Sin[*x*]**

Solution

First, find the critical numbers in the range:

In[ ]:= **crit = Solve[D[f[x], x] == 0 && 1 ≤ x ≤ 6, x] // N**

Out[ ]= **{{x → 2.02876}, {x → 4.91318}}**

Then evaluate the function at the critical numbers and the endpoints:

In[ ]:= **f /@ {1, 6, crit〚1, 1, 2〛, crit〚2, 1, 2〛} // N**

Out[ ]= **{0.841471, −1.67649, 1.81971, −4.81447}**

You can see the function has an absolute maximum at roughly (2.03, 1.82) and an absolute minimum at roughly (4.91, −4.81).

The function and its extreme values are plotted in a graph:

In[ ]:= **Plot[f[x], {x, 1, 6}, Epilog → {PointSize[Large], Point[{x, f[x]} /. crit]}]**

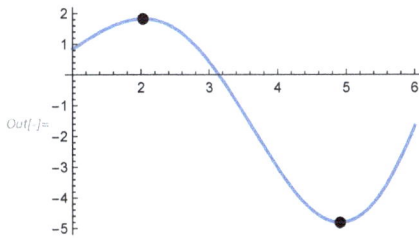

Exercise 2—Critical Numbers

Find the critical numbers of the following function:

In[ ]:= **f[*x*_] := RealAbs[*x*^3 − 4 *x*^2 + 3]**

Solution

First, look at its derivative:

$In[\cdot]:=$ **f '[x]**

$$Out[\cdot]= \frac{\left(-8x+3x^2\right)\left(3-4x^2+x^3\right)}{RealAbs\left[3-4x^2+x^3\right]}$$

The critical numbers will be wherever $f\,'[x]$ equals 0 or does not exist.

Set the numerator and denominator equal to 0 and solve to find the critical numbers:

$In[\cdot]:=$ **numerator = Solve[(−8 x + 3 x²) (3 − 4 x² + x³) == 0, x];**
denominator = Solve[RealAbs[x^3 − 4 x^2 + 3] == 0, x];

$In[\cdot]:=$ **crit = Union[numerator, denominator]**

$$Out[\cdot]= \left\{\{x \to 0\}, \{x \to 1\}, \left\{x \to \frac{8}{3}\right\}, \left\{x \to \frac{1}{2}\left(3-\sqrt{21}\right)\right\}, \left\{x \to \frac{1}{2}\left(3+\sqrt{21}\right)\right\}\right\}$$

The critical numbers lie at 0, 1, $\frac{8}{3}$, $\frac{1}{2}\left(3-\sqrt{21}\right)$ and $\frac{1}{2}\left(3+\sqrt{21}\right)$.

Plot them and the function:

$In[\cdot]:=$ **Plot[f[x], {x, −2, 5}, Epilog → {PointSize[Large], Point[{x, f[x]} /. crit]}] // Quiet**

$Out[\cdot]=$

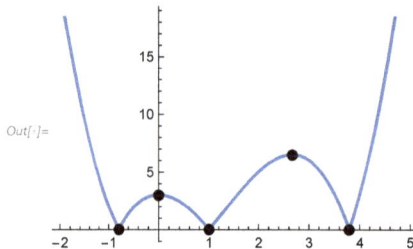

Exercise 3—Max and Min of a Polynomial

Find the absolute maximum and minimum for the following function in the range $[-5, 5]$:

$In[\cdot]:=$ **f[x_] := x^3 − 3 x^2 + 3 x − 1**

Solution

Use the closed interval method and find the critical numbers of the function.

Since f is a polynomial, its derivative is defined everywhere, so the values exist only where the derivative is 0:

In[ ]:= **crit = Solve[f '[x] == 0, x]**

Out[ ]= **{{x → 1}, {x → 1}}**

Then evaluate the function at the critical number and the endpoints:

In[ ]:= **f /@ {1, −5, 5}**

Out[ ]= **{0, −216, 64}**

The function has an absolute maximum at $(5, 64)$ and an absolute minimum at $(-5, -216)$ in the range $[-5, 5]$.

Here is a plot of the function, its critical number and its extreme points:

In[ ]:= **Plot[f[x], {x, −5, 5}, ⋯ ✦]**

Out[ ]=

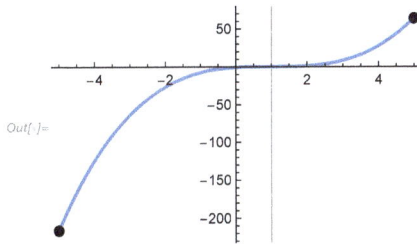

Exercise 4—Business

A business with a new product has the profit function:

In[ ]:= **profit[x_] := 10.9 + 12.5219 x − 0.004865 x^2 + 5.3538*^−7 x^3**

If the business can only afford to produce 5000 units, how many units should it produce to maximize its profit?

Solution

Use the closed interval method. Find the critical values of the function in the range $[0, 5000]$:

In[ ]:= **sol = Solve[profit '[x] == 0 && 0 ≤ x ≤ 5000, x]**

Out[ ]= **{{x → 1854.87}, {x → 4203.13}}**

Then evaluate the function at the critical numbers and the endpoints:

In[ ]:= **profit /@ {0, 5000, sol[[1, 1, 2]], sol[[2, 1, 2]]}**

Out[ ]= **{10.9, 7917.9, 9915.82, 6449.49}**

The function has a maximum at about 1855, so the business should produce 1855 units to maximize profits.

Here are the function and its maximum in a graph:

In[·]:= `Plot[profit[x], {x, 0, 5000}, Epilog → {PointSize[Large], Point[{x, profit[x]} /. sol[[1]]]}]`

Out[·]=

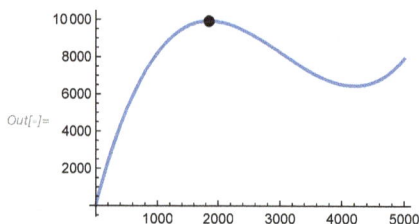

Exercise 5—Friction Revisited

An object with weight w is dragged along a horizontal plane by a force acting along a rope attached to the object. If the rope makes an angle θ with the plane, recall that the magnitude of the force is given by:

In[·]:= `Clear[w, μ];`
`force[θ_] := μ w / (μ Sin[θ] + Cos[θ])`

where μ is a positive constant known as the **coefficient of friction**.

In terms of μ, find when the force is minimized in the range $[0, \pi/2]$:

Solution

Simply solve for μ in the equation force'$[\theta]$ = 0 with the constraints:

In[·]:= `sol = Solve[force'[θ] == 0 && 0 ≤ θ ≤ π/2 && μ > 0, μ]`

Out[·]= $\left\{\left\{\mu \to \boxed{\text{Tan}[\theta] \ \text{if} \ \ 0 < \theta < \dfrac{\pi}{2}}\right\}\right\}$

So if the angle is 45° ($\pi/4$ radians) and μ is 1, then the force will be minimized:

In[·]:= `force'[θ] /. {θ → π/4, μ → 1}`

Out[·]= `0`

Plot the function with the minimum when μ is 1 and the weight is 50 in the range $[0, \pi / 2]$:

```
In[ ]:= Plot[force[θ] /. {μ → 1, w → 50}, {θ, 0, π/2}, PlotRange → {25, 50},
        Epilog → {PointSize[Large], Point[{π/4, force[π/4] /. {μ → 1, w → 50}}]}]
```

Out[ ]=

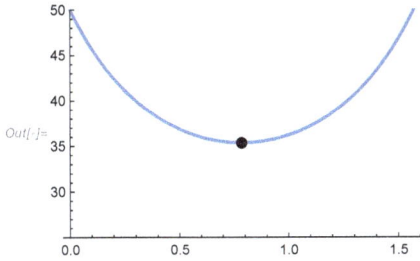

17 | The Mean Value Theorem

Overview

The **mean value theorem** relates the slope of an interval (i.e. the slope of the secant line connecting the two endpoints of the interval) with the instantaneous slope of some point in the interval:

$In[\ ]:=$ `Plot[{x^2, 1/4 - (x + 1/2)}, {x, -2, 1}, ⋯ ÷ ]`

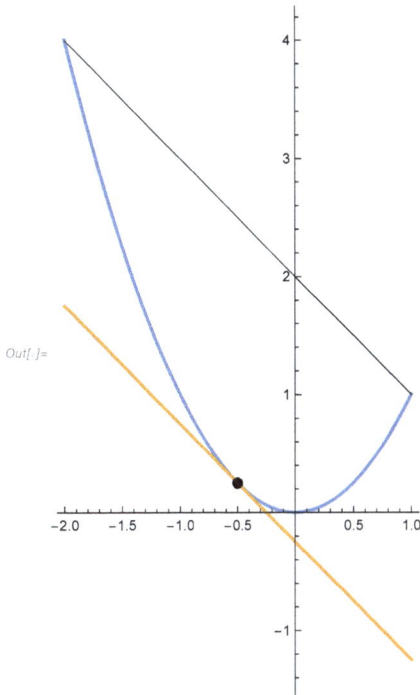

$Out[\ ]=$

It has many uses, some of which will be covered in later lessons.

The goal of this lesson is to develop a good understanding of the mean value theorem and provide a few applications. It begins with a special version of the mean value theorem, called **Rolle's theorem**.

Observation

To start, look at the function x^2 in the range $[-2, 2]$:

In[·]:= **Plot[x^2, {x, -2, 2}]**

Out[·]=

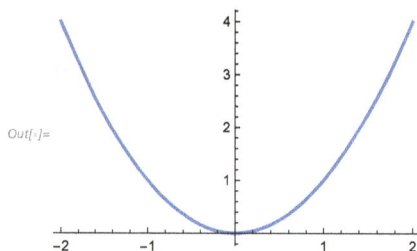

Notice that at the endpoints -2 and 2, the function has the same value:

In[·]:= **# ^2 &/@ {-2, 2}**

Out[·]= **{4, 4}**

Notice also that the slope at 0 is 0:

In[·]:= **f[x_] := x^2**

In[·]:= **f'[0]**

Out[·]= **0**

Rolle's Theorem

The following theorem is by Michel Rolle:

Let f be a function that satisfies the following three hypotheses:

1. f is continuous on the closed interval $[a, b]$.

2. f is differentiable on the open interval (a, b).

3. $f[a] = f[b]$.

Then there is a number c in (a, b) such that $f'[c] = 0$.

The previous function in the interval $[-2, 2]$ satisfies all three hypotheses, so at some point the slope is zero.

In that case, the point was (0, 0). Here is the plot of the function again:

In[ ]:= **Plot[x^2, {x, −2, 2}]**

Out[ ]=

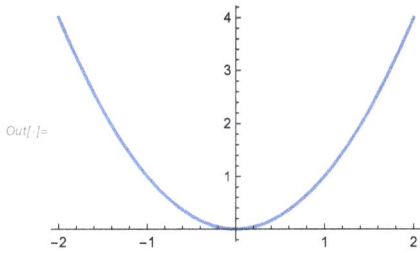

Observation 2

Now look at the function x^{-2} in the range $[-1, 1]$:

In[ ]:= **Plot[1/x^2, {x, −1, 1}]**

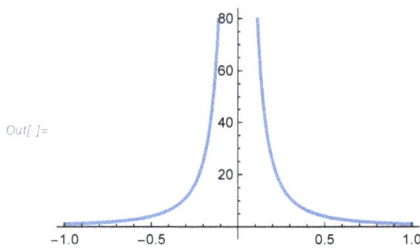

Out[ ]=

The function has a discontinuity in the interval $[-1, 1]$ at $x = 0$. It is therefore not differentiable at $x = 0$ either.

At the endpoints -1 and 1, this function has the same value:

In[ ]:= **1/#^2 &/@ {−1, 1}**

Out[ ]= **{1, 1}**

Because it does not satisfy all three hypotheses of Rolle's theorem, it cannot be guaranteed that it has any number in its range where the slope is zero.

Solve in fact shows it does not:

In[ ]:= **Solve[D[1/x^2, x] == 0 && −1 ≤ x ≤ 1, x]**

Out[ ]= **{}**

Observation 3

Finally, look at the function $|x|$ in the range $[-1, 1]$:

In[ ]:= **Plot[RealAbs[x], {x, −1, 1}]**

Out[ ]=

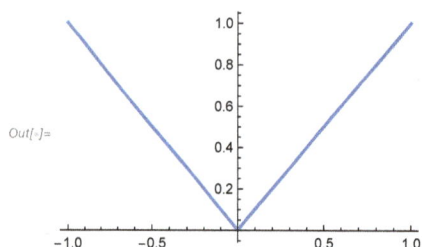

The function is continuous in the interval $[-1, 1]$. However it is not differentiable at $x = 0$.

At the endpoints -1 and 1, this function also has the same value:

In[ ]:= **RealAbs[#] & /@ {−1, 1}**

Out[ ]= **{1, 1}**

However, it does not have any number in its range where the slope is zero. Because it does not satisfy all three hypotheses of Rolle's theorem, it cannot be guaranteed that has any number in its range where the slope is zero.

Solve again shows it does not:

In[ ]:= **Solve[RealAbs'[x] == 0 && −1 ≤ x ≤ 1, x]**

Out[ ]= **{}**

Motivation for Mean Value Theorem

In the first example, notice that the secant line between $x = -2$ and $x = 2$ and the tangent line at $x = 0$ have the same slope:

In[ ]:= **Plot[{x^2, 0}, {x, −2, 2}, Axes → {False, True}, Epilog → {Line[{{−2, (−2)^2}, {2, 2^2}}]}]**

Out[ ]=

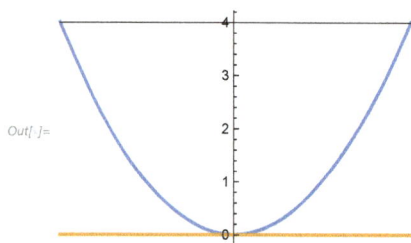

Now consider the same function in the range $[-0.5, 1]$:

In[ ]:= **Plot[{x^2, 1/16 + 1/2 (x − 1/4)}, {x, −0.5, 1}, AspectRatio → Automatic,**
 Epilog → {PointSize[Large], Point[{1/4, 1/16}], Line[{{−0.5, 0.25}, {1, 1}}]}]

Out[ ]=

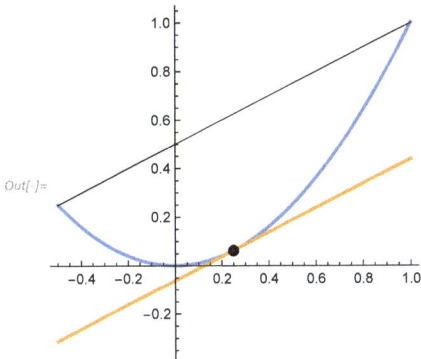

The secant line between $x = -0.5$ and $x = 1$ has slope $1/2$. Notice that at $x = 1/4$, the tangent line also has slope $1/2$.

Mean Value Theorem

The previous two observations can be summarized by the mean value theorem:

Let f be a function that satisfies the following hypotheses:

  1. f is continuous on the closed interval $[a, b]$.

  2. f is differentiable on the open interval (a, b).

Then there is a number c in (a, b) such that the slope of the tangent line at c equals the slope of the secant line between a and b.

Put in equation form:

$$f'(c) = \frac{f(b) - f(a)}{b - a}$$

$$f(b) - f(a) = f'(c)(b - a)$$

Notice that Rolle's theorem follows from the mean value theorem: if $f[b] = f[a]$, then $f[b] - f[a] = 0 = f'[c]$.

The following plot illustrates the theorem for the sine function in the range $[0, \pi]$:

In[·]:= **Plot[{Sin[x], 1}, {x, 0, π}, Axes → {False, True}, Epilog → {PointSize[Large], Point[{{0, Sin[0]}, {π, Sin[π]}}], Line[{{0, Sin[0]}, {2 π, Sin[2 π]}}]}]**

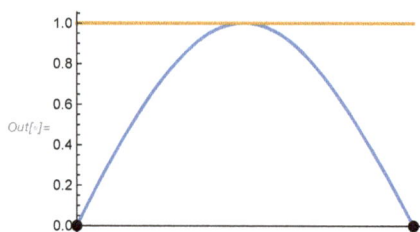

Out[·]=

Mean Value Theorem Subtleties

Note that there may be more than one c with tangent parallel to the secant line:

In[·]:= **Plot[{Sin[x], Sin[ArcCos[-$\dfrac{3}{5\pi}$]] + -$\dfrac{3}{5\pi}$ (x - ArcCos[-$\dfrac{3}{5\pi}$]),**

Sin[-ArcCos[-$\dfrac{3}{5\pi}$]] + -$\dfrac{3}{5\pi}$ (x + ArcCos[-$\dfrac{3}{5\pi}$] - 2 π)}, {x, π/6, 11 π/6}, ⋯ ⊕]

Out[·]=

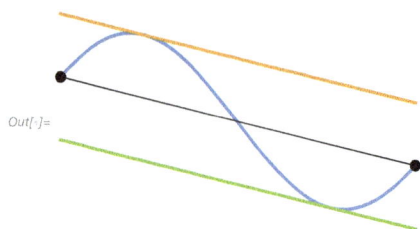

In particular, every function with slope 0 everywhere must be constant. In this case, $f'[c] (b - a) = 0 = f[b] - f[a]$, so $f[b] = f[a]$ for any a and b.

You can see this in the plot for the constant function $f[x] = 5$:

In[·]:= **f[x_] := 5**

In[·]:= **Plot[{f[x], f'[x]}, {x, −10, 10}, ⋯ ⊕]**

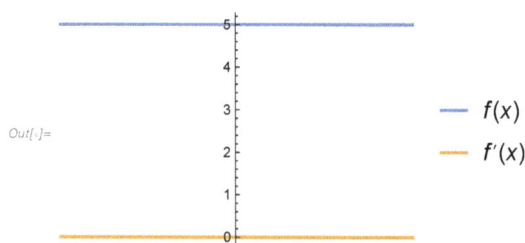

Out[·]=

— $f(x)$

— $f'(x)$

Application

Consider the following function with $a = 0$, $b = 4$:

In[ ]:= **f[x_] := x^3 – 8 x**

Find c in the range (a, b) that has tangent line parallel to the secant line going through $(a, f[a])$ and $(b, f[b])$. By the mean value theorem, you know this is possible because the function is a polynomial (therefore continuous and differentiable).

Solve for c using the mean value theorem:

In[ ]:= **sol = Solve[f[4] – f[0] == f'[x] (4 – 0) && 0 ≤ x ≤ 4, x]**

Out[ ]= $\left\{\left\{x \rightarrow \dfrac{4}{\sqrt{3}}\right\}\right\}$

So at $c = \dfrac{4}{\sqrt{3}}$, there is a tangent line that is parallel to the secant line going through $(0, 0)$ and $(4, 32)$.

The tangent line is given from point-slope form:

In[ ]:= **tangent = f[4 / Sqrt[3]] + f'[4 / Sqrt[3]] (x – 4 / Sqrt[3]);**

Plot the function, tangent line and secant line:

In[ ]:= **Plot[{f[x], tangent}, {x, 0, 5}, ⋯ ⊹]**

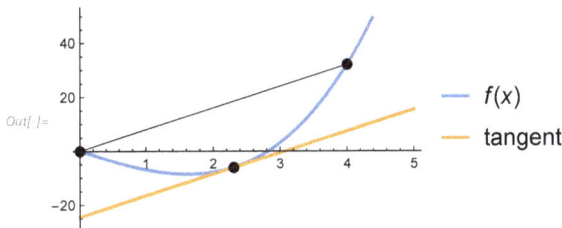

Maximum Value

Suppose that $f[0] = 4$ and $f'[x] \le 6$ for all values of x. How large can $f[3]$ possibly be? Since f is differentiable, it is also continuous, so you can apply the mean value theorem on the interval $[0, 3]$.

In this case, the mean value theorem states that for some c:

$$f(3) - f(0) = f'(c) (3 + 0)$$

Since $f'[x] \leq 6$ for all x, there is a maximum for $f'[c]$ and a maximum for $f[3]$:

In[·]:= **Clear[f]**

In[·]:= **Reduce[f[3] − f[0] ≤ f'[c] (3 − 0) /. {f[0] → 4, f'[c] → 6}, f[3]]**

Out[·]= **f[3] ≤ 22**

$f[3]$ can be 22 at maximum.

Roots of a Polynomial

Show that the following function has only one root:

In[·]:= **f[x_] := x^3 + 8 x − 2**

The function is continuous, so first use the intermediate value theorem to show that a root exists.

Evaluate the function at 0 and 1:

In[·]:= **f /@ {0, 1}**

Out[·]= **{−2, 7}**

By the intermediate value theorem, a root exists in the interval [0, 1].

Now use the mean value theorem: assuming roots a and b exist, the slope between them must be zero. From the mean value theorem, there must be a point c between a and b with slope zero.

Look at the derivative:

In[·]:= **f'[x]**

Out[·]= $8 + 3 x^2$

Since $x^2 \geq 0$, the derivative is always positive and there is a contradiction. Therefore no other roots exist:

In[·]:= **Plot[f[x], {x, −5, 5}]**

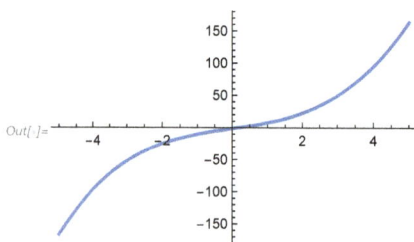

Speed Limit

The mean value theorem can be used to give people speeding tickets.

Suppose two points A and B on a road are 90 miles apart and you start from point A at noon. Assuming you do not take any breaks at any point in your journey, suppose you reach point B at 1pm. If the speed limit is 80 miles per hour, should you receive a speeding ticket?

If you represent A by the coordinates (0, 0) and B by the coordinates (1, 90) (i.e. letting the first coordinate be time and the second coordinate be distance), you can find the average speed for the whole trip by finding the slope of the line between the two points:

In[ ]:= **avg = (90 – 0)/(1 – 0)**

Out[ ]= 90

The mean value theorem states that at some point in the interval [0, 1] you had an instantaneous speed equal to your average speed (90 mph). Therefore you should receive a speeding ticket.

Summary

The mean value theorem relates the average slope of an interval to the instantaneous slope in the interval.

The mean value theorem will prove to be a very useful theorem.

The next lesson will use the mean value theorem to help draw the graphs of functions.

Exercises

Exercise 1—Rolle's Theorem

A ball is thrown vertically in the air and falls back into your hand three seconds later. What can you say about the ball's velocity at some point during the trajectory?

Solution

At some time t, the ball's height was s feet off the ground:

In[·]:= **height[t] = s**

Out[·]= **s**

Three seconds later, the ball has the same height s:

In[·]:= **height[t + 3] = s**

Out[·]= **s**

Assuming the ball did not teleport magically back to your hand, you can use Rolle's theorem.

Since the endpoints have equal value in the interval $[t, t + 3]$, at some point the ball had velocity equal to 0.

Exercise 2—Mean Value Theorem

Find the value of c in the interval $[4, 8]$ that has slope equal to the slope of the interval for the following function:

In[·]:= **f[x_] := x^3 − 6 x^2 − 3 x − 3**

Solution

The function is a polynomial (continuous and differentiable), so solve for c using the mean value theorem:

In[·]:= **sol = Solve[{f[8] − f[4] == f'[c] (8 − 4), 4 ≤ c ≤ 8}, c]**

Out[·]= $\left\{\left\{c \to \frac{1}{3}\left(6 + 2\sqrt{39}\right)\right\}\right\}$

So at $c = \frac{1}{3}\left(6 + 2\sqrt{39}\right)$, there is a tangent line that is parallel to the secant line going through $(4, -47)$ and $(8, 101)$:

In[ ]:= **c = $\frac{1}{3}$ (6 + 2 $\sqrt{39}$);**

The tangent line is given from point-slope form:

In[ ]:= **tangent = f[c] + f'[c] (x − c);**

Plot the function, tangent line and secant line:

In[ ]:= **Plot[{f[x], tangent}, {x, 4, 8}, ⋯ +]**

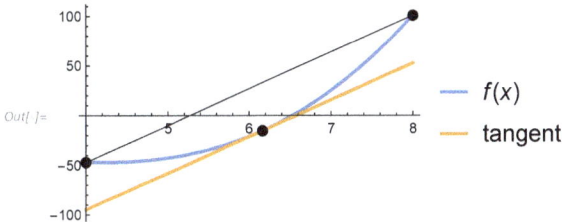

Exercise 3—Roots of a Polynomial

Show that the following function only has one root:

In[ ]:= **f[x_] := x^3 + x − 4**

Solution

The function is a polynomial, so you can use the intermediate value theorem.

Evaluate the function at the points 0 and 2:

In[ ]:= **f /@ {0, 2}**

Out[ ]= **{−4, 6}**

Therefore the function has a root in the interval $[0, 2]$.

To show it does not have any other roots, use the mean value theorem and look at the function's derivative:

In[ ]:= **f'[x]**

Out[ ]= $1 + 3 x^2$

Since $x^2 \geq 0$, the derivative is always positive. If there were any other roots (say a and b), the slope between them would be zero and there would be a point between them with derivative zero. Therefore the function only has one root.

Here is the function's graph:

In[·]:= **Plot[f[x], {x, −2, 2}]**

Out[·]=

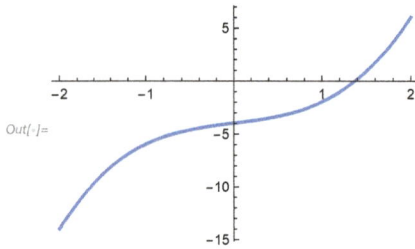

Exercise 4—Minimum Value

If $g[4] = 5$ and $g'[x] \geq 3$ for all x, find the least possible value for $g[10]$.

Solution

The derivative is defined for all x, so you can use the mean value theorem.

Using the constraints in the problem, you can find the minimum value for $g[10]$:

In[·]:= **Reduce[g[10] − g[4] ≥ g'[x] (10 − 4) /. {g[4] → 5, g'[x] → 3}, g[10]]**

Out[·]= **g[10] ≥ 23**

The least possible value for $g[10]$ is 23.

Exercise 5—Speed Limit

A driver travels from San Diego, California at 3pm. He arrives at El Cajon, California (15.8 miles away) at 3:12pm. If the speed limit was 75 mph for the whole trip, and the driver never took any stops, should he get a speeding ticket?

Solution

It is given that the cities are 15.8 miles apart:

In[·]:= **dist = 15.8**

Out[·]= **15.8**

The time for the whole trip is also given (12 minutes).

Convert minutes to hours and divide the distance by time:

In[·]:= **avg = dist / (12 / 60)**

Out[·]= 79.

The average speed for the whole trip was 79 miles per hour. By the mean value theorem, at some point during the trip the driver had instantaneous speed 79 miles per hour. Therefore, the driver should get a ticket.

18 | Derivatives and the Shape of Graphs

Overview

The previous lesson hinted at the extreme power of the mean value theorem. Now you get to see that power in action as you use the derivative of a function to sketch its graph!

Consider the plot of this trig function and its derivative, for example:

In[ ]:= `f[x_] := x^2`

In[ ]:= `Plot[f'[x], {x, −5, 5}, PlotLegends → "Expressions"]`

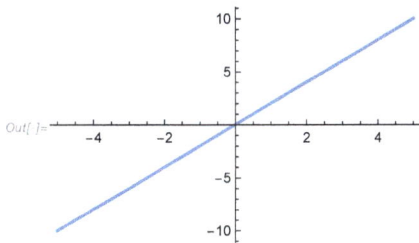

Out[ ]=

From this, you can plot the original function.

The goal of this lesson is to develop a way for sketching such graphs.

Intuition

The way a graph is sketched from its derivative is not hard to follow.

Use the following simple facts:

- If the derivative is positive at a point, then the function is increasing at that point.
- If the derivative is negative at a point, then the function is decreasing at that point.
- If the derivative is zero at a point, then the function will be horizontal at that point.

Consider the graph of the derivative of x^2 and x^2 as shown in the previous slide:

In[·]:= **{Plot[f '[x], {x, −2, 2}], Plot[f[x], {x, −2, 2}]}**

Out[·]:=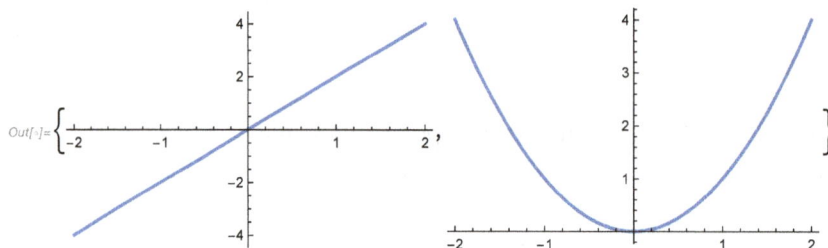

When $x < 0$, the derivative is negative, so the function should be decreasing.

When $x > 0$, the derivative is positive, so the function should be increasing.

When $x = 0$, the derivative is zero, so the function should be horizontal.

Increasing/Decreasing Test

The simple facts shown in the previous test are true because of the mean value theorem.

Recall that the mean value theorem states that for a continuous, differentiable function on an interval $[a, b]$, there is a point c in the interval (a, b) with tangent line parallel to the secant line connecting the endpoints of the interval.

The equation for the theorem is:

$$f(b) - f(a) = (b - a) f'(c)$$

If you assume $b > a$, then $b - a$ is positive. Let c be any number in the interval $[a, b]$.

If $f'[c]$ is **positive** for all c, then $f[b] = f[a] + (b - a) f'[c] > f[a]$, so the function is **increasing** on that interval.

If $f'[c]$ is **negative** for all c, then $f[b] = f[a] + (b - a) f'[c] < f[a]$, so the function is **decreasing** on that interval.

These two facts are collectively called the **increasing/decreasing test**.

Here is a plot describing what was just said for the curve x^3:

In[ ]:= **f[x_] := x^3**

In[ ]:= **{Plot[f[x], {x, −1, 1}], Plot[f'[x], {x, −1, 1}]}**

Out[ ]:=

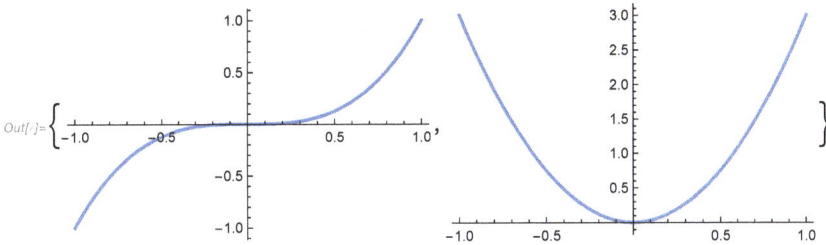

The curve is always increasing except at 0, where it is horizontal, and this is reflected in the graph of the derivative.

Example 1

Find the places where the following function is increasing or decreasing without graphing:

In[ ]:= **f[x_] := 2 x^3 + 3 x^2 − 12 x**

Use the increasing/decreasing test. First, find where the derivative is 0:

In[ ]:= **sol = Solve[f'[x] == 0, x]**

Out[ ]:= **{{x → −2}, {x → 1}}**

Based on the derivative, you know the function is horizontal at 1 and −2.

Now evaluate the derivative at points in the intervals $(-\infty, -2)$, $(-2, 1)$ and $(1, \infty)$:

In[ ]:= **f'/@{−3, 0, 2}**

Out[ ]:= **{24, −12, 24}**

The function is increasing on the intervals $(-\infty, -2)$, $(-2, 1)$ and $(1, \infty)$, and decreasing on the interval $(-2, 1)$.

Here is the graph of the derivative and the function:

In[ ]:= `Plot[{f '[x], f[x]}, {x, −3, 3}, GridLines → {{−2, 1}, None}, PlotLegends → "Expressions"]`

Out[ ]=

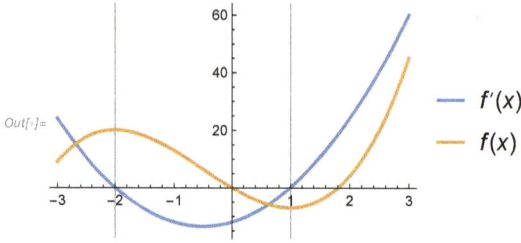

— $f'(x)$
— $f(x)$

Finding Local Max and Min

You can also use the derivative of a function to find its local maxima and minima.

Recall that a **local maximum** of a function f is a point c where $f[c] \geq f[x]$ for x near c. A **local minimum** of a function f is a point c where $f[c] \leq f[x]$ for x near c.

Look at the previous function and note the local maximum and minimum:

In[ ]:= `Plot[f[x], {x, −3, 3}]`

Out[ ]=

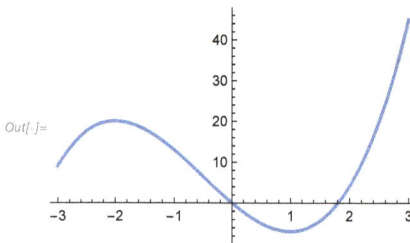

At $x = -2$ there is a local maximum, and the derivative goes from positive to negative. At $x = 1$ there is a local minimum, and the derivative goes from negative to positive.

Now look at the graph of x^3. It is horizontal at 0, but is not a local max or min. The derivative does not change sign as it goes across 0:

In[ ]:= `Plot[x^3, {x, −1, 1}]`

Out[ ]=

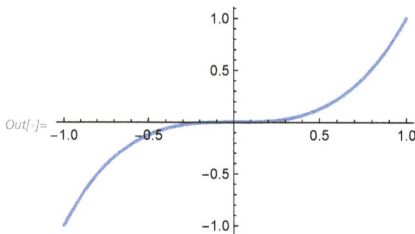

First Derivative Test

The **first derivative test** summarizes what was just covered and is based on the increasing/decreasing test (which is based on the mean value theorem).

Suppose that c is a critical number of a continuous function f. Then the first derivative test states:

- If f' changes from **positive** to **negative** at c, then f has a **local maximum** at c.
- If f' changes **negative** to **positive** at c, then f has a **local minimum** at c.
- If f' **does not change sign** at c (i.e. goes from positive to positive or negative to negative), then f **has no local maximum or minimum** at c.

For example, the sine function has a local maximum at $\pi/2$ and a local minimum at $3\pi/2$:

In[ ]:= **Plot[{Sin[x], Sin'[x]}, {x, 0, 2 π}]**

Out[ ]=

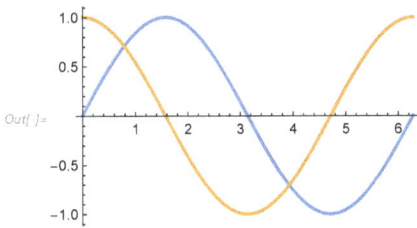

At $\pi/2$, the slope is zero and the derivative (cosine) goes from positive to negative. At $3\pi/2$, the slope is zero and the derivative goes from negative to positive.

Second Derivative

What about the second derivative f''? What can it tell you about the function f?

Look at the graph of the function x^3 and its second derivative $6x$:

In[ ]:= **{Plot[x^3, {x, −1, 1}], Plot[6 x, {x, −1, 1}]}**

Out[ ]=

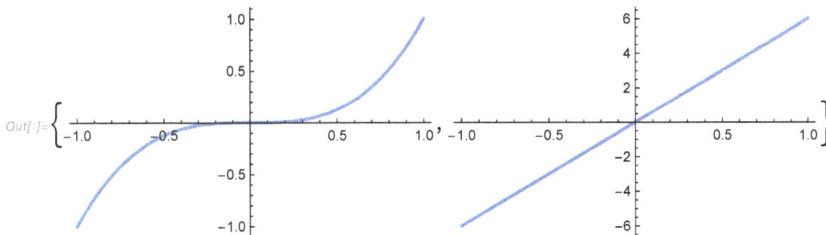

When the second derivative is negative, the slope gets progressively smaller (starting positive and becoming zero). When the second derivative is positive, the slope gets progressively bigger (starting at zero and becoming more positive).

In the first case, the curve is said to be **concave downward**. In the second case, the curve is said to be **concave upward**.

Look at the sine function and its second derivative:

In[·]:= **Plot[{Sin[x], Sin''[x]}, {x, 0, 2 π}]**

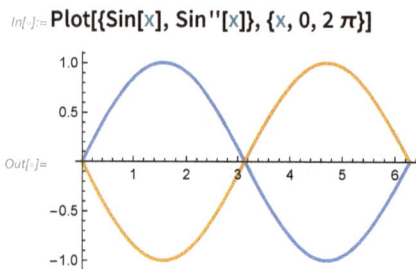

The curve is concave downward on $[0, \pi]$, and concave upward on $[\pi, 2\pi]$.

Concavity Test

An easy way to remember whether a function is concave upward or downward at a point is as follows:

- If the function is concave **up** at c, then it looks like part of a **cup** (it holds water).
- If the function is concave **down** at c, then it looks like part of a **frown** (it loses water).

The **concavity test** sums up what has been covered:

- If $f''[x] > 0$ for all x in the interval I, then the graph of f is concave upward on I.
- If $f''[x] < 0$ for all x in the interval I, then the graph of f is concave downward on I.

Another way to look at the second derivative is to look at tangent lines.

Look at x^2, its second derivative 2 and some tangent lines on x^2:

In[·]:= **Plot[{x^2, 2, 0, .25 + (x − .5), .25 − (x + .5)}, {x, −1, 1}, Axes → {False, True},**
Epilog → {PointSize[Large], Point[{{0, 0}, {−0.5, 0.25}, {0.5, 0.25}}]}]

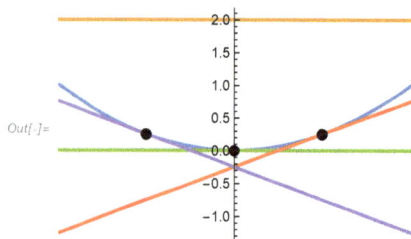

Its second derivative is always positive, so the curve is concave upward. All the tangent lines lie **below** the curve. If the curve was concave downward, then all the tangent lines would lie **above** the curve.

Inflection Points

An **inflection point** on a curve is a place where the second derivative goes from positive to negative or from negative to positive. In other words, the curve switches from being concave upward to concave downward or from being concave downward to concave upward.

The following graph has an inflection point at 1:

In[ ]:= **Plot[x^3 − 3 x^2 , {x, −1, 3}, Epilog → {PointSize[Large], Point[{1, −2}]}]**

Out[ ]=

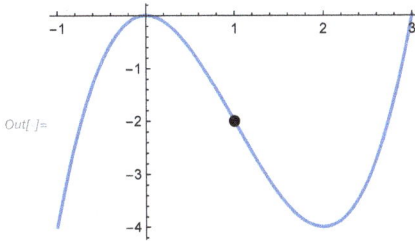

Notice the graph changes from being concave downward to concave upward at 1.

Generally a function's inflection points are where the function has the maximum or minimum slope. To find a function's inflection points, find places where the second derivative is equal to 0 or not defined and solve for the point.

Confirm 1 is an inflection point for the preceding curve:

In[ ]:= **Solve[D[x^3 − 3 x^2, {x, 2}] == 0, x]**

Out[ ]= **{{x → 1}}**

Second Derivative Test

You can also use second derivatives to find local maxima and minima.

The **second derivative test** states that if f'' is continuous near c:

If $f'[c] = 0$ and $f''[c] > 0$, then f has a **local minimum** at c.

If $f'[c] = 0$ and $f''[c] < 0$, then f has a **local maximum** at c.

Intuitively this makes sense: when the $f'[c]$ is zero, the function is flat.

If it is concave upward there too, then it looks like a cup and should be a local minimum. If it is concave downward there, then it looks like a frown and should be a local maximum.

Look at $x^3 - 3x^2$ again:

In[·]:= **Plot[x^3 − 3 x^2, {x, −1, 3}]**

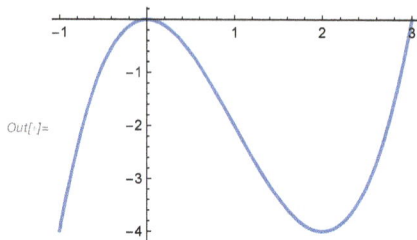

Out[·]=

At 0, the slope is zero and the graph is concave downward, so it is a local maximum. At 2, the slope is zero and the graph is concave upward, so it is a local minimum.

Summary

From the mean value theorem, you can do a lot just from the graph of a derivative.

The increasing/decreasing test lets you know where a function is increasing or decreasing.

The first derivative test uses the derivative of a function to find local maxima and minima.

The concavity test tells where a function is concave upward or downward using the second derivative.

Inflection points tell where the rate of change is at a maximum or minimum for a function.

The second derivative test uses the first and second derivatives to find local maxima and minima.

The next lesson will cover limits at infinity and horizontal asymptotes.

Exercises

Exercise 1—Increasing or Decreasing?

Find where the following function is increasing and decreasing:

In[ ]:= **f[x_] := 2160 x − 270 x^2 − 560 x^3 − 15 x^4 + 24 x^5**

Solution

First, factor the function's derivative:

In[ ]:= **Factor[f '[x]]**

Out[ ]= $60 (−4 + x) (−1 + x) (3 + x) (3 + 2 x)$

The function is horizontal at 4, 1, $−3$, and $−\frac{3}{2}$. Evaluate the derivative in the intervals $(−\infty, −3)$, $\left(−3, −\frac{3}{2}\right)$, $\left(−\frac{3}{2}, 1\right)$, $(1, 4)$ and $(4, \infty)$:

In[ ]:= **f '/@{−4, −2, 0, 2, 5}**

Out[ ]= $\{12\,000, −1080, 2160, −4200, 24\,960\}$

The function is increasing on $(−\infty, −3)$, $\left(−\frac{3}{2}, 1\right)$ and $(4, \infty)$. The function is decreasing on $\left(−3, −\frac{3}{2}\right)$ and $(1, 4)$.

Here is the function's graph:

In[ ]:= **Plot[f[x], {x, −4, 5}, GridLines → {{−3/2, −3, 1, 4}, None}]**

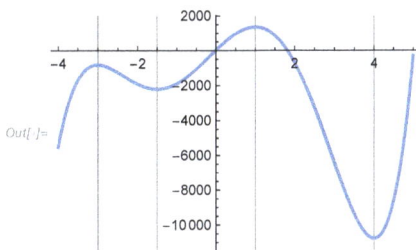

Exercise 2—First Derivative Test

Use the first derivative test to find the following function's local maxima and minima:

In[ ]:= **f[x_] := 3 x^5 − 5 x^3**

Solution

Find the places where the derivative is zero:

In[·]:= **sol = Solve[f'[x] == 0, x]**

Out[·]:= $\{\{x \to -1\}, \{x \to 0\}, \{x \to 0\}, \{x \to 1\}\}$

The derivative is 0 at -1, 0 and 1. Evaluate the derivative in the intervals $(-\infty, -1)$, $(-1, 0)$, $(0, 1)$ and $(1, \infty)$:

In[·]:= **f' /@ {−2, −1/2, 1/2, 2}**

Out[·]:= $\left\{180, -\dfrac{45}{16}, -\dfrac{45}{16}, 180\right\}$

The function is increasing on $(-\infty, -1)$ and $(1, \infty)$ and decreasing on $(-1, 0)$ and $(0, 1)$. Therefore the function has a local maximum at -1 and a local minimum at 1.

Here is a plot of the function:

In[·]:= **Plot[f[x], {x, −2, 2}, GridLines → {{1, −1}, None}]**

Out[·]:=

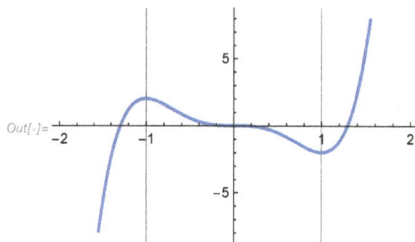

Exercise 3—Inflection Points

Find the inflection points of the following function:

In[·]:= **f[x_] := −90 x^2 − 4 x^3 + x^4**

Solution

Find the second derivative of the function:

In[·]:= **f''[x]**

Out[·]:= $-180 - 24x + 12x^2$

Then factor it:

In[·]:= **Factor[%]**

Out[·]:= $12(-5+x)(3+x)$

The function has inflection points at -3 and 5.

Here is the function and its inflection points in a graph:

In[ ]:= **Plot[f[x], {x, −10, 10}, Epilog → {PointSize[Large], Point[{{5, f[5]}, {−3, f[−3]}}]}]**

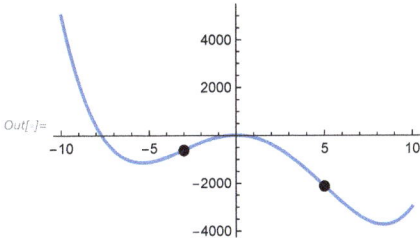

Out[ ]=

Exercise 4—Second Derivative Test

Use the second derivative test to find the following function's local maxima and minima:

In[ ]:= **f[x_] := 60 x + 21 x^2 + 2 x^3**

Solution

First, find the places where the first derivative is zero:

In[ ]:= **sol = Solve[f'[x] == 0, x]**

Out[ ]= **{{x → −5}, {x → −2}}**

Then evaluate the second derivative at −5 and −2:

In[ ]:= **f'' /@ {−5, −2}**

Out[ ]= **{−18, 18}**

By the second derivative test, you can conclude there is a local maximum at −5 and a local minimum at −2.

Here is a graph of the function and its local maximum and minimum:

In[ ]:= **Plot[f[x], {x, −6, −1}, Epilog → {PointSize[Large], Point[{{−2, f[−2]}, {−5, f[−5]}}]}]**

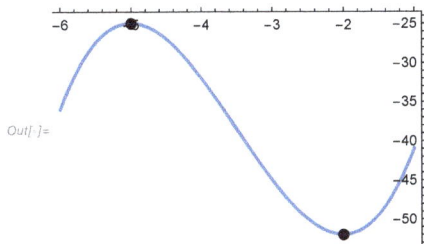

Out[ ]=

Exercise 5—Graphs of Derivatives

The first and second derivatives of a function are plotted here:

```
In[·]:= f[x_] := x^3 + 12 x^2;
    {Plot[f'[x], {x, -15, 5}, ImageSize → Medium],
      Plot[f''[x], {x, -15, 5}, ImageSize → Medium]}
```

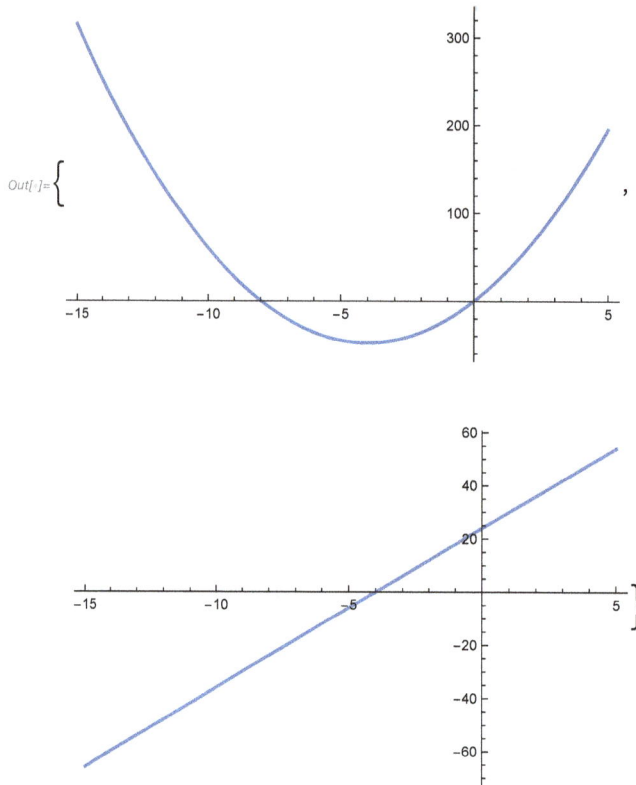

What can you conclude about the original function?

Solution

From the first graph, you know the function is decreasing in the range $(-8, 0)$. It is increasing in the range $(-\infty, -8)$ and $(0, \infty)$.

From the second graph, you know the function has an inflection point at -4 and is concave downward for values less than -4 and concave upward for values greater than -4.

From the first and second derivative tests, you know there is a local maximum at -8 and a local minimum at 0.

Here is the graph of the original function:

In[ ]:= **Plot[f[x], {x, −15, 5}, ImageSize → Medium]**

Out[ ]=

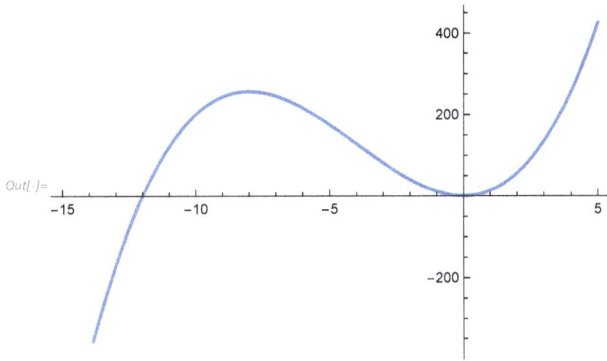

19 | Asymptotes

Overview

In the past, limits have been calculated at points. For example, the limit of $3\,x$ as x approaches 2 is 6:

In[ ]:= **Limit[3 x, x → 2]**

Out[ ]= 6

You can also take limits at positive and negative infinity:

In[ ]:= **Limit[3 x, x → Infinity]**

Out[ ]= ∞

In[ ]:= **Limit[3 x, x → −Infinity]**

Out[ ]= −∞

Sometimes these limits approach finite numbers:

In[ ]:= **Limit[1 / x, x → Infinity]**

Out[ ]= 0

This lesson covers this so-called **end behavior** of functions, which will be helpful when sketching graphs.

Horizontal Asymptotes

To see why x^{-1} approaches 0 as x approaches ∞, make a table of values:

1	1
10	$\frac{1}{10}$
100	$\frac{1}{100}$
1000	$\frac{1}{1000}$
10 000	$\frac{1}{10\,000}$
100 000	$\frac{1}{100\,000}$

As x approaches ∞, x^{-1} gets closer and closer to 0. Zero is a **horizontal asymptote** of x^{-1}.

You can also see the horizontal asymptote in the graph of x^{-1}:

In[·]:= **Plot[{1 / x, 0}, {x, 1, 100 000}, Axes → {False, True}, PlotLegends → "Expressions"]**

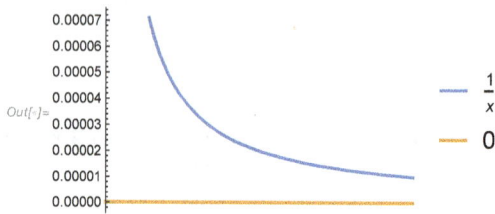

Definition

In general, the following definition is used for limits at positive and negative infinity:

Let f be a function defined on some (a, ∞).

Then **$\text{Limit}_{x \to \infty} f[x] = L$** means that the values of $f[x]$ can be made arbitrarily close to L by taking x sufficiently large.

Let f be a function defined on some interval $(-\infty, a)$.

Then **$\text{Limit}_{x \to \infty} f[x] = L$** means that the values of $f[x]$ can be made arbitrarily close to L by taking x sufficiently large negative.

The line $y = L$ is called a **horizontal asymptote** of the curve $y = f[x]$ if either $\text{Limit}_{x \to \infty} f[x] = L$ or $\text{Limit}_{x \to -\infty} f[x] = L$.

The following function has $y = 1$ as a horizontal asymptote:

In[·]:= **Plot[{(x^2 + 1)/(x^2 − 1), 1}, {x, −1000, 1000}, PlotLegends → "Expressions"]**

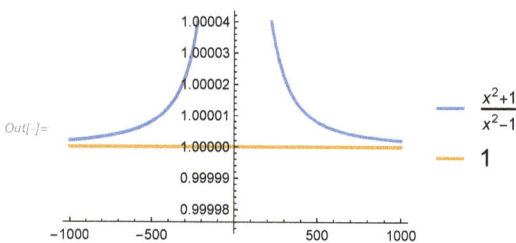

Infinite Limits at Infinity

Consider the function given in the introduction, $3\,x$:

In[·]:= **f[x_] := 3 x**

The limit of f as x approaches ∞ is ∞:

In[ ]:= **Limit[3 x, x → ∞]**

Out[ ]= ∞

This means that the function gets bigger and bigger as x gets bigger and bigger.

The limit of f as x approaches $-\infty$ is $-\infty$:

In[ ]:= **Limit[3 x, x → −∞]**

Out[ ]= −∞

This means that the function gets smaller and smaller (i.e. more negative) as x gets smaller and smaller.

You can see this in the graph of $3\,x$:

In[ ]:= **Plot[3 x, {x, −1000, 1000}]**

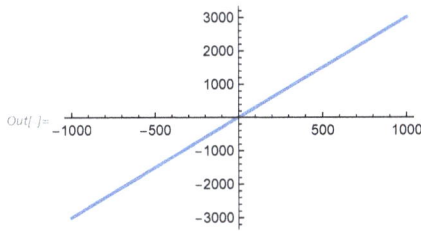

Out[ ]=

Limit Laws

A previous lesson covered the mathematical laws that limits obey. Most of these laws still apply for limits at infinity.

To recall:

- The limit of a sum is the sum of the limits (**sum law**).
- The limit of a difference is the difference of the limits (**difference law**).
- The limit of a constant is the constant (**constant law**).
- The limit of a product is the product of the limits (**product law**).
- The limit of a quotient is the quotient of the limits, so long as the denominator is not 0 (**quotient law**).

You must be careful with these limit laws, however, since you are now dealing with infinities. For example, the limit of $x^3 - x$ as x approaches ∞ is **not** 0, because $\infty - \infty$ is not well defined.

Large numbers minus large numbers can give you 0, large numbers or even negative large numbers:

In[-]:= **Limit[x^3 − x, x → ∞] == 0**

Out[-]= **False**

If you reframe it as $x(x^2 − 1)$ and use the product law, then you can see that the limit is ∞.

Large numbers multiplied by large numbers do give you large numbers:

In[-]:= **Limit[x^3 − x, x → ∞]**

Out[-]= **∞**

Special Law

For limits at infinity, an important rule can be derived from the limit laws.

The rule states that if $r > 0$ is a rational number, then $\text{Limit}_{x \to \infty} 1/x^r = 0$:

In[-]:= **Limit[(1/x)^r, x → Infinity, Assumptions → r > 0]**

Out[-]= **0**

If $r > 0$ is a rational number such that x^r is defined for all x, then $\text{Limit}_{x \to -\infty} 1/x^r = 0$:

In[-]:= **Limit[(1/x)^r, x → −Infinity, Assumptions → r > 0]**

Out[-]= **0**

So the limit of x^{-2} as x approaches ∞ or − ∞ is 0:

In[-]:= **Limit[x^−2, x → Infinity]**

Out[-]= **0**

In[-]:= **Limit[x^−2, x → −Infinity]**

Out[-]= **0**

Rational

Find the limit at ∞ for the following function:

In[-]:= **f[x_] := (2 x^2 − 3 x + 4)/(3 x^2 + 9 x + 10)**

First, divide the numerator and denominator by x^2:

In[ ]:= **{num = (2 x^2 − 3 x + 4)/x^2, den = (3 x^2 + 9 x + 10)/x^2}**

Out[ ]= $\left\{ \dfrac{4 - 3x + 2x^2}{x^2}, \dfrac{10 + 9x + 3x^2}{x^2} \right\}$

By the sum rule, split each of these into sums and use the special law to conclude the limit of the new numerator is 2:

In[ ]:= **Limit[num, x → ∞]**

Out[ ]= 2

... and the limit of the new denominator is 3:

In[ ]:= **Limit[den, x → ∞]**

Out[ ]= 3

So by the quotient law, the limit is $\frac{2}{3}$. Limit agrees:

In[ ]:= **Limit[f[x], x → ∞]**

Out[ ]= $\dfrac{2}{3}$

Radical Rational

Find the limit at infinity for the following function:

In[ ]:= **f[x_] := Sqrt[5 x^2 + 3]/(2 x − 3)**

First, divide the numerator and denominator by x.

Since $x = \sqrt{x^2}$ for $x > 0$, the new numerator is:

In[ ]:= **{num = Sqrt[5 + 3/x^2], den = (2 x − 3)/x}**

Out[ ]= $\left\{ \sqrt{5 + \dfrac{3}{x^2}}, \dfrac{-3 + 2x}{x} \right\}$

From the special law, you know the limits at ∞ for the numerator and denominator are 5 and 2, respectively:

In[ ]:= **{Limit[num, x → ∞], Limit[den, x → ∞]}**

Out[ ]= $\left\{ \sqrt{5}, 2 \right\}$

From the quotient law, the limit is 52. Limit agrees:

In[·]:= **Limit[f[x], x → ∞]**

Out[·]= $\dfrac{\sqrt{5}}{2}$

Radical

Find the limit at ∞ for the following function:

In[·]:= **f[x_] := Sqrt[x^2 + 4] − x**

In this case, the difference law will not work because you will get ∞ − ∞, which is undefined.

Use algebra and multiply and divide by the conjugate radical:

In[·]:= **f[x] (Sqrt[x^2 + 4] + x) // Simplify**

Out[·]= **4**

The new numerator is 4, so the function can be rewritten as $\dfrac{4}{x + \sqrt{x^2 + 4}}$.

From the special law, you can conclude the limit is 0. Limit agrees:

In[·]:= **Limit[f[x], x → ∞]**

Out[·]= **0**

Here is the graph of the function:

In[·]:= **Plot[f[x], {x, 0, 10 000}]**

Out[·]=

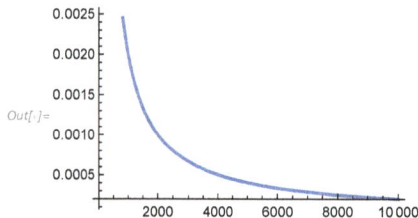

Compositions

Recall that if f is continuous at b and the limit of g at a is b, then the limit of $f[g[x]]$ at a equals f applied to the limit of g at a (i.e. $f[b]$). So $\text{Limit}_{x \to a} f[g[x]] = f[\text{Limit}_{x \to a} g[x]]$.

You can use this property to solve the following example.

Find the limit at ∞ for the following function:

In[ ]:= **f[x_] := Cos[1 / x]**

Cosine is a continuous function, so find the limit at ∞ for x^{-1} and apply it to cosine to get the answer.

The limit at ∞ for x^{-1} is 0:

In[ ]:= **Limit[1 / x, x → Infinity]**

Out[ ]= 0

Cosine at 0 is 1, so the answer is 1. Limit agrees:

In[ ]:= **Limit[Cos[1 / x], x → ∞]**

Out[ ]= 1

Here is a plot of the function:

In[ ]:= **Plot[f[x], {x, 0, 1000}]**

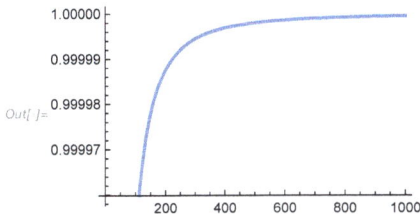

Shortcut 1

For rational functions, there is a special trick for finding limits at ∞ and −∞.

Recall that the highest power of a polynomial is called its **degree**, so the degree of $x^5 - 3 x^3$ is 5.

For rational functions, if the degree of the numerator is less than the degree of the denominator, then the limits at ∞ and −∞ are both 0:

In[ ]:= **{Limit[$\dfrac{3 x}{4 x \char94 2 + 1}$, x → Infinity], Limit[$\dfrac{3 x}{4 x \char94 2 + 1}$, x → −Infinity]}**

Out[ ]= {0, 0}

If the degree of the numerator equals the degree of the denominator, then the limits at ∞ and −∞ are both the leading coefficient of the numerator divided by the leading coefficient of the denominator:

$$\textit{In[·]:=} \left\{ \text{Limit}\left[\frac{2\,x^{\wedge}6+7\,x^{\wedge}4}{7\,x^{\wedge}6-3\,x^{\wedge}2}, x \rightarrow \text{Infinity}\right], \text{Limit}\left[\frac{2\,x^{\wedge}6+7\,x^{\wedge}4}{7\,x^{\wedge}6-3\,x^{\wedge}2}, x \rightarrow -\text{Infinity}\right] \right\}$$

$$\textit{Out[·]=} \left\{ \frac{2}{7}, \frac{2}{7} \right\}$$

Shortcut 2

If the degree of the numerator is greater than the degree of the denominator, then one of four things can happen:

1. If the function is positive and the degree of the numerator minus the degree of the denominator is even, then the limits at ∞ and −∞ are both ∞:

$$\textit{In[·]:=} \left\{ \text{Limit}\left[\frac{x^{\wedge}6+2\,x}{x^{\wedge}2+1}, x \rightarrow \text{Infinity}\right], \text{Limit}\left[\frac{x^{\wedge}6+2\,x}{x^{\wedge}2+1}, x \rightarrow -\text{Infinity}\right] \right\}$$

$$\textit{Out[·]=} \{\infty, \infty\}$$

2. If the function is negative and the degree of the numerator minus the degree of the denominator is even, then the limits at ∞ and −∞ are both −∞:

$$\textit{In[·]:=} \left\{ \text{Limit}\left[\frac{-2\,x^{\wedge}4+2\,x}{3\,x^{\wedge}2+7\,x}, x \rightarrow \text{Infinity}\right], \text{Limit}\left[\frac{-2\,x^{\wedge}4+2\,x}{3\,x^{\wedge}2+7\,x}, x \rightarrow -\text{Infinity}\right] \right\}$$

$$\textit{Out[·]=} \{-\infty, -\infty\}$$

3. If the function is positive and the degree of the numerator minus the degree of the denominator is odd, then the limits at ∞ and −∞ are ∞ and −∞, respectively:

$$\textit{In[·]:=} \left\{ \text{Limit}\left[\frac{x^{\wedge}9+2\,x}{x^{\wedge}6+10\,x}, x \rightarrow \text{Infinity}\right], \text{Limit}\left[\frac{x^{\wedge}9+2\,x}{x^{\wedge}6+10\,x}, x \rightarrow -\text{Infinity}\right] \right\}$$

$$\textit{Out[·]=} \{\infty, -\infty\}$$

4. If the function is negative and the degree of the numerator minus the degree of the denominator is odd, then the limits at ∞ and −∞ are −∞ and ∞, respectively:

$$\textit{In[·]:=} \left\{ \text{Limit}\left[\frac{-x^{\wedge}6+5\,x}{2\,x^{\wedge}3+8}, x \rightarrow \text{Infinity}\right], \text{Limit}\left[\frac{-x^{\wedge}6+5\,x}{2\,x^{\wedge}3+8}, x \rightarrow -\text{Infinity}\right] \right\}$$

$$\textit{Out[·]=} \{-\infty, \infty\}$$

Summary

Limits at infinity follow the same laws as limits at individual points.

There is a special rule that helps evaluate limits at infinity for rational functions.

Using algebraic techniques such as multiplying by conjugates can help.

The next lesson will use what has been learned here and in previous lessons to help sketch curves.

Exercises

Exercise 1—Rational Functions

Find the limits at ∞ and $-\infty$ for the following functions:

In[ ]:= $f[x\_] := x^2/(x^6 + 1)$

In[ ]:= $g[x\_] := (5\,x^3 + x)/(9\,x^3 + 4\,x^2)$

In[ ]:= $h[x\_] := 3\,x^7/x + 2$

Solution

f's numerator has a degree that is smaller than the degree of its denominator, so its limits at ∞ and $-\infty$ are 0:

In[ ]:= {Limit[f[x], x → Infinity], Limit[f[x], x → −Infinity]}

Out[ ]= {0, 0}

g's numerator has a degree that is equal to the degree of its denominator, so its limits at ∞ and $-\infty$ are $5/9$:

In[ ]:= {Limit[g[x], x → Infinity], Limit[g[x], x → −Infinity]}

Out[ ]= $\left\{\dfrac{5}{9}, \dfrac{5}{9}\right\}$

h is positive, its numerator has a degree greater than the degree of its denominator, and their difference is even, so its limits at ∞ and $-\infty$ are both ∞:

In[ ]:= {Limit[h[x], x → ∞], Limit[h[x], x → −∞]}

Out[ ]= {∞, ∞}

Here are plots of all three functions:

In[ ]:= {Plot[f[x], {x, 0, 100}, ⋯ ⬦], Plot[g[x], {x, 0, 100}, ⋯ ⬦], Plot[h[x], {x, 0, 100}, ⋯ ⬦]}

Out[ ]:= {

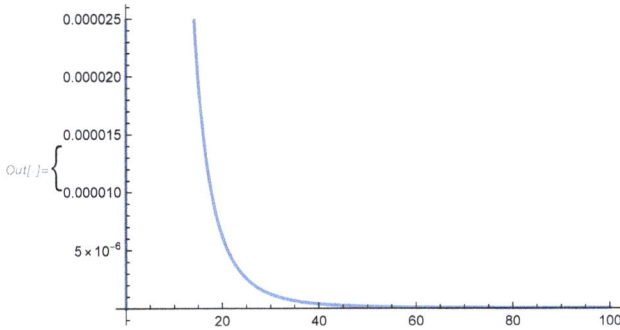

,

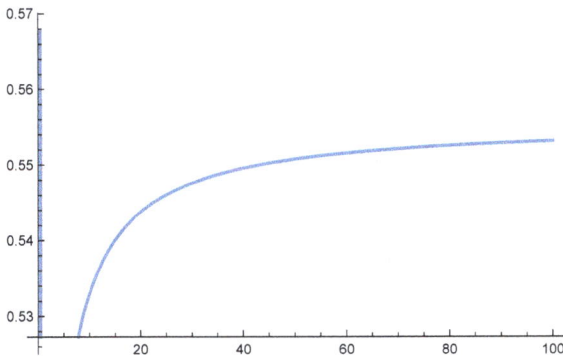

,

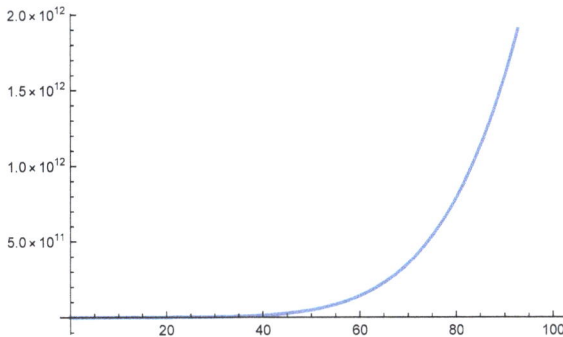

}

Exercise 2—Radical Function

Find the limit at ∞ for the following function:

In[ ]:= f[x_] := Sqrt[2 x^2 - 3 x + 9] - Sqrt[2 x^2 + 3 x + 3]

Solution

First, rationalize by multiplying and dividing by $\sqrt{2\,x^2 - 3\,x + 9} + \sqrt{2\,x^2 + 3\,x + 3}$:

In[·]:= **f[x] (Sqrt[2 x^2 – 3 x + 9] + Sqrt[2 x^2 + 3 x + 3]) // Simplify**

Out[·]= **6 – 6 x**

The numerator is $-6\,x + 6$, so an equivalent form is $\dfrac{-6\,x+6}{\sqrt{2\,x^2-3\,x+9} + \sqrt{2\,x^2+3\,x+3}}$.

Since $\sqrt{x^2} = x$ for $x > 0$, divide the top by x and the bottom by $\sqrt{x^2}$:

In[·]:= **{num = (–6 x + 6) / x, den = Sqrt[2 – 3 / x + 9 / x^2] + Sqrt[2 + 3 / x + 3 / x^2]}**

Out[·]= $\left\{ \dfrac{6-6\,x}{x},\ \sqrt{2 + \dfrac{9}{x^2} - \dfrac{3}{x}} + \sqrt{2 + \dfrac{3}{x^2} + \dfrac{3}{x}} \right\}$

From the sum law, the new numerator approaches -6 and the new denominator approaches $2\,\sqrt{2}$:

In[·]:= **{Limit[num, x → Infinity], Limit[den, x → Infinity]}**

Out[·]= $\left\{ -6,\ 2\,\sqrt{2} \right\}$

By the quotient law, the limit is $-\dfrac{3}{\sqrt{2}}$:

In[·]:= **Limit[f[x], x → Infinity]**

Out[·]= $-\dfrac{3}{\sqrt{2}}$

Exercise 3—Composition

Find the limit at ∞ for the following function:

In[·]:= **f[x_] := x Cos[1 / x]**

Solution

To find the limit, use the product rule on the functions x and $\text{Cos}[x^{-1}]$.

You know the limit at ∞ for x; it is just ∞:

In[·]:= **Limit[x, x → ∞]**

Out[·]= **∞**

Since cosine is a continuous function, find the limit for $\text{Cos}[x^{-1}]$ by using the law for compositions.

The limit for x^{-1} as x approaches ∞ is 0, so the limit for $\text{Cos}[x^{-1}]$ is $\text{Cos}[0] = 1$:

In[·]:= **Limit[Cos[1/x], x → Infinity]**

Out[·]= 1

Therefore the limit is $1 * \infty = \infty$. Limit agrees:

In[·]:= **Limit[f[x], x → ∞]**

Out[·]= ∞

Here is a plot of the function:

In[·]:= **Plot[f[x], {x, 0, 100}]**

Out[·]=

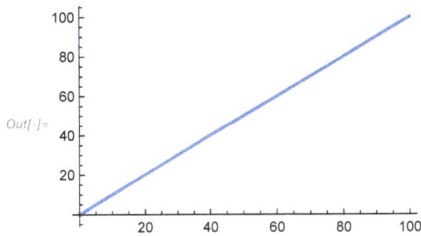

Exercise 4—Asymptotes

Find the vertical and horizontal asymptotes for the following function:

In[·]:= **f[x_] := (3 x^2 + 1)/(x^2 − 5 x − 6)**

Solution

Since the degree of the numerator equals the degree of the denominator, $y = 3$ is a horizontal asymptote.

Limit agrees:

In[·]:= **{Limit[f[x], x → ∞], Limit[f[x], x → −Infinity]}**

Out[·]= {3, 3}

To find the vertical asymptotes, find the places where the denominator is 0:

In[·]:= **sol = Solve[x^2 − 5 x − 6 == 0, x]**

Out[·]= {{x → −1}, {x → 6}}

So there are vertical asymptotes at − 1 and 6.

Here is its graph with its asymptotes:

In[·]:= **Plot[f[x], {x, −15, 15}, GridLines → {{−1, 6}, {3}}, ImageSize → Medium]**

Out[·]=

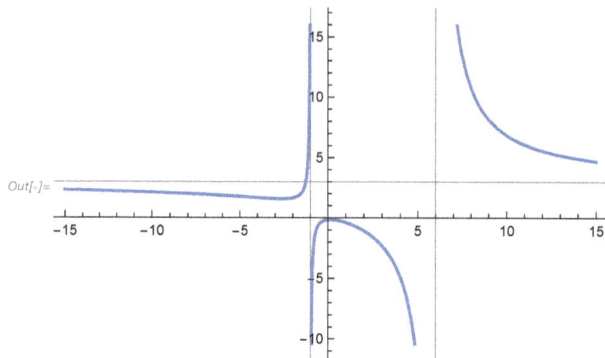

Exercise 5—Squeeze Theorem

Use the squeeze theorem to find the limit at ∞ for the following function:

In[·]:= **f[x_] := Cos[x] / x**

Solution

Since cosine lies between 1 and − 1, you can bound f by the function x^{-1} and $-x^{-1}$:

In[·]:= **g[x_] := 1 / x**

In[·]:= **h[x_] := −1 / x**

The limits as x approaches ∞ for g and h are both 0:

In[·]:= **{Limit[1 / x, x → ∞], Limit[−1 / x, x → Infinity]}**

Out[·]= **{0, 0}**

By the squeeze theorem, the limit at ∞ for f should be 0. Limit agrees:

In[·]:= **Limit[f[x], x → Infinity]**

Out[·]= **0**

Here is a plot of all three functions:

In[·]:= **Plot[{f[x], g[x], h[x]}, {x, 0, 100}, PlotLegends → "Expressions"]**

Out[·]=

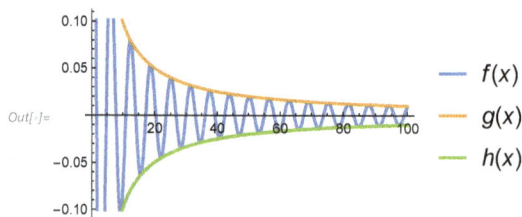

20 | Curve Sketching

Overview

Given a function f:

$\textit{In[ ]:=}$ **f[x_] := x^2/(x^2 − 4)**

you can calculate its zeros:

$\textit{In[ ]:=}$ **zeroes = Solve[f[x] == 0, x]**

$\textit{Out[ ]=}$ **{{x → 0}, {x → 0}}**

You can find its vertical asymptotes:

$\textit{In[ ]:=}$ **vert = Solve[x^2 − 4 == 0, x]**

$\textit{Out[ ]=}$ **{{x → −2}, {x → 2}}**

You can find its horizontal asymptotes:

$\textit{In[ ]:=}$ **{Limit[f[x], x → −Infinity], Limit[f[x], x → Infinity]}**

$\textit{Out[ ]=}$ **{1, 1}**

You can do all this without plotting the function. Here is a plot of the function with its asymptotes of zero:

$\textit{In[ ]:=}$ **Plot[f[x], {x, −5, 5}, ⋯ +]**

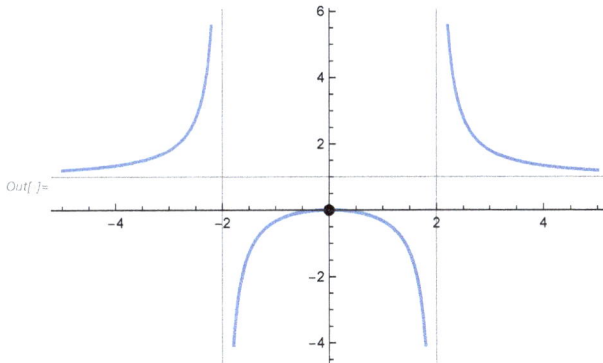

The goal of this lesson is to develop a way for sketching functions using what you have learned so far.

General Guidelines 1

Here is a list of things to do when sketching a curve by hand.

1. Find the domain of the function. This is the set of values where the function is defined.

2. Find the intercepts. Evaluate the function at 0 and find the places where the function is 0.

3. Check symmetry.

    i. An **even** function is symmetric about the y axis (so $f[-x] = f[x]$).

    ii. An **odd** function is symmetric about the origin (so $f[-x] = -f[x]$).

    iii. A **periodic** function can be translated by some amount to produce the same function (so $f[x + k] = f[k]$ for some k). k is called the **period**.

4. Find asymptotes.

    i. Vertical asymptotes are points where the function has limit positive or negative infinity on either side, and this is usually where the denominator is 0.

    ii. Horizontal asymptotes are found by calculating the limits of the function at positive and negative infinity.

    iii. Slant asymptotes are also important (these will be discussed later).

5. Find the intervals of increase and decrease. Compute $f'[x]$ and use the increasing/decreasing test.

    i. Find the intervals where $f'[x]$ is positive (where f is increasing) and negative (where f is decreasing).

General Guidelines 2

6. Find local maxima and minima. Find the critical numbers. Use the first derivative test or the second derivative test. The critical numbers are where $f'[x]$ is 0 or undefined.

    i. For the first derivative test, compute $f'[x]$ and evaluate the derivative on both sides of the critical numbers. If it goes from positive to negative, then it is a local maximum. If it goes from negative to positive, then it is a local minimum.

    ii. For the second derivative test, find the critical values and evaluate the second derivative at each one. If $f''[x]$ is positive, then it is a local minimum. If $f''[x]$ is negative, then it is a local maximum.

7. Compute $f''[x]$ and use the concavity test.

    i. Find the intervals where $f''[x]$ is positive (concave upward) and negative (concave downward).

    ii. Find the inflection points (the places where $f''[x]$ is zero and switches sign).

8. With all the preceding information, sketch the curve by hand.

Rational Function 1

Use the guidelines to sketch the following function:

In[ ]:= **f[x_] := 3 x^2 / (x^2 − 4)**

1. The function is defined everywhere except at 2 and − 2. So the domain is $(-\infty, -2) \cup (-2, 2) \cup (2, \infty)$.

2. The x- and y-intercepts of the function are both at 0:

In[ ]:= **{zeroes = Solve[f[x] == 0, x], f[0]}**

Out[ ]= **{{{x → 0}, {x → 0}}, 0}**

3. Since $f[-x] = f[x]$, the function is even:

In[ ]:= **f[−x] == f[x]**

Out[ ]= **True**

4. The function has vertical asymptotes at 2 and − 2:

In[ ]:= **{{Limit[f[x], x → 2, Direction → "FromBelow"],**
    Limit[f[x], x → 2, Direction → "FromAbove"]},
    {Limit[f[x], x → −2, Direction → "FromBelow"],
    Limit[f[x], x → −2, Direction → "FromAbove"]}}

Out[ ]= **{{−∞, ∞}, {∞, −∞}}**

The function has horizontal asymptote $y = 3$:

In[ ]:= **{Limit[f[x], x → −Infinity], Limit[f[x], x → Infinity]}**

Out[ ]= **{3, 3}**

The function has no slant asymptotes.

Rational Function 2

Calculate the first and second derivatives:

In[ ]:= **{f'[x], f''[x]}**

Out[ ]= $\left\{-\dfrac{6 x^3}{\left(-4 + x^2\right)^2} + \dfrac{6 x}{-4 + x^2}, \dfrac{24 x^4}{\left(-4 + x^2\right)^3} - \dfrac{30 x^2}{\left(-4 + x^2\right)^2} + \dfrac{6}{-4 + x^2}\right\}$

5. The derivative is 0 at 0. In this case, $f'[x]$ and $f''[x]$ are both undefined at 2 and -2:

In[·]:= **Solve[f'[x] == 0, x]**

Out[·]= $\{\{x \to 0\}\}$

In[·]:= **crit = {2, 0, –2};**

By the increasing/decreasing test, the function is increasing on $(-\infty, -2)$ and $(-2, 0)$ and decreasing on $(0, 2)$ and $(2, \infty)$:

In[·]:= **f' /@ {–3, –1, 1, 3}**

Out[·]= $\left\{\dfrac{72}{25}, \dfrac{8}{3}, -\dfrac{8}{3}, -\dfrac{72}{25}\right\}$

6. By the first derivative test, there is a local maximum at 0. The second derivative test agrees:

In[·]:= **f''[0]**

Out[·]= $-\dfrac{3}{2}$

7. The function has no inflection points. By the concavity test, it is concave upward on $(-\infty, -2)$ and $(2, \infty)$ and concave downward on $(-2, 2)$:

In[·]:= **{infl = Solve[f''[x] == 0, x, Reals], f'' /@ {–3, 0, 3}}**

Out[·]= $\left\{\{\}, \left\{\dfrac{744}{125}, -\dfrac{3}{2}, \dfrac{744}{125}\right\}\right\}$

The Plot

You can see all the previous information in the following plot:

In[·]:= **Plot[{f[x], 0}, {x, –5, 5}, GridLines → {{–2, 2}, {3}},**
　　　　Epilog → {PointSize[Large], Point[{0, f[0]}]}]

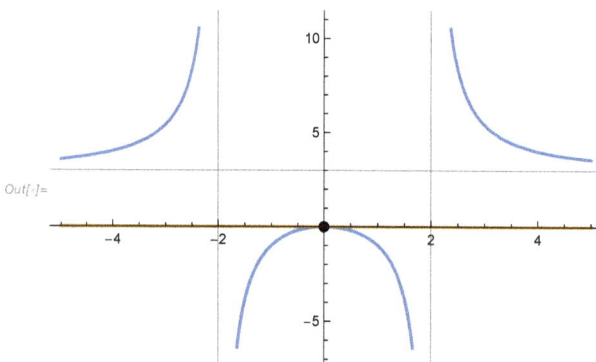

Trig Function 1

Use the guidelines to sketch the following function:

In[ ]:= **f[x_] := Sin[x] / (2 + Cos[x])**

1. The function is defined everywhere. So the domain is $(-\infty, \infty)$.

2. The x-intercepts occur wherever sine is 0, so they are at multiples of π. The y-intercept of the function is at 0:

In[ ]:= **{zeroes = Solve[f[x] == 0, x], f[0]}**

Out[ ]= $\left\{\left\{\left\{x \to \boxed{2\pi c_1 \;\; \text{if} \;\; c_1 \in \mathbb{Z}}\right\}, \left\{x \to \boxed{\pi + 2\pi c_1 \;\; \text{if} \;\; c_1 \in \mathbb{Z}}\right\}\right\}, 0\right\}$

3. Since $f[-x] = -f[x]$, the function is odd. Since $f[x] = f[x + 2\pi]$, the function is periodic with period 2π. Therefore, look at the range $[0, \; 2\pi]$:

In[ ]:= **{f[-x] == -f[x], f[x] == f[x + 2 π]}**

Out[ ]= {True, True}

4. The function has no vertical asymptotes, horizontal asymptotes or slant asymptotes:

In[ ]:= **{Limit[f[x], x → -Infinity], Limit[f[x], x → Infinity]}**

Out[ ]= {Indeterminate, Indeterminate}

Trig Function 2

Calculate the first and second derivatives:

In[ ]:= **{f'[x], f''[x]}**

Out[ ]= $\left\{\dfrac{Cos[x]}{2 + Cos[x]} + \dfrac{Sin[x]^2}{(2 + Cos[x])^2}, \dfrac{3\,Cos[x]\,Sin[x]}{(2 + Cos[x])^2} - \dfrac{Sin[x]}{2 + Cos[x]} + \dfrac{2\,Sin[x]^3}{(2 + Cos[x])^3}\right\}$

5. The derivative is 0 at $2\pi/3$ and $4\pi/3$. In this case, $f'[x]$ and $f''[x]$ are never undefined:

In[ ]:= **Solve[f'[x] == 0 && 0 ≤ x ≤ 2 π, x]**

Out[ ]= $\left\{\left\{x \to \dfrac{2\pi}{3}\right\}, \left\{x \to \dfrac{4\pi}{3}\right\}\right\}$

By the increasing/decreasing test, the function is increasing on $(0, 2\pi/3)$ and $(4\pi/3, 2\pi)$ and decreasing on $(2\pi/3, 4\pi/3)$:

In[ ]:= **f'/@{π/3, π, 5π/3}**

Out[ ]= $\left\{\dfrac{8}{25}, -1, \dfrac{8}{25}\right\}$

6. By the first derivative test, there is a local maximum at $2\pi/3$ and a local minimum at $4\pi/3$. The second derivative test agrees:

In[ ]:= **f''/@{2π/3, 4π/3}**

Out[ ]= $\left\{-\dfrac{4}{3\sqrt{3}}, \dfrac{4}{3\sqrt{3}}\right\}$

7. The function has inflection points at 0, π and 2π. By the concavity test, it is concave upward on $(\pi, 2\pi)$ and concave downward on $(0, \pi)$:

In[ ]:= **{infl = Solve[f''[x] == 0 && 0 ≤ x ≤ 2π, x], f''/@{π/2, 3π/2}}**

Out[ ]= $\left\{\{\{x \to 0\}, \{x \to 0\}, \{x \to 0\}, \{x \to \pi\}, \{x \to 2\pi\}, \{x \to 2\pi\}, \{x \to 2\pi\}\}, \left\{-\dfrac{1}{4}, \dfrac{1}{4}\right\}\right\}$

The Plot

You can see all the previous information in the following plot:

In[ ]:= **Plot[{f[x], $\dfrac{1}{\sqrt{3}}$, $-\dfrac{1}{\sqrt{3}}$}, {x, −π, 3π},**

Epilog → {PointSize[Large], Point[{{0, f[0]}, {π, f[π]}, {2π, f[2π]}}]}]

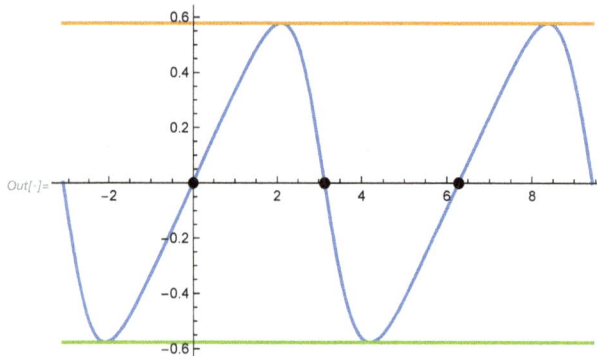

Slant Asymptotes

Recall that for a rational function, when the degree of the numerator is greater than the degree of the denominator, the limits at positive and negative infinity are positive or negative infinity, depending on several factors.

This does not tell the whole story, however. The function actually does have asymptotes, but they are **oblique**, i.e. neither horizontal nor vertical.

The line $y = m x + b$ is called a **slant asymptote** if $\text{Limit}_{x \to \infty}(f[x] - (m x + b)) = 0$.

For rational functions, slant asymptotes only occur when the degree of the numerator is one more than the degree of the denominator.

In such cases, the slant asymptote can be found by polynomial long division (the Wolfram Language uses PolynomialQuotient).

For example, the function $\frac{(x^2 + 1)}{x}$ has $y = x$ as a slant asymptote:

In[ ]:= **PolynomialQuotient[x^2 + 1, x, x]**

Out[ ]= **x**

Here is a plot of the function and its slant asymptote:

In[ ]:= **Plot[{(x^2 + 1)/x, x}, {x, −10, 10}, PlotStyle → {Red, Dashed},**
      PlotRange → {−10, 10}, AspectRatio → 1, PlotLegends → "Expressions"]

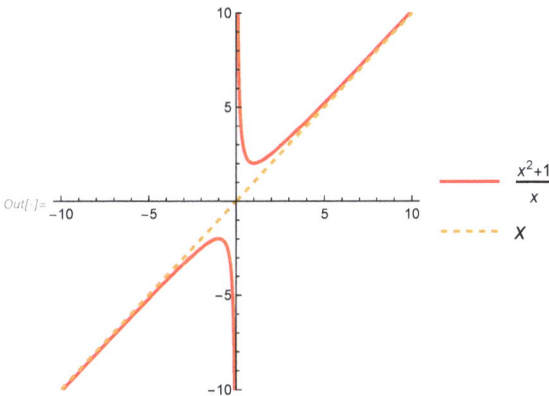

Slant 1

Use the general guidelines to sketch the graph of following function:

In[ ]:= **f[x_] := 2 x^3 / (x^2 + 2)**

1. The function is defined everywhere, so the domain is $(-\infty, \infty)$.

2. The *x*- and *y*-intercepts both occur at 0:

In[·]:= **{zeroes = Solve[f[x] == 0, x], f[0]}**

Out[·]= $\{\{\{x \to 0\}, \{x \to 0\}, \{x \to 0\}\}, 0\}$

3. Since $f[-x] = -f[x]$, the function is odd:

In[·]:= **f[−x] == f[x]**

Out[·]= $-\dfrac{2x^3}{2+x^2} == \dfrac{2x^3}{2+x^2}$

4. The function has no vertical asymptotes or horizontal asymptotes:

In[·]:= **{Limit[f[x], x → −Infinity], Limit[f[x], x → Infinity]}**

Out[·]= $\{-\infty, \infty\}$

The function has $y = 2\,x$ as a slant asymptote:

In[·]:= **PolynomialQuotient[2 x^3, x^2 + 2, x]**

Out[·]= **2 x**

In[·]:= **Limit[f[x] − 2 x, x → Infinity]**

Out[·]= **0**

Slant 2

Calculate the first and second derivatives:

In[·]:= **{f'[x], f''[x]}**

Out[·]= $\left\{-\dfrac{4x^4}{(2+x^2)^2} + \dfrac{6x^2}{2+x^2}, \dfrac{16x^5}{(2+x^2)^3} - \dfrac{28x^3}{(2+x^2)^2} + \dfrac{12x}{2+x^2}\right\}$

5. The derivative is 0 at 0. $f'[x]$ and $f''[x]$ are never undefined:

In[·]:= **Solve[f'[x] == 0, x, Reals]**

Out[·]= $\{\{x \to 0\}, \{x \to 0\}\}$

By the increasing/decreasing test, the function is increasing on $(-\infty, 0)$ and $(0, \infty)$:

In[·]:= **f' /@ {−1, 1}**

Out[·]= $\left\{\dfrac{14}{9}, \dfrac{14}{9}\right\}$

6. By the first derivative test, the function has no local maxima or minima. The second derivative test is inconclusive:

In[ ]:= **f''[0]**

Out[ ]= 0

7. The function has inflection points at 0, $\sqrt{6}$ and $-\sqrt{6}$. By the concavity test, it is concave upward on $\left(-\infty, \ -\sqrt{6}\right)$ and $\left(0, \ \sqrt{6}\right)$ and concave downward on $\left(-\sqrt{6}, \ 0\right)$ and $\left(\sqrt{6}, \ \infty\right)$:

In[ ]:= **{infl = Solve[f''[x] == 0, x], f'' /@ {−3, −1, 1, 3}}**

Out[ ]= $\left\{\left\{\{x \rightarrow 0\}, \left\{x \rightarrow -\sqrt{6}\right\}, \left\{x \rightarrow \sqrt{6}\right\}\right\}, \left\{\dfrac{72}{1331}, -\dfrac{40}{27}, \dfrac{40}{27}, -\dfrac{72}{1331}\right\}\right\}$

The Plot

You can see all the previous information in the following plot:

In[ ]:= **Plot[{f[x], 2 x}, {x, −10, 10}, PlotStyle → {Red, Dashed},**
        Epilog → {PointSize[Large], Point[{{0, f[0]}, {−Sqrt[6], f[−Sqrt[6]]}, {Sqrt[6], f[Sqrt[6]]}}]}]

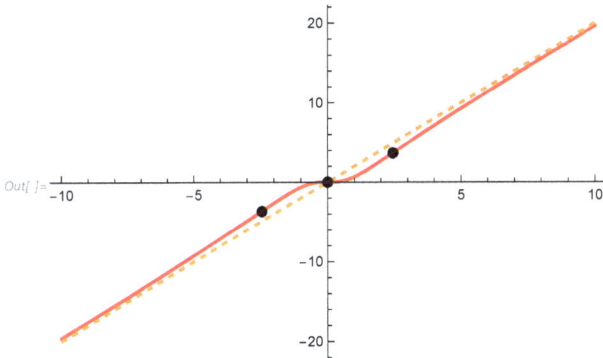

Summary

You can use many of the properties of a function to sketch its graph.

Intercepts determine where the function is 0 or crosses the y axis.

Limits can help find asymptotes.

First and second derivatives can help find local maxima and minima, inflection points and intervals of increase/decrease.

The next lesson will delve further into optimization problems.

Exercises

Exercise 1—Polynomial

Sketch the following polynomial:

In[·]:= **f[x_] := x^3 − 6 x^2 + 9 x**

Solution

Since the function is a polynomial, it is defined everywhere and has no asymptotes.

The function has x- and y-intercepts at 0 and another x-intercept at 3:

In[·]:= **{f[0], Solve[f[x] == 0, x]}**

Out[·]= **{0, {{x → 0}, {x → 3}, {x → 3}}}**

Calculate the first and second derivatives:

In[·]:= **{f'[x], f''[x]}**

Out[·]= $\{9 - 12x + 3x^2, -12 + 6x\}$

There are critical numbers at 1 and 3:

In[·]:= **Solve[f'[x] == 0, x]**

Out[·]= **{{x → 1}, {x → 3}}**

By the increasing/decreasing test, the function is increasing on $(-\infty, 1)$ and $(3, \infty)$ and decreasing on $(1, 3)$:

In[·]:= **f' /@ {0, 2, 4}**

Out[·]= **{9, −3, 9}**

Therefore, there is a local maximum at 1 and a local minimum at 3. The second derivative test agrees:

In[·]:= **f'' /@ {1, 3}**

Out[·]= **{−6, 6}**

The function has an inflection point at 2 and is concave upward on $(-\infty, 2)$ and concave downward on $(2, \infty)$:

In[·]:= **{Solve[f''[x] == 0, x], f'' /@ {1, 3}}; f[1]**

Out[·]= **4**

Here is all the previous information in a graph:

In[ ]:=**Plot[{f[x], 4, 0}, {x, −2, 5}, Axes → {False, True},**
Epilog → {PointSize[Large], Point[{{0, f[0]}, {2, f[2]}, {3, f[3]}}]}]

Out[ ]=

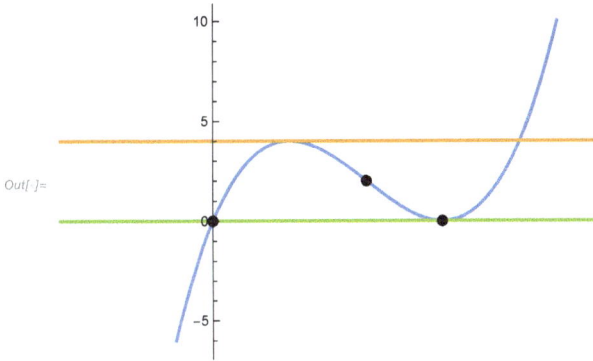

Exercise 2—Rational Function

Sketch the following rational function:

In[ ]:=**f[x_] := (2 x − 6) / (x − 4)**

Solution

1. The function is defined everywhere except at 4. So the domain is $(-\infty, 4) \cup (4, \infty)$.

2. The function has a y-intercept at $\frac{3}{2}$ and an x-intercept at 3:

In[ ]:=**{zeroes = Solve[f[x] == 0, x], f[0]}**

Out[ ]=$\left\{\{\{x \to 3\}\}, \dfrac{3}{2}\right\}$

3. The function has no symmetry.

4. The function has a vertical asymptote at 4:

In[ ]:=**{Limit[f[x], x → 4, Direction → "FromBelow"], Limit[f[x], x → 4, Direction → "FromAbove"]}**

Out[ ]=$\{-\infty, \infty\}$

The function has horizontal asymptote $y = 2$:

In[ ]:=**{Limit[f[x], x → −Infinity], Limit[f[x], x → Infinity]}**

Out[ ]=$\{2, 2\}$

The function has no slant asymptotes.

Calculate the first and second derivatives:

In[·]:= **{f'[x], f''[x]}**

$$Out[·]=\left\{\frac{2}{-4+x}-\frac{-6+2x}{(-4+x)^2}, -\frac{4}{(-4+x)^2}+\frac{2(-6+2x)}{(-4+x)^3}\right\}$$

5. The derivative is 0 nowhere. In this case, $f'[x]$ and $f''[x]$ are both undefined at 4:

In[·]:= **Solve[f'[x] == 0, x]**

Out[·]= **{}**

In[·]:= **crit = 4;**

By the increasing/decreasing test, the function is decreasing on $(-\infty, 4)$ and $(4, \infty)$:

In[·]:= **f' /@ {0, 5}**

$$Out[·]=\left\{-\frac{1}{8}, -2\right\}$$

6. By the first derivative test, there are no local maxima or minima. The second derivative test is inconclusive:

In[·]:= **f''[4] // Quiet**

Out[·]= Indeterminate

7. The function has no inflection points. By the concavity test, it is concave downward on $(-\infty, -4)$ and concave upward on $(4, \infty)$:

In[·]:= **{infl = Solve[f''[x] == 0, x], f'' /@ {0, 5}}**

$$Out[·]=\left\{\{\}, \left\{-\frac{1}{16}, 4\right\}\right\}$$

You can see all the previous information in the following plot:

In[·]:= **Plot[f[x], {x, 0, 8}, GridLines → {{4}, {2}}]**

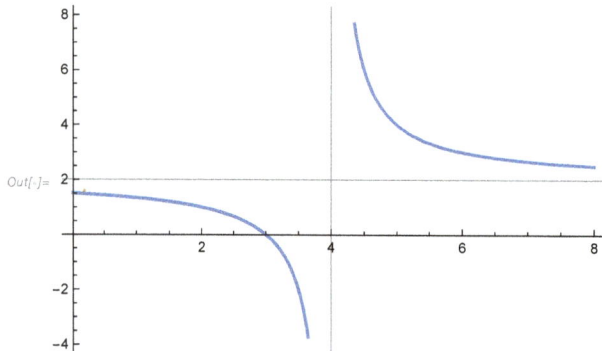

Exercise 3—Slant

Sketch the following rational function:

$_{In[\cdot]:=}$ **f[x_] := (2 x^3 + x^2 + 1)/(x^2 − 4)**

Solution

1. The function is defined everywhere except at 2 and −2. So the domain is $(-\infty, -2) \cup (-2, 2) \cup (2, \infty)$.

2. The function has a y-intercept at $-1/4$ and has an x-intercept at -1:

$_{In[\cdot]:=}$ **{zeroes = Solve[f[x] == 0, x, Reals], f[0]}**

$_{Out[\cdot]=}\left\{\{\{x \rightarrow -1\}\}, -\dfrac{1}{4}\right\}$

3. The function has no symmetry.

4. The function has vertical asymptotes at 2 and −2:

$_{In[\cdot]:=}$ **{{Limit[f[x], x → 2, Direction → "FromBelow"],**
　　　Limit[f[x], x → 2, Direction → "FromAbove"]},
　　{Limit[f[x], x → −2, Direction → "FromBelow"],
　　　Limit[f[x], x → −2, Direction → "FromAbove"]}}

$_{Out[\cdot]=}\{\{-\infty, \infty\}, \{-\infty, \infty\}\}$

The function has no horizontal asymptotes:

$_{In[\cdot]:=}$ **{Limit[f[x], x → −Infinity], Limit[f[x], x → Infinity]}**

$_{Out[\cdot]=}\{-\infty, \infty\}$

The function has $y = 2 x + 1$ as a slant asymptote:

$_{In[\cdot]:=}$ **PolynomialQuotient[2 x^3 + x^2 + 1, x^2 − 4, x]**

$_{Out[\cdot]=}$ **1 + 2 x**

$_{In[\cdot]:=}$ **Limit[f[x] − (2 x + 1), x → Infinity]**

$_{Out[\cdot]=}$ **0**

Calculate the first and second derivatives:

$_{In[\cdot]:=}$ **{f'[x], f''[x]}**

$_{Out[\cdot]=}\left\{\dfrac{2x+6x^2}{-4+x^2} - \dfrac{2x\left(1+x^2+2x^3\right)}{\left(-4+x^2\right)^2}, \dfrac{2+12x}{-4+x^2} - \dfrac{4x\left(2x+6x^2\right)}{\left(-4+x^2\right)^2} + \dfrac{8x^2\left(1+x^2+2x^3\right)}{\left(-4+x^2\right)^3} - \dfrac{2\left(1+x^2+2x^3\right)}{\left(-4+x^2\right)^2}\right\}$

5. The derivative is 0 at 0 and roughly -3.23, -0.42 and 3.66. In this case, $f'[x]$ and $f''[x]$ are both undefined at 2 and -2:

In[·]:= **sol = Solve[f '[x] == 0, x] // FullSimplify // N**

Out[·]= {{x → 0.}, {x → −3.23319}, {x → −0.422973}, {x → 3.65617}}

By the increasing/decreasing test, the function is increasing on $(-\infty, -3.23)$ and $(3.66, \infty)$ and decreasing on all other intervals:

In[·]:= **f ' /@ {−4, −3, −1, −1/2, 1, 4}**

Out[·]= $\left\{\dfrac{7}{6}, -\dfrac{24}{25}, -\dfrac{4}{3}, -\dfrac{14}{225}, -\dfrac{32}{9}, \dfrac{11}{18}\right\}$

6. By the first derivative test, there is a local minimum at roughly 3.66 and a local maximum at roughly -3.23. The second derivative test agrees:

In[·]:= **f '' /@ {sol[[2, 1, 2]], sol[[3, 1, 2]]}**

Out[·]= {−3.00597, 0.664165}

7. The function has an inflection point at roughly -0.21. By the concavity test, it is concave downward on $(-\infty, -2)$ and $(-0.21, 2)$ and concave upward on $(-2, -0.21)$ and $(2, \infty)$:

In[·]:= **{infl = Solve[f ''[x] == 0, x, Reals], f '' /@ {−3, −1, 1, 3}} // N**

Out[·]= {{{x → −0.214712}}, {−5.584, 5.11111, −10.2963, 10.544}}

You can see the previous information in the following plot:

In[·]:= **Plot[{f[x], 2 x + 1}, {x, −5, 5}, ⋯ +]**

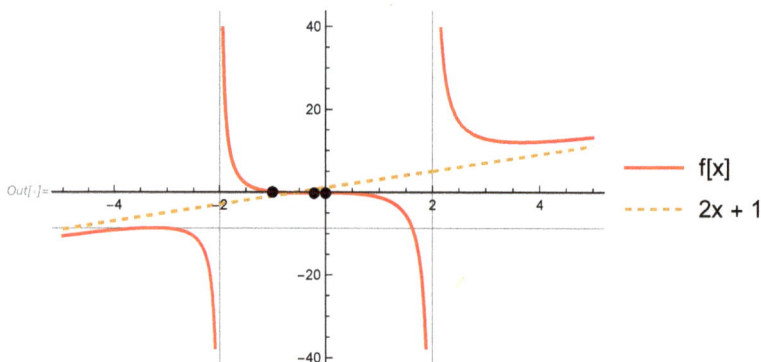

21 | Optimization

Overview

The ways that extreme values have been found so far are very useful in applications.

You have already seen how derivatives can be used to maximize profits and minimize costs.

This lesson continues this approach to solve more optimization problems.

The goal of this lesson is to develop a way for solving optimization problems.

Guidelines

When solving related rates problems, a methodology was introduced to make them easier to solve. A similar methodology is used for optimization problems.

1. Understand the problem.

2. Make a picture.

3. Assign a symbol to the quantity (call it M) being maximized or minimized. Also, select symbols (like a, b, x or y) for other unknown quantities.

4. Use the other symbols to find an equivalent expression for M.

5. If M is in terms of multiple symbols, try to find relationships between them to express M in terms of only one variable, so that $M = f[x]$ for some x. Find the domain of f.

6. Find the absolute maximum or minimum of f using the techniques covered, such the closed interval method, the first derivative test or the second derivative test.

Area

Using 3600 feet of fencing, a farmer wants to fence off a field that borders a river. If the farmer wants to make a rectangular field that runs parallel to the river and wants no fencing on the side touching the river, find the dimensions of the field that maximize its area.

1. Use an animation to better understand the problem:

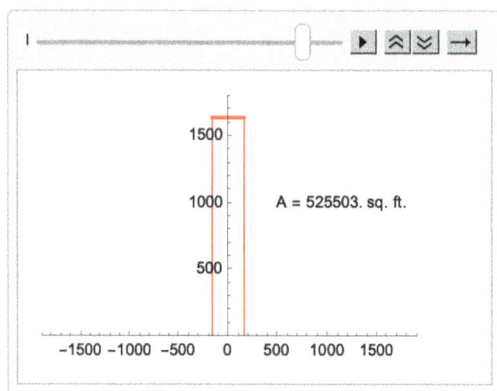

Based on the animation, you want a configuration that is not too thin and not too tall.

2. (and 3.) In general, you have the following picture:

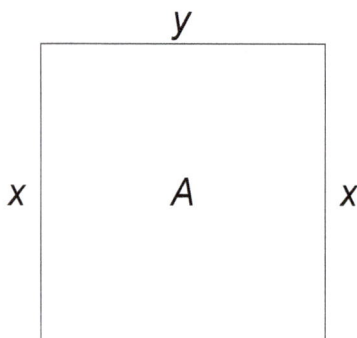

where A is the area, x is the length, and y is the width.

Area 2

4. From geometry, you know the formula for A:

$A = x y$

5. To express A in terms of one variable, use the given information that there are 3600 feet of fencing. Therefore:

$2 x + y = 3600$

$y = 3600 - 2 x$

$A = x (3600 - 2 x) = 3600 x - 2 x^2$

Note that $0 \le A \le 1800$, otherwise A is less than 0, so the domain of A is [0, 1800].

6. Use the closed interval method to find the absolute maximum. Since the area function can be expressed as a polynomial, the critical numbers are where the derivative of A is 0:

In[ ]:= **area[x_] := 3600 x − 2 x^2**

In[ ]:= **Solve[area '[x] == 0 && 0 ≤ x ≤ 1800]**

Out[ ]= **{{x → 900}}**

In[ ]:= **area[900]**

Out[ ]= **1 620 000**

$A[0] = 0$, $A[1800] = 0$ and $A[900] = 1\,620\,000$. So there is an absolute maximum at $(900, 1\,620\,000)$. Therefore, the rectangle should be 900 feet long and 1800 feet wide.

Cylindrical Can 1

A company wants to make a cylindrical can that holds 1 liter of water. What are the dimensions that will minimize the total cost of plastic to make the can?

1. To minimize the cost of plastic, you should minimize the surface area of the can.

2. (and 3.) Draw a picture and label the unknowns:

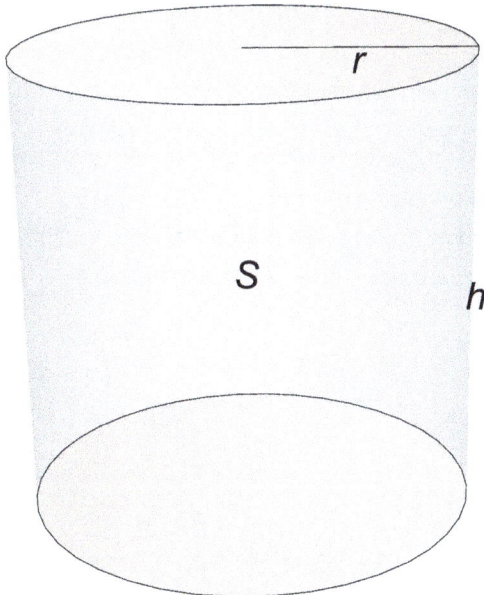

Here S is the surface area, r is the radius, and h is the height.

4. From geometry, you know the formula for S:

$$S = 2\pi r h + 2\pi r^2$$

5. You want to get rid of h in the formula. To do so, note that the volume of the can is 1000 cm^3 and use the formula for volume:

$$\pi r^2 h = 1000$$

$$h = \frac{1000}{\pi r^2}$$

Cylindrical Can 2

The new equation for S is:

In[ ]:= **surfacearea[r_] := 2 π r^2 + 2 π r (1000/(π r^2))**

The domain of S is $(0, \infty)$.

6. You cannot use the closed interval method since the domain is $(0, \infty)$, but you want to find the critical numbers in the domain:

In[ ]:= **sol = Solve[surfacearea '[r] == 0 && r > 0, r] // N**

Out[ ]= **{{r → 5.41926}}**

Using the first derivative test, you can see that the function is decreasing on $(0, 5.42)$ and increasing on $(5.42, \infty)$, so the function has an absolute maximum at roughly 5.42.

You can also see this in the plot:

In[ ]:= **Plot[surfacearea[r], {r, 0, 20}, PlotRange → {0, 2000},**
      Epilog → {PointSize[Large], Point[{sol[[1, 1, 2]], surfacearea[sol[[1, 1, 2]]]}]}]

Out[ ]=

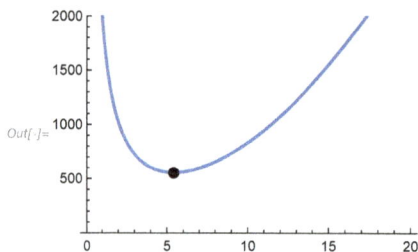

With that radius, the height will be roughly 10.84, or twice the radius:

In[ ]:= **1000/(π sol[[1, 1, 2]]^2)**

Out[ ]= **10.8385**

So the can should have radius ~ 5.42 cm and height ~ 10.84 cm.

First Derivative Test for Absolute Extreme Values

To find the absolute extreme values of a function, use what was learned in the previous example.

Suppose that c is a critical number of a continuous function f defined on an interval.

   i. If $f'[x] > 0$ for **all** $x < c$ and $f'[x] < 0$ for **all** $x > c$, then $f[c]$ is the **absolute maximum value** of f.

   ii. If $f'[x] < 0$ for **all** $x < c$ and $f'[x] > 0$ for **all** $x > c$, then $f[c]$ is the **absolute minimum value** of f.

Notice that this is just the first derivative test, but tweaked slightly to find absolute maxima and minima instead of local maxima and minima.

For example, x^2 has an absolute minimum at (0, 0).

The derivative of x^2, $2x$ is 0 at 0, and is less than the derivative for all values of $x > 0$.

It is also greater than the derivative for all values of $x < 0$.

You can see this is in the graph of x^2 too:

In[ ]:= **Plot[x^2, {x, −1000, 1000}]**

Out[ ]=

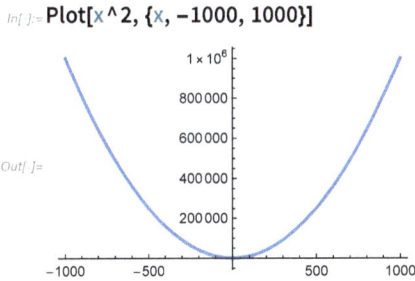

Implicit Differentiation

In the cylindrical can example, implicit differentiation could have been used instead.

The same equations are used, but h is expressed as a function of r:

In[ ]:= **eqn1 = s[r] == 2 π r^2 + 2 π r h[r]**

Out[ ]= $s[r] == 2\pi r^2 + 2\pi r\, h[r]$

In[ ]:= **eqn2 = π r^2 h[r] == 1000**

Out[ ]= $\pi r^2 h[r] == 1000$

Now differentiate both equations with respect to r **implicitly**:

In[·]:= **D[eqn1, r]**

Out[·]= s′[r] == 4 π r + 2 π h[r] + 2 π r h′[r]

In[·]:= **D[eqn2, r]**

Out[·]= 2 π r h[r] + π r² h′[r] == 0

If you substitute 0 for $s'[r]$, note that $r > 0$ and eliminate $h'[r]$ from both equations, you find that the height should be twice the radius:

In[·]:= **{new1 = Simplify[D[eqn1, r] /. s'[r] → 0], new2 = Simplify[D[eqn2, r], r > 0]}**

Out[·]= {h[r] + r (2 + h′[r]) == 0, 2 h[r] + r h′[r] == 0}

In[·]:= **Eliminate[{new1, new2}, h'[r]]**

Out[·]= h[r] == 2 r

In the preceding equations, Eliminate and Simplify were used. Now you can easily solve for the radius given the desired volume:

In[·]:= **Solve[π r^2 h == 1000 && r > 0 /. h → 2 r, r] // N**

Out[·]= {{r → 5.41926}}

Distance

Find the point on the parabola $y^2 = 2 x$ that is closest to the point $(1, 32)$.

The distance between the point $(1, 32)$ and the point (x, y) is given by:

$$d = \sqrt{(x - 1)^2 + (y - 32)^2}$$

If (x, y) lies on the parabola, then $x = y^2 / 2$.

Minimize the following function:

In[·]:= **d[y_] := (y^2/2 − 1)^2 + (y − 32)^2**

You can minimize the square of the distance because the numbers are easier to deal with, and you will still get the same answer.

Find the critical numbers of the function:

In[·]:= **sol = Solve[d'[y] == 0, y, Reals]**

Out[·]= {{y → 4}}

By the first derivative test, 4 is an absolute minimum:

In[ ]:= **d'/@{3, 5}**

Out[ ]= **{−37, 61}**

So $x = 4^2 / 2 = 8$ and the closest point is $(8, 4)$.

Business and Economics

Recall that if $C[x]$ is the cost of producing x units, then $C'[x]$ is the **marginal cost**, the cost of producing one more unit.

If a company sells x units, let $p[x]$ be the price per unit. It is usually called the **demand function** or **price function**.

If x units are sold, then the total revenue is $R[x] = x\, p[x]$ and R is called the **revenue function**.

The **marginal revenue function** $R'[x]$ tells how much money is gained for producing one more unit.

The **profit function** $P[x] = R[x] - C[x]$ gives the total profit for producing x units.

The **marginal profit function** $P'[x]$ tells how much extra profit is earned for producing one more unit.

Businesses want to **maximize** their profits and revenue and **minimize** their costs.

You can see that the profits are maximized when the marginal revenue equals the marginal cost.

i. If the marginal revenue is greater than the marginal cost, then clearly more units should be made to make more money.

ii. If the marginal revenue is less than the marginal cost, then clearly fewer units should be made to save money.

Business Example

A store has been selling 300 monitors a week at $200 each. For each $10 rebate that the store offers to buyers, the number of monitors sold will increase by 30 a week. What are the demand function and revenue function? What rebate should the store offer to maximize its revenue?

Let x be the number of units sold. The weekly increase in sales is $x - 300$. For each increase of 30 units sold, the price goes down by \$10. So for each additional unit sold, the decrease in price will be $\frac{1}{30} * 10$, and the demand function is:

In[-]:= $p[x\_] := 200 - \frac{10}{30}(x - 300)$

The revenue function is:

In[-]:= $r[x\_] := x\, p[x]$

To maximize revenue, find the critical numbers of the function. The revenue function is a polynomial, so the critical numbers are where the derivative is 0:

In[-]:= **Solve[r'[x] == 0]**

Out[-]= $\{\{x \to 450\}\}$

By the first derivative test, this is an absolute maximum:

In[-]:= **r'/@{400, 500}**

Out[-]= $\left\{\dfrac{100}{3}, -\dfrac{100}{3}\right\}$

The corresponding price is:

In[-]:= **p[450]**

Out[-]= 150

Therefore the rebate should be \$200 – \$150 = \$50.

Maximize and Minimize

The Wolfram Language has two special functions to find the absolute maximum or minimum for a given range. They are called Maximize and Minimize, respectively.

Find the absolute maximum for the previous revenue function:

In[-]:= **Maximize[r[x], x]**

Out[-]= $\{67\,500, \{x \to 450\}\}$

You can also specify it on a range:

In[-]:= **Maximize[{r[x], 0 ≤ x ≤ 400}, x]**

Out[-]= $\left\{\dfrac{200\,000}{3}, \{x \to 400\}\right\}$

Find the absolute minimum for the following function in the range $[0, 2\pi]$:

In[ ]:= **g[*x*_] := *x* Sin[*x*] – Cos[*x*]**

Use Minimize:

In[ ]:= **min = Minimize[{g[*x*], 0 ≤ *x* ≤ 2 *π*}, *x*] // N**

Out[ ]= **{–5.10013, {*x* → 5.08699}}**

Here is the function and its minimum in the range:

In[ ]:= **Plot[g[*x*], {*x*, 0, 2 *π*}, Epilog → {PointSize[Large], Point[{min[[2, 1, 2]], min[[1]]}]}]**

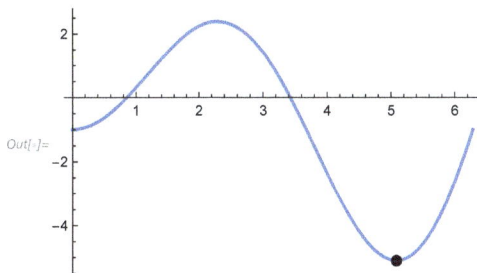

Summary

Optimization is important in many areas.

To solve optimization problems, it is important to understand the problem being solved, draw pictures, assign variables, find relations and use differentiation techniques to find absolute maxima and minima.

Optimization is especially useful in business, where companies want to maximize profits.

The Wolfram Language's Maximize and Minimize are especially useful for optimization problems.

The next lesson will cover antiderivatives.

Exercises

Exercise 1—Minimum Product

Find two numbers whose difference is 200 and whose product is a minimum.

Solution

You have two numbers x and y with the following property:

$$x - y = 200$$

You want to minimize the following expression:

$$x y$$

You express y in terms of x and want to minimize the following function:

In[·]:= **f[x_] := x (x − 200)**

Find the critical numbers of the function:

In[·]:= **crit = Solve[f'[x] == 0, x]**

Out[·]= **{{x → 100}}**

By the first derivative test, this is an absolute minimum:

In[·]:= **f' /@ {99, 101}**

Out[·]= **{−2, 2}**

So the two numbers you want are 100 and $100 - 200 = -100$.

Confirm with Minimize:

In[·]:= **Minimize[f[x], x]**

Out[·]= **{−10 000, {x → 100}}**

Exercise 2—Closest Point

Find the point on the line $y = 5 x + 3$ that is closest to the origin.

Solution

The distance to the origin from the line is given by the following expression:

$$\sqrt{x^2 + (5x + 3)^2}$$

Square the expression and find its minimum, since the numbers are easier to deal with:

In[ ]:= **d[x_] := x^2 + (5 x + 3)^2**

Find the critical number(s) of the function:

In[ ]:= **sol = Solve[{d'[x] == 0}, x]**

Out[ ]= $\left\{\left\{x \to -\dfrac{15}{26}\right\}\right\}$

By the first derivative test, this is an absolute minimum:

In[ ]:= **d' /@ {−1, 0}**

Out[ ]= **{−22, 30}**

$y = 5\,(-15/26) + 3 = 3/26$, so the point $(-15/26,\ 3/26)$ is closest to the origin.

Exercise 3—Souvenirs

Jake sells souvenirs in the city. On an average day he sells 15 for \$30. If he raises the price by \$1, then he loses on average two sales. Find the demand function.

Solution

If x is the number of souvenirs that Jake sells per day, then his weekly increase is $(x − 15)$. Therefore his demand function is:

In[ ]:= **p[x_] := 30 − 2 (x − 15)**

If it costs \$10 to produce each souvenir, find the price that maximizes his profits.

The cost of producing x souvenirs is $10\,x$:

In[ ]:= **c[x_] := 10 x**

Therefore, the profit function is $R[x] − C[x] = x\,p[x] − C[x]$:

In[ ]:= **profit[x_] := x p[x] − c[x]**

Maximize the profit function by first finding its critical number(s):

In[·]:= **sol = Solve[profit'[x] == 0, x]**

Out[·]= $\left\{\left\{x \to \dfrac{25}{2}\right\}\right\}$

By the first derivative test, this is a maximum:

In[·]:= **profit'/@{12, 13}**

Out[·]= **{2, −2}**

This corresponds to a price of $35:

In[·]:= **p[sol[[1, 1, 2]]]**

Out[·]= **35**

Exercise 4—Poster

A poster is to have an area of 200 in^2 with 1-inch margins at the bottom and sides and a 2-inch margin at the top. What dimensions will give the largest printed area?

Solution

Here is a general picture to illustrate the problem:

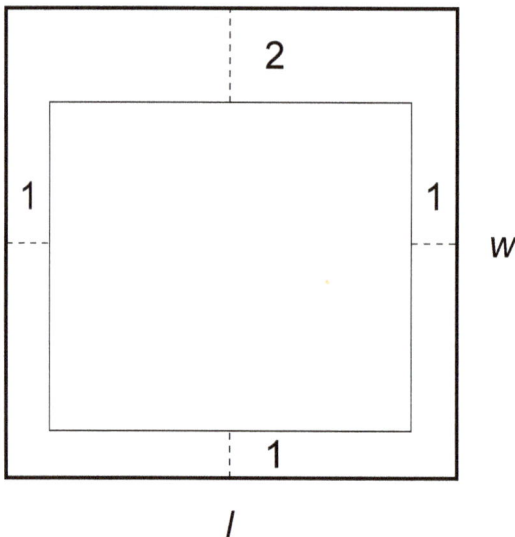

You want to maximize $(l - 2)\,(w - 3)$.

The problem states that there is a relationship between l and w:

In[ ]:= **l * w == 200**

Out[ ]= **l w == 200**

Therefore, $w = 200/l$. So maximize $(l - 2)(200/l - 3)$:

In[ ]:= **area[l_] := (l − 2) (200/l − 3)**

Find the critical number(s):

In[ ]:= **sol = Solve[area'[l] == 0 && l > 0, l]**

$$Out[ ]= \left\{\left\{l \rightarrow \frac{20}{\sqrt{3}}\right\}\right\}$$

By the first derivative test, there is an absolute maximum:

In[ ]:= **area' /@ {11, 12}**

$$Out[ ]= \left\{\frac{37}{121}, -\frac{2}{9}\right\}$$

Calculate the width:

In[ ]:= **200/sol[[1, 1, 2]]**

Out[ ]= $10 \sqrt{3}$

So the poster should have dimensions $20 \big/ \sqrt{3}$ in. by $10 \sqrt{3}$ in.

22 | Antiderivatives

Overview

So far, you have been taking derivatives of many functions.

For example, the derivative of $3\,x$ is 3:

In[ ]:= **D[3 x, x]**

Out[ ]= 3

Say you wanted to go *backward* to **find a function whose derivative is 3**.

In this case, you already know that $3\,x$ is such a function.

This problem has various applications: a physicist may want to find the position of a particle given the particle's velocity function.

This lesson will go over how to do this for various functions.

Antiderivative

The function with a given derivative f has a special name: the **antiderivative of f**.

Consider the following function:

In[ ]:= **f[x_] := x**

To find its antiderivative, just use the power rule, except in reverse.

Instead of **multiplying** by the exponent and **reducing** the exponent by one, **increase** the exponent by one and **divide** by the new exponent.

So an antiderivative of x is $x^2/2$.

Confirm with D:

In[ ]:= **D[x^2/2, x]**

Out[ ]= x

Here is the graph of x and the antiderivative $x^2/2$:

In[ ]:= **Plot[{x, x^2/2}, {x, -1, 1}, PlotLegends → {"function", "antiderivative"}]**

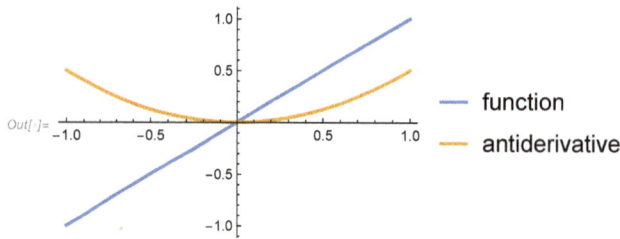

Families of Antiderivatives

Notice before that it was said that $x^2/2$ is *an* antiderivative of x.

That is because a function has more than one antiderivative. In the previous case, $x^2/2 + 1$ is also an antiderivative:

In[ ]:= **D[x^2/2 + 1, x]**

Out[ ]= **x**

In general, for the function $F'[x]$, its antiderivatives have the general form $F[x] + C$, where C is a constant. The functions $F[x] + C$ form a **family** of antiderivatives.

Here is a plot of x and some of its antiderivatives:

In[ ]:= **anti = Table[x^2/2 + i, {i, -5, 5, 2}];**

In[ ]:= **Plot[Evaluate[Flatten[{x, anti}]], {x, -5, 5}, PlotLegends → "Expressions"]**

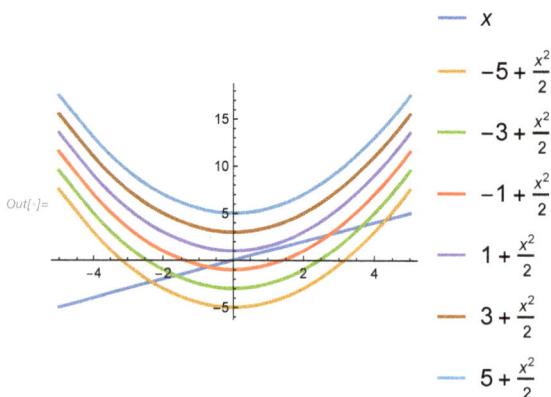

Notice that the antiderivatives are all vertical shifts of $x^2/2$.

Sum and Difference

Finding the antiderivative of a sum is the same as taking each antiderivative in the sum and adding them together.

So the general antiderivative of $x^2 + x^3$ is $x^3/3 + x^4/4 + C$:

In[ ]:= **D[x^3/3 + x^4/4 + c, x]**

Out[ ]= $x^2 + x^3$

Finding the antiderivative of a difference is the same as taking each antiderivative in the difference and subtracting the second from the first.

So the general antiderivative of $5 - x^4$ is $5\,x - x^5/5 + C$:

In[ ]:= **D[5 x − x^5/5 + c, x]**

Out[ ]= $5 - x^4$

Notice that these correspond to the sum and difference laws for differentiation.

The constant multiple law also applies: the general antiderivative of $k\,F'[x]$ for k a constant is $k\,F[x] + C$.

There is also a correspondence for products and quotients, but those are generally more difficult to find, so they will not be covered.

Trig Functions

Recall the derivatives of the following trigonometric functions:

In[ ]:= **trig = {Sin[x], Cos[x], Tan[x], Sec[x], Csc[x], Cot[x]};**

In[ ]:= **deriv = Table[D[i, x], {i, trig}]**

Out[ ]= $\{Cos[x], -Sin[x], Sec[x]^2, Sec[x]\,Tan[x], -Cot[x]\,Csc[x], -Csc[x]^2\}$

So the antiderivative of $Sec[x]\,Tan[x]$ is $Sec[x] + C$.

Here is a table of each function and its antiderivative (do not forget the $+C$ for each one):

Function	Antiderivative
Cos[x]	Sin[x]
−Sin[x]	Cos[x]
Sec[x]^2	Tan[x]
Sec[x] Tan[x]	Sec[x]
−Cot[x] Csc[x]	Csc[x]
−Csc[x]^2	Cot[x]

Example

Find the general antiderivative of the following function:

In[ ]:= **f[x_] := 3 Sin[x] + (4 x^5 − Sqrt[x]) / x**

Since it is a sum, find the antiderivative of each constituent and add them together.

The antiderivative of $3\,\mathrm{Sin}[x]$ is $-3\,\mathrm{Cos}[x] + C$:

In[ ]:= **D[−3 Cos[x], x]**

Out[ ]= 3 Sin[x]

To find the antiderivative of the quotient, first rewrite it as a difference:

In[ ]:= **(4 x^5 − Sqrt[x]) / x == 4 x^5 / x − Sqrt[x] / x;**

Note that $\sqrt{x} = x^{1/2}$, so you can use the power rules to get rid of the x in the denominator. The Wolfram Language already does that for you:

In[ ]:= **4 x^5 / x − Sqrt[x] / x**

Out[ ]= $-\dfrac{1}{\sqrt{x}} + 4\,x^4$

The antiderivative of $4\,x^4$ is $4\,x^5/5 + C$ and the antiderivative of $-x^{-1/2}$ is $-x^{1/2}/(-1/2) + C = 2\,x^{1/2} + C$:

In[ ]:= **{D[4 x^5 / 5, x], D[2 Sqrt[x], x]}**

Out[ ]= $\left\{4\,x^4,\ \dfrac{1}{\sqrt{x}}\right\}$

So the total antiderivative is $-3\,\mathrm{Cos}[x] + 4\,x^5/5 - 2\,\sqrt{x} + C$. Confirm with **D**:

In[ ]:= **D[−3 Cos[x] + 4 x^5 / 5 − 2 Sqrt[x] + c, x] == f[x] // Simplify**

Out[ ]= **True**

First-Order Differential Equation

Antiderivatives are used in **differential equations**: equations that involve derivatives.

Differential equations in general can be very difficult to solve, but you can solve some very basic ones.

A differential equation's general solution includes some number of arbitrary constants, but with some extra constraints you can find a particular solution.

For example, find f when $f'[x] = x \sqrt{x}$ and $f[1] = 3$.

In this case $f'[x] = x^{3/2}$, so the general antiderivative is $f[x] = x^{5/2} / (5/2) + C = 2\, x^{5/2} / 5 + C$:

In[ ]:= **D[2 x^(5/2)/5, x]**

Out[ ]= $x^{3/2}$

Given the constraint $f[1] = 3$, you can solve for C:

In[ ]:= **Solve[3 == 2*x^(5/2)/5 + c /. x → 1, c]**

Out[ ]= $\left\{\left\{c \to \dfrac{13}{5}\right\}\right\}$

So the particular solution is $(2\, x^{5/2} + 13) / 5$.

You can also solve this with the built-in function DSolve:

In[ ]:= **Clear[f]**

In[ ]:= **DSolve[{f '[x] == x Sqrt[x], f[1] == 3}, f[x], x]**

Out[ ]= $\left\{\left\{f[x] \to \dfrac{1}{5}\left(13 + 2\, x^{5/2}\right)\right\}\right\}$

Second-Order Differential Equation

Find f when $f''[x] = 12\, x^2 - 6\, x - 8$, $f[0] = 1$, $f[1] = 2$.

The general antiderivative for the function $f''[x]$ is $f'[x] = 4\, x^3 - 3\, x^2 - 8\, x + C$:

In[ ]:= **D[4 x^3 − 3 x^2 − 8 x + c, x]**

Out[ ]= $-8 - 6\, x + 12\, x^2$

The general antiderivative for the function $f'[x]$ is $f[x] = x^4 - x^3 - 4x^2 + Cx + D$:

In[·]:= **D[x^4 − x^3 − 4 x^2 + c x + d, x]**

Out[·]= $c - 8x - 3x^2 + 4x^3$

To solve for C and D, use the given constraints $f[0] = 1$, $f[1] = 2$:

In[·]:= **Solve[(1 == x^4 − x^3 − 4 x^2 + c x + d /. {x → 0}) &&**
(2 == x^4 − x^3 − 4 x^2 + c x + d /. {x → 1}), {c, d}]

Out[·]= $\{\{c \rightarrow 5, d \rightarrow 1\}\}$

So the particular solution is $f[x] = x^4 - x^3 - 4x^2 + 5x + 1$.

Confirm with DSolve:

In[·]:= **DSolve[{f ''[x] == 12 x^2 − 6 x − 8, f[0] == 1, f[1] == 2}, f[x], x]**

Out[·]= $\{\{f[x] \rightarrow 1 + 5x - 4x^2 - x^3 + x^4\}\}$

Rectilinear Motion

Antidifferentiation can help find the position of a particle that moves in a straight line.

Recall that for a particle with position function **s[t]**, its velocity is given by **v[t] = s'[t]** and its acceleration is given by **a[t] = s''[t]**.

So if you are given a particle's velocity or acceleration and some initial conditions, you can use antidifferentiation to find its position!

For example, a particle moving in a straight line has acceleration $a[t]$ given by the following function:

In[·]:= **a[t_] := 6 t^2 − 10**

Given that its initial velocity is $v[0] = -4$ cm / sec. and its initial displacement is $s[0] = 3$ cm, you can find its position function $s[t]$.

First find the velocity, which by antidifferentiation is $v[t] = 2t^3 - 10t + C$ for some constant C:

In[·]:= **D[2 t^3 − 10 t + c, t]**

Out[·]= $-10 + 6t^2$

Its position function is therefore $s[t] = t^4 / 2 - 5t^2 + Ct + D$ for some constants C, D:

In[·]:= **D[t^4 / 2 − 5 t^2 + c t + d, t]**

Out[·]= $c - 10t + 2t^3$

To solve for C and D, use the given constraints $v[0] = -4$, $s[0] = 3$:

In[ ]:= **Solve[(−4 == 2 t^3 − 10 t + c /. {t → 0}) && (3 == t^4/2 − 5 t^2 + c t + d /. {t → 0}), {c, d}]**

Out[ ]= {{c → −4, d → 3}}

So the particle has position function $s[t] = t^4/2 - 5\,t^2 - 4\,t + 3$. Confirm with **DSolve**:

In[ ]:= **DSolve[{s ''[t] == a[t], s '[0] == −4, s[0] == 3}, s[t], t] // Simplify**

Out[ ]= $\left\{\left\{s[t] \rightarrow 3 - 4\,t - 5\,t^2 + \dfrac{t^4}{2}\right\}\right\}$

Summary

The antiderivative of a function f is a function g such that $g'[x] = f[x]$.

Antiderivatives come in families, so the general antiderivative of $F'[x]$ is $F[x] + C$ for C an arbitrary constant.

Antiderivatives can be found by using the usual differentiation rules and going in reverse.

Antidifferentiation is used in differential equations to solve for functions.

With extra constraints, a particular solution of a differential equation can be found.

Antidifferentiation is very useful when analyzing the one-dimensional motion of a particle.

The next lesson will start covering integral calculus.

Exercises

Exercise 1—Find the Antiderivative

Find the general antiderivative of the following function:

In[·]:= **f[*x*_] := 4 Sec[*x*] Tan[*x*] + 6 Sqrt[*x*]**

Solution

The function is a sum, so first break it up into its constituents:

In[·]:= **g[*x*_] := 4 Sec[*x*] Tan[*x*]**
     h[*x*_] := 6 Sqrt[*x*]

$4\,\mathrm{Sec}[x]\,\mathrm{Tan}[x]$ is a trigonometric function, and from the trigonometric table you know its antiderivative is $4\,\mathrm{Sec}[x] + C$:

In[·]:= **D[4 Sec[x], x]**

Out[·]= 4 Sec[x] Tan[x]

$6\sqrt{x} = 6\,x^{1/2}$, so by the power rule its antiderivative is $6\,x^{3/2}/(3/2) + C = 4\,x^{3/2} + C$:

In[·]:= **D[4 x^(3/2), x]**

Out[·]= 6 $\sqrt{x}$

So the general antiderivative is $4\,\mathrm{Sec}[x] + 4\,x^{3/2} + C$:

In[·]:= **D[4 Sec[x] + 4 x^(3/2), x] == f[x]**

Out[·]= True

Exercise 2—First-Order Differential Equation

If $f'[x] = 5\,x^4 + 3\,\mathrm{Cos}[x]$ and $f[0] = -5$, find f.

Solution

The function is a sum, so first break it up into constituents:

In[·]:= **g[*x*_] := 5 x^4**
     h[*x*_] := 3 Cos[*x*]

By the power rule, the general antiderivative of $5\,x^4$ is $5\,x^5/(5) + C = x^5 + C$:

In[ ]:= **D[x^5, x]**

Out[ ]= $5\,x^4$

$3\,\mathrm{Cos}[x]$ is a trigonometric function, and from the trig table you know its antiderivative is $3\,\mathrm{Sin}[x] + C$:

In[ ]:= **D[3 Sin[x], x]**

Out[ ]= $3\,\mathrm{Cos}[x]$

So the general antiderivative is $x^5 + 3\,\mathrm{Sin}[x] + C$. Given the constraint $f[0] = -5$, solve for C:

In[ ]:= **Solve[−5 == 0^5 + 3 Sin[0] + c, c]**

Out[ ]= $\{\{c \rightarrow -5\}\}$

So $f[x] = x^5 + 3\,\mathrm{Sin}[x] - 5$:

In[ ]:= **D[x^5 + 3 Sin[x] − 5, x] == g[x] + h[x]**

Out[ ]= **True**

Exercise 3—Second-Order Differential Equation

If $f''[x] = 4/\sqrt{x}$, $f[9] = -5$ and $f'[9] = 2$, find f.

Solution

Since $4/\sqrt{x} = 4\,x^{-1/2}$, $f'[x] = 4\,x^{1/2}/(1/2) + C = 8\,x^{1/2} + C$:

In[ ]:= **D[8 x^(1/2) + c, x]**

Out[ ]= $\dfrac{4}{\sqrt{x}}$

So $f[x] = 8\,x^{3/2}/(3/2) + C\,x + D = 16\,x^{3/2}/3 + C\,x + D$:

In[ ]:= **D[16 x^(3/2)/3 + c x + d, x]**

Out[ ]= $c + 8\,\sqrt{x}$

Given the constraints $f[9] = -5$ and $f'[9] = 2$, find C and D:

In[ ]:= **Solve[(−5 == 16 x^(3/2)/3 + c x + d /. {x → 9}) && (2 == 8 x^(1/2) + c /. {x → 9}), {c, d}]**

Out[ ]= $\{\{c \rightarrow -22, d \rightarrow 49\}\}$

So $f[x] = 16\,x^{3/2}/3 - 22\,x + 49$.

Confirm with DSolve:

In[·]:= **Clear[f];**
DSolve[{f''[x] == 4 / Sqrt[x], f[9] == −5, f'[9] == 2}, f[x], x] // Simplify

Out[·]= $\left\{\left\{f[x] \rightarrow 49 - 22x + \dfrac{16\,x^{3/2}}{3}\right\}\right\}$

Exercise 4—Falling Stones

A stone is dropped from a tower 400 m above the ground. When does it hit the ground? Assume the acceleration due to gravity is $-9.8\,\text{m}/\sec^2$.

Solution

The problem tells us:

1. The acceleration function $a[t] = -9.8$.

2. The initial displacement $s[0] = 400$.

3. The initial velocity $v[0] = 0$.

With these, first solve for $s[t]$:

$a[t] = s''[t] = -9.8$, so $s'[t] = v[t] = -9.8\,t + C$.

Therefore, $s[t] = -9.8\,t^2/2 + C\,t + D = -4.9\,t^2 + C\,t + D$:

In[·]:= **D[−4.9 t^2 + c t + d, t]**

Out[·]= c − 9.8 t

From the initial conditions, solve for C and D:

In[·]:= **Solve[(400 == −4.9 t^2 + c t + d /. {t → 0}) && (0 == −9.8 t + c /. {t → 0}), {c, d}]**

Out[·]= {{c → 0., d → 400.}}

So $s[t] = -4.9\,t^2 + 400$. Now solve for t when $s[t] = 0$:

In[·]:= **Solve[0 == −4.9 ∗ t^2 + 400 && t > 0, t]**

Out[·]= {{t → 9.03508}}

The stone hits the ground approximately 9.04 seconds later.

Exercise 5—Cost

A company wants to make a new product. It calculates the marginal cost (in dollars per item) of producing x items to be given by the following function:

In[ ]:= **mcost[x_] := 3.76 – 0.006 x**

How much would it cost to produce 50 items if the cost of producing one item is $425? Since the marginal cost is the derivative of the cost, you can find the cost by finding the antiderivative of the marginal cost.

Solution

In this case the function is $3.76 - 0.006\,x$, so its general antiderivative is $3.76\,x - 0.006\,x^2 / 2 + C = 3.76\,x - 0.003\,x^2 + C$:

In[ ]:= **D[3.76 x – 0.003 x^2 + c, x] == mcost[x]**

Out[ ]= **True**

To solve for C, use the fact that it costs $425 to produce one item:

In[ ]:= **sol = Solve[(425 == 3.76 x – 0.003 x^2 + c /. {x → 1}), c]**

Out[ ]= **{{c → 421.243}}**

So the general cost function is $3.76\,x - 0.003\,x^2 + 421.243$. Now plug in 50:

In[ ]:= **3.76 x – 0.003 x^2 + sol[[1, 1, 2]] /. x → 50**

Out[ ]= **601.743**

You can see that it costs approximately $601.74 to produce 50 items.

23 | Riemann Sums

Overview

Every calculus course covers the two fundamental aspects of calculus. The first aspect, differential calculus, has already been covered. Its prime motivation was finding the slope of a curve at a point.

The second aspect, integral calculus, is what will soon be covered. The prime motivation of integral calculus is simple to understand.

Consider the following plot:

In[ ]:= **Plot[Sin[x] + 1, {x, 0, 3}, Filling → Bottom]**

Out[ ]=

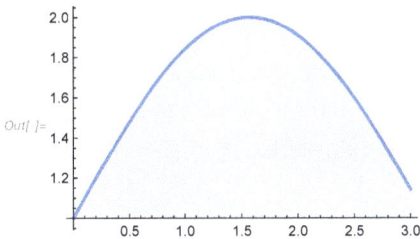

You would like to find the area under this curve.

The goal of this lesson is to develop a precise way of expressing this idea mathematically.

Area Problem 1

Finding the area of some objects is easy.

The area of a rectangle is its length times its width:

In[ ]:= **rectanglearea[l_, w_] := l ∗ w**

The area of a triangle is its base times its height divided by 2:

In[ ]:= **trianglearea[b_, h_] := b ∗ h / 2**

To find the area of a general convex polygon, you divide the polygon into triangles, find the area of each one, and add them all together.

Conceptually speaking, these are all easy to understand too.

Finding the area of a circle is harder to comprehend, but its equation is π times the radius of the circle squared:

In[ ]:= **circlearea[r_] := π r ^ 2**

How do you find the area under a general curve, though? Conceptually speaking, it is a little difficult to understand, and may seem impossible to calculate.

What is the general formula for the area under this curve?

In[·]:= **Plot[x^2, {x, 0, 4}, Filling → Bottom]**

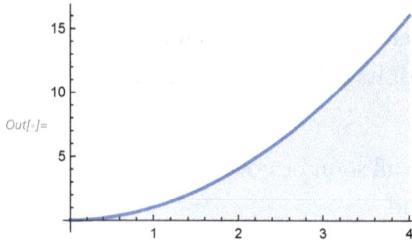

Out[·]=

Slope at a Point

To start, recall what you did when you started out finding the slope of a function at a point. You took secant lines. One point (P) intersected the desired point on the curve, and another point intersected the curve at some other point Q.

You calculated the slope and then took another secant line with P and another point Q' on the curve closer to P than Q. Then you calculated the slope again and kept repeating in this fashion.

If the function was differentiable, the slopes of the secant lines eventually approached some value m that you took to be the slope at that point.

Here is an interactive example for the function $f[x] = x^2$ at the point $P = (1, 1)$:

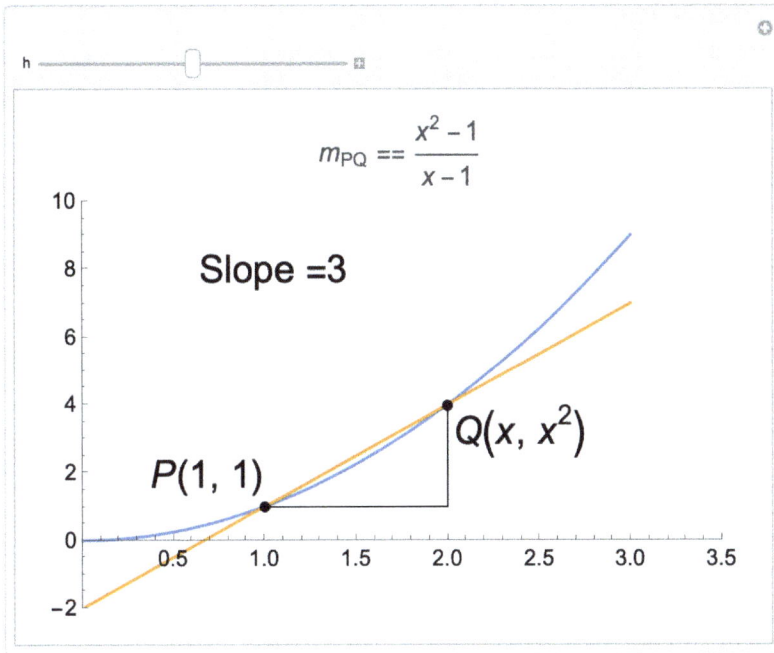

Area Problem 2

In a similar way of doing things, divide the given plot region into separate rectangles. Then calculate the area of each rectangle and add them all together to get the approximate area of the region. Then increase the number of rectangles by reducing the base lengths to better approximate the region.

You can visualize this better with the built-in function DiscretePlot. Use the function $f[x] = x^2$ in the range $[0, 1]$ and divide the plot region into four rectangles:

```
In[ ]:= Show[Plot[x^2, {x, 0, 1}],
         DiscretePlot[x^2, {x, 0, 1, 1/4}, ExtentSize → Right, ExtentMarkers → {"Filled", None}]]
```

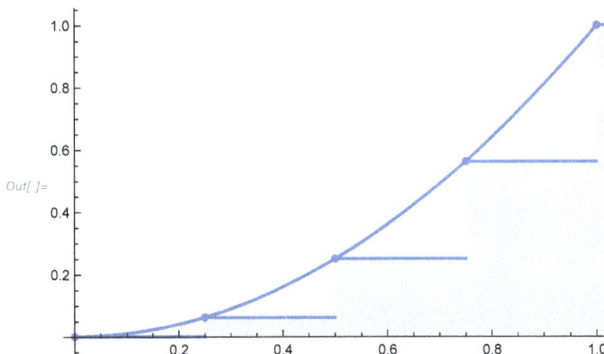

By dividing the region into 10 rectangles, you get a better approximation:

In[·]:= **Show[Plot[x^2, {x, 0, 1}],**

DiscretePlot[x^2, {x, 0, 1, 1/10}, ExtentSize → Right, ExtentMarkers → {"Filled", None}]]

Out[·]=

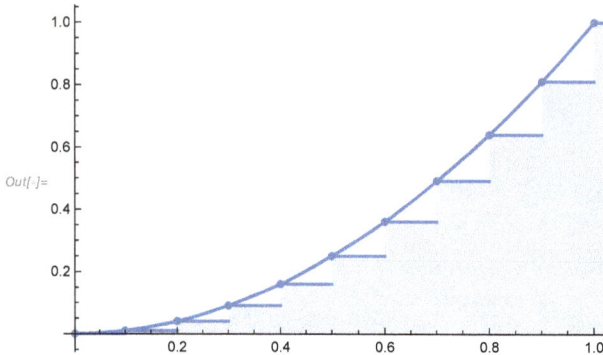

Area under a Parabola 1

When the top-left corner of each rectangle is the point on the curve, you can see that the overall area should be less than the total area under the curve:

In[·]:= **Show[Plot[x^2, {x, 0, 1}],**

DiscretePlot[x^2, {x, 0, 1, 1/4}, ExtentSize → Right, ExtentMarkers → {"Filled", None}]]

Out[·]=

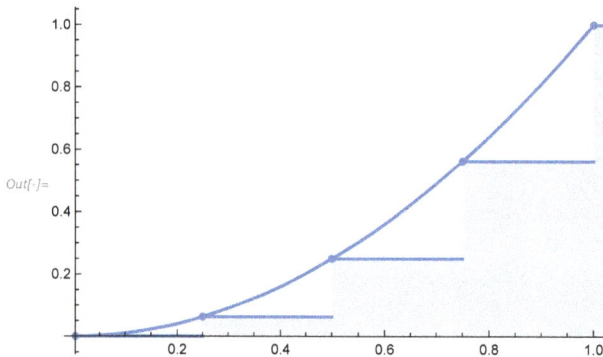

In this case, the calculated area would be the base of each rectangle $(1/4)$ times the height of the left corner (x^2).

The corners are at 0, 0.25, 0.5 and 0.75, so the sum should be:

In[·]:= **1/4 (0^2 + 0.25^2 + 0.5^2 + 0.75^2)**

Out[·]= **0.21875**

When the top-right corner is the point on the curve, you can see that the overall area should be more than the total area under the curve:

In[ ]:= **Show[Plot[x^2, {x, 0, 1}],**
DiscretePlot[x^2, {x, 0, 1, 1/4}, ExtentSize → Left, ExtentMarkers → {None, "Filled"}]]

Out[ ]=

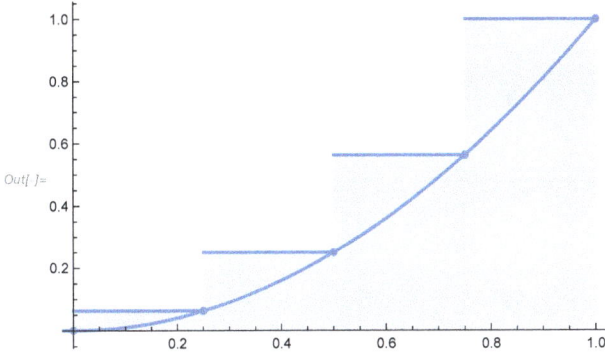

The corners are at 0.25, 0.5, 0.75 and 1, so the sum should be:

In[ ]:= **1/4 (0.25^2 + 0.5^2 + 0.75^2 + 1^2)**

Out[ ]= **0.46875**

Area under a Parabola 2

The total area under the parabola from 0 to 1 is between 0.21875 and 0.46875. Now increase the number of rectangles by reducing the base lengths. If you use 10 rectangles, the lower limit is 0.285 and the upper limit is 0.385.

Here is a table showing the lower and upper limits of the area for various numbers of rectangles:

Number of rectangles	Lower	Upper
10.	0.285	0.385
20.	0.30875	0.35875
50.	0.3234	0.3434
100.	0.32835	0.33835
1000.	0.332834	0.333834

Here is a list of plots showing the rectangles approximating the plot as more rectangles are used:

In[ ]:= **Table[DiscretePlot[x^2, {x, 0, 1, 1/i}, ExtentSize → Right], {i, {10, 100, 1000}}]**

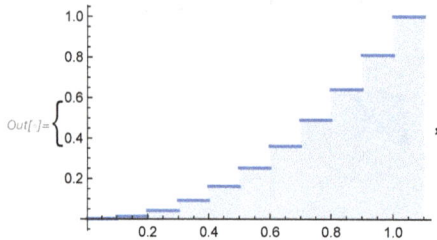

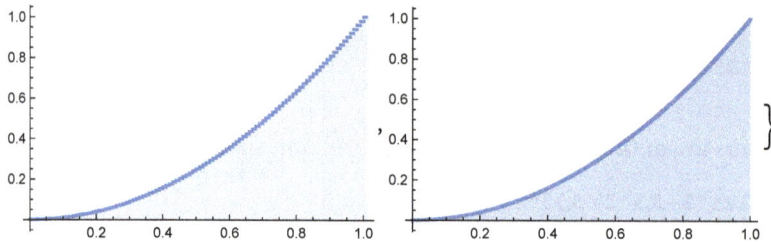

Taking the Limit

As you use more rectangles, the value approaches the number $1/3$.

You can show this with Limit. Use width $1/n$ and calculate the upper limit as $n \to \infty$:

In[ ]:= **Limit[1/n Sum[(i/n)^2, {i, 1, n}], n → Infinity]**

Out[ ]= $\dfrac{1}{3}$

The lower limit also approaches $1/3$ as $n \to \infty$:

In[ ]:= **Limit[1/n Sum[((i − 1)/n)^2, {i, 1, n}], n → Infinity]**

Out[ ]= $\dfrac{1}{3}$

Here the built-in function Sum is used. It takes the expression and adds the values for i between i_{min} and i_{max} in steps of 1. The mathematical symbol used for sums is Σ, the Greek letter capital sigma.

When you divide the following expression by n, you get the value for the area of n rectangles approximating the region under the parabola from 0 to 1.

The $i = 1$ indicates you start at 1; the n above it indicates you stop at n. The $\left(\frac{i}{n}\right)^2$ to the side indicates you add $\left(\frac{i}{n}\right)^2$ to the sum:

$$\sum_{i=1}^{n} \left(\frac{i}{n}\right)^2$$

Riemann Sums

Now use a general function $f[x]$ and find the area under the curve in the interval $[a, b]$. If you use n rectangles, then the width of each rectangle should be $\frac{b-a}{n}$. Denote the width by Δx.

You now have n subintervals of the form $[x_0, x_1]$, $[x_1, x_2]$, ..., $[x_{n-1}, x_n]$ with $x_0 = a$ and $x_n = b$. The right endpoints have the form $x_1 = a + \Delta x$, $x_2 = a + \Delta x + \Delta x = a + 2 \Delta x$, $x_3 = a + 3 \Delta x$, ... etc.

If you use the right-aligned triangles, then the height of the ith strip is $f[x_i]$. Therefore, the approximate area is $\sum_{i=1}^{n} f[x_i] \Delta x$.

For left-aligned triangles, the approximate area is $\sum_{i=1}^{n} f[x_{i-1}] \Delta x$.

In general, such sums are called **Riemann sums**, in honor of the German mathematician Bernhard Riemann.

Here is an animation for the function $f[x] = \text{Sin}[x] + 2$. Notice the widths get smaller, but the rectangles better approach the area under the curve as n increases:

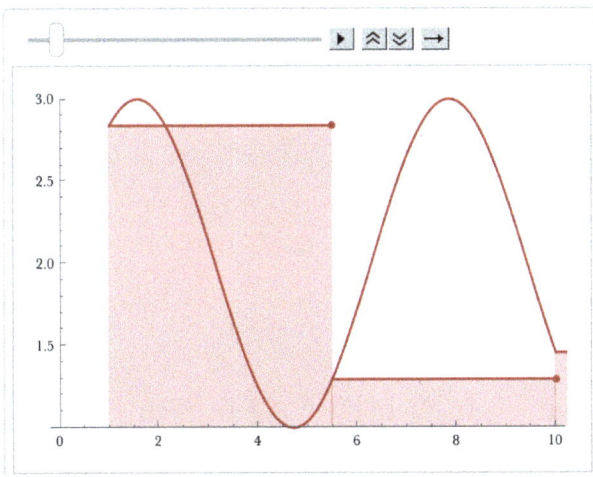

Definition of Area

Let f be a continuous function. Define the **area** A of the region S lying under the graph of f to be the limit of the sum of the area of approximating rectangles:

$$A = \lim_{n \to \infty} \left(\sum_{i=1}^{n} f(x_i) \Delta x \right)$$

This is the limit for right-aligned triangles. For left-aligned triangles the limit is:

$$A = \lim_{n \to \infty} \left(\sum_{i=0}^{n-1} f(x_i) \Delta x \right)$$

It can be shown that both limits exist, and you can even let the height of the ith rectangle in the subinterval $[x_{i-1}, x_i]$ be the value of f at *any* number x_i^* in the interval.

The numbers x_1^*, x_2^*, x_3^*, ..., x_n^* are called **sample points**. For right-aligned rectangles $x_i^* = x_i$, while for left-aligned rectangles $x_i^* = x_{i-1}$.

A more general formula for the area A is:

$$A = \lim_{n \to \infty} \left(\sum_{i=1}^{n} f((x_i)^*) \Delta x \right)$$

For example, you can let the sample points be the midpoint of each rectangle, as shown in the plot below:

In[ ]:= **Show[Plot[x^2, {x, 0, 1}, ⋯ → ⋯ +], DiscretePlot[x^2, {x, 0, 1, 1/10}, ExtentSize → Full]]**

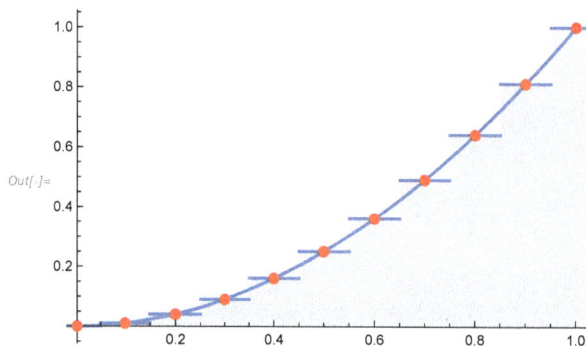

Example

Find the area of the region under the function Sin[x] from 0 to b where $0 \leq b \leq \pi$.

Here is the plot of the region from 0 to π:

In[ ]:= **Plot[Sin[x], {x, 0, π}, Filling → Bottom]**

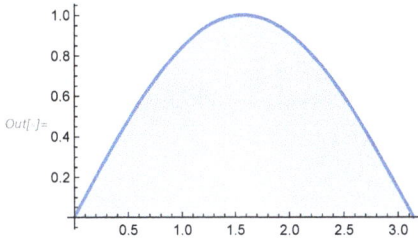

Out[ ]=

Divide the region into rectangles, each with base length $\frac{b-0}{n} = \frac{b}{n}$, and use right-aligned rectangles, so $f[x_i^*] = f[x_i]$.

The expression for the area is:

$$\lim_{n \to \infty} \frac{b \sum_{i=1}^{n} \sin\left(\frac{i\,b}{n}\right)}{n}$$

This equals:

In[ ]:= **Limit[b/n Sum[Sin[i b/n], {i, 1, n}], n → Infinity]**

Out[ ]= **1 − Cos[b]**

So when $b = \pi$, the sum is 2:

In[ ]:= **Limit[π/n Sum[Sin[i π/n], {i, 1, n}], n → Infinity]**

Out[ ]= **2**

Distance Problem

A car is moving on a highway and its velocity is given by the following table:

time (minutes)	0	10	20	30	40	50	60	70	80	90	100
velocity (mph)	63	62	61	64	67	70	65	67	63	66	65

Suppose you wanted to find the total distance covered for the full 100 minutes.

If you assume the velocity during each 10-minute interval is constant, you can calculate the distance in each interval and add them all together:

```
In[·]:= 10/60 (63 + 62 + 61 + 64 + 67 + 70 + 65 + 67 + 63 + 66)
```

```
Out[·]= 108
```

Notice that the velocities are added at the **beginning** of each interval, and not the last value, 65.

However, you could have added the velocities at the **end** of each interval, and not have included the first value, 63. The total distance would then be:

```
In[·]:= 10/60 (62 + 61 + 64 + 67 + 70 + 65 + 67 + 63 + 66 + 65) // N
```

```
Out[·]= 108.333
```

Plot the velocities on a graph:

```
In[·]:= time = {0, 10, 20, 30, 40, 50, 60, 70, 80, 90, 100};
        velocity = {63, 62, 61, 64, 67, 70, 65, 67, 63, 66, 65};
```

```
In[·]:= ListPlot[Transpose[{time, velocity}], PlotRange → {0, 75}]
```

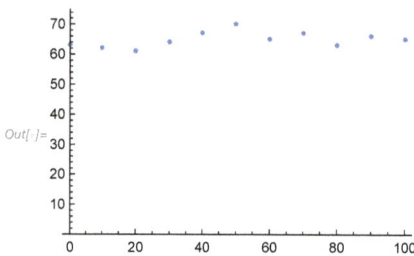

Distance Problem 2

In the previous problem, you can assume the velocity of the car was actually continuous, so the distances calculated were just approximations. You could get better approximations by calculating the velocity every five minutes instead of every 10 minutes.

In fact, if you were given the velocity as a **function**, you could calculate the total distance by taking the limit of a Riemann sum!

You actually already used Riemann sums in the previous example. The width of each rectangle was $10/60$ and the height was the beginning or end velocity in each interval.

If you model the velocity of a car by the following function:

```
In[·]:= Clear[time, velocity]
```

```
In[·]:= velocity[t_] := 65 + 5 Sin[π t / 20]
```

the area under the curve from 0 to 100 will give the total distance covered during the time period:

In[ ]:= **Limit[100 / (60 n) Sum[velocity[100 i / n], {i, 1, n}], n → Infinity] // N**

Out[ ]= 109.394

Here is a plot of the velocity function and the region below the curve:

In[ ]:= **Plot[velocity[t], {t, 0, 100}, PlotRange → {0, 80}, Filling → Bottom]**

Out[ ]=

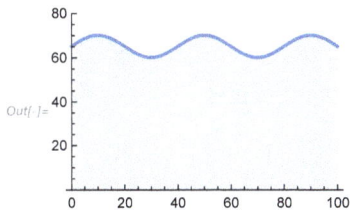

Summary

Finding the area of an object is a very practical endeavor, and you already know how to find the area of various shapes.

To approximate the area under a curve, divide the curve into rectangles, find the area of each rectangle, and then add the areas together. If you take the limit as the number of rectangles goes to infinity, you get the area under the curve.

Such sums are called Riemann sums. Riemann sums can be used to find the total distance traveled by an object given only its velocity.

The next lesson will cover Riemann sums in more detail.

Exercises

Exercise 1—Approximating Rectangles

Using 4, 8 and 16 rectangles with their right endpoints on the curve, approximate the area under the following curve in the range [1, 9]:

In[·]:= **f[x_] := Sqrt[x]**

Solution

For four rectangles, the width of each rectangle is $\frac{9-1}{4} = 2 = \Delta x$.

Take the sample points $x_i^* = x_i = x_0 + i \Delta x = 1 + 2i$. Using Sum the area is:

In[·]:= **2 Sum[f[1 + 2 i], {i, 1, 4}] // N**

Out[·]= **19.2277**

For eight rectangles, the width of each rectangle is $\frac{9-1}{8} = 1 = \Delta x$.

Take the sample points $x_i^* = x_i = x_0 + i \Delta x = 1 + i$. Using Sum the area is:

In[·]:= **Sum[f[1 + i], {i, 1, 8}] // N**

Out[·]= **18.306**

For 16 rectangles, the width of each rectangle is $\frac{9-1}{16} = \frac{1}{2} = \Delta x$.

Take the sample points $x_i^* = x_i = x_0 + i \Delta x = 1 + \frac{i}{2}$. Using Sum the area is:

In[·]:= **1/2 Sum[f[1 + i/2], {i, 1, 16}] // N**

Out[·]= **17.8264**

Exercise 2—Area under a Curve

Using limits and right-aligned rectangles, find the area under the curve for the following function in the range [2, 5]:

In[·]:= **f[x_] := x^5**

Solution

The width of each rectangle is $\Delta x = \frac{5-2}{n} = \frac{3}{n}$.

Take the sample points $x_i^* = x_i = 2 + i \Delta x = 2 + \frac{3i}{n}$.

When you approximate the appropriate Riemann sum with n rectangles, you get the following expression:

In[ ]:= **3/n Sum[f[2 + 3 i/n], {i, 1, n}]**

Out[ ]= $\dfrac{3\left(-189 + 3045\,n^2 + 6186\,n^3 + 3458\,n^4\right)}{4\,n^4}$

Using Limit, you can find the answer to be:

In[ ]:= **Limit[3/n Sum[f[2 + 3 i/n], {i, 1, n}], n → Infinity]**

Out[ ]= $\dfrac{5187}{2}$

Left-aligned rectangles give the same answer:

In[ ]:= **Limit[3/n Sum[f[2 + 3 (i − 1)/n], {i, 1, n}], n → Infinity]**

Out[ ]= $\dfrac{5187}{2}$

Exercise 3—Runner

A runner speeds up in the first six seconds of a race. The runner's speed is taken every second and is given in the following table:

time (seconds)	0	1	2	3	4	5	6
speed (feet / second)	0	9.8	17.5	21.3	22.3	22.9	23.1

Calculate the lower and upper estimates for the distance the runner travels during the first six seconds.

Solution

To find the lower estimate, add the speeds at the beginning of each one-second interval and multiply by one second:

In[ ]:= **1 (0 + 9.8 + 17.5 + 21.3 + 22.3 + 22.9)**

Out[ ]= **93.8**

To find the upper estimate, add the speeds at the end of each one-second interval and multiply by one second:

In[ ]:= **1 (9.8 + 17.5 + 21.3 + 22.3 + 22.9 + 23.1)**

Out[ ]= **116.9**

So the runner ran between 93.8 feet and 116.9 feet in the first six seconds of the race.

Exercise 4—Oil Tanker

An oil tanker crashes into a massive coral reef and leaks oil into the ocean at a rate of $r[t]$ liters per hour (in thousands of liters). For the first 20 hours, there is the following table:

time (hours)	0	2	4	6	8	10	12	14	16	18	20
r[t] (liters / hour)	118.9	88.9	67.6	50.	36.4	28.3	22.7	17.5	13.9	9.8	7.4

Calculate the lower and upper estimates for the total amount of oil that spills into the ocean after the 20 hours.

Solution

To find the upper estimate, add the rates at the beginning of each two-hour interval and multiply by two hours:

In[·]:= **2 (118.9 + 88.9 + 67.6 + 50 + 36.4 + 28.3 + 22.7 + 17.5 + 13.9 + 9.8)**

Out[·]= **908.**

To find the lower estimate, add the rates at the end of each two-hour interval and multiply by two hours:

In[·]:= **2 (88.9 + 67.6 + 50 + 36.4 + 28.3 + 22.7 + 17.5 + 13.9 + 9.8 + 7.4)**

Out[·]= **685.**

So the tanker leaked between 908,000 and 685,000 liters in the first 20 hours.

Exercise 5—Car Brake

The driver of a car moving at 35 mph sees a stop sign and puts on the brakes. The car's speed while it brakes in mph is given by the following function as a function of t seconds:

In[·]:= **brakespeed[$t\_$] := 35. − 18.6948 t + 3.31493 t^2 − 0.195225 t^3**

Find the total distance (in feet) the car travels from the time the driver sees the stop sign to the time it completely stops.

Solution

Since the function gives the speed in mph, it is clear that the driver sees the stop sign at $t = 0$. Find the time the car completely stops by setting the function equal to 0 and solving for t:

In[ ]:= **Solve[brakespeed[t] == 0, t]**

Out[ ]= {{t → 4.98014}, {t → 5.97991}, {t → 6.02}}

Take the first time the car stops (4.98 seconds).

To find the total distance covered, first convert the function from mph to ft./sec. That is just the following expression:

In[ ]:= **5280 brakespeed[t] / 3600**

Out[ ]= $\dfrac{22}{15} \left(35. - 18.6948\,t + 3.31493\,t^2 - 0.195225\,t^3\right)$

Then find the area under the curve from $t = 0$ to $t = 4.98$ using the altered function:

In[ ]:= **Limit[4.98 / n Sum[5280 brakespeed[4.98 t / n] / 3600, {t, 1, n}], n → Infinity]**

Out[ ]= 71.7687

The car goes about 71.8 feet before it stops.

24 | The Definite Integral

Overview

In the previous lesson, the Riemann sum $\text{Limit}_{n \to \infty}(\Sigma_{i=1}^{n} f[x_i] \Delta x)$ was used for the function $f[x]$. Usually f was positive, and the sum could be interpreted as the area under the curve corresponding to f.

The Riemann sum also has practical applications. For example, the area under the following velocity curve:

In[ ]:= **v[t_] := Sin[3 t] + 5**

In[ ]:= **Plot[v[t], {t, 0, 5}, PlotRange → {0, 6}, Filling → Bottom]**

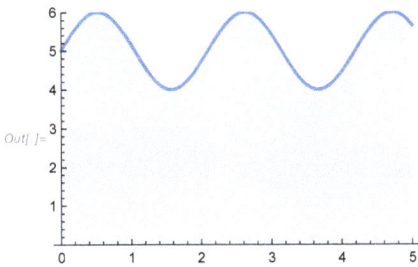

Out[ ]=

could be interpreted as the total distance traveled by a particle from $t = 0$ to 5.

What if f is negative, though? By definition, the velocity of a particle can be positive or negative.

This lesson will answer this question and go over some properties of the Riemann sum.

Definite Integral

Since the Riemann sum is used often, mathematicians give it a special name: the **definite integral**.

The definite integral also has a special notation, given here:

$$\lim_{n \to \infty} \left(\sum_{i=1}^{n} f(x_i) \Delta x \right) = \int_{a}^{b} f(x) \, dx$$

Here the $\int$ is an elongated S standing for "sum," $f[x]$ is the function, and $\int_{a}^{b} f[x] \, dx$ signifies the area from $x = a$ to $x = b$. Δx becomes dx.

When you find the area under a function from a to b, you are said to be **integrating the function from *a* to *b***. The Wolfram Language also has a built-in function that integrates functions. It is called Integrate.

For example, to find the area from 0 to 1 for the function $f[x] = x^2$, you can take the limit of a Riemann sum as usual:

In[·]:= **Limit[Sum[1 / n (i / n)^2, {i, 1, n}], n → Infinity]**

Out[·]= $\dfrac{1}{3}$

Or you can use the easier-to-use Integrate:

In[·]:= **Integrate[x^2, {x, 0, 1}]**

Out[·]= $\dfrac{1}{3}$

Integrable Functions

Not all functions can be integrated. This usually occurs when the limit does not converge to a finite value.

For example, the following function cannot be integrated in the range [0, 1]:

In[·]:= **f[x_] := 1 / x**

In[·]:= **Integrate[f[x], {x, 0, 1}] // Quiet**

Out[·]= $\displaystyle\int_0^1 \frac{1}{x} \, dx$

To see why, look at the function's graph:

In[·]:= **Plot[f[x], {x, 0, 1}, Filling → Bottom]**

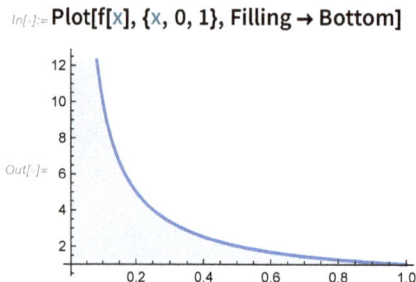

As you can see, at 0 the function approaches ∞ from the right, so the area under the curve should approach ∞ as well, and is therefore not finite.

Luckily, the functions the lesson will cover can be integrated so long as you are careful. In general, a function is considered **integrable** if it is continuous or has a finite number of jump discontinuities.

Negative Area

What about functions with negative values, as was mentioned at the beginning? How do you interpret integrals of negative functions?

Since taking the area **under** the curve no longer makes sense (it would go to $-\infty$), instead take the area of the curve from the curve to the x axis.

In this case, it would be the area **above** the curve. For negative functions, the area above the curve is calculated by taking the area as usual and then negating it.

Consider the following function. It is negative from 0 to 1:

In[ ]:= $f[x\_] := -x$

You can see this in its graph:

In[ ]:= Plot[f[x], {x, 0, 1}, PlotRange → {-1, 1}, Filling → Axis]

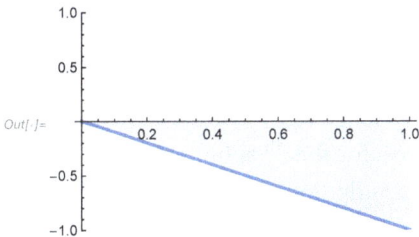

Since the region is a triangle with base 1 and height 1, the area is just $-1/2$.

Integrate agrees:

In[ ]:= Integrate[f[x], {x, 0, 1}]

Out[ ]= $-\dfrac{1}{2}$

Net Area

What about functions that are positive and negative? Consider the following function:

In[ ]:= $f[x\_] := Sin[x]$

From its plot, you can see that it is positive from 0 to π and negative from π to 2π:

In[ ]:= **Plot[f[x], {x, 0, 2 π}, Filling → Axis]**

Out[ ]=

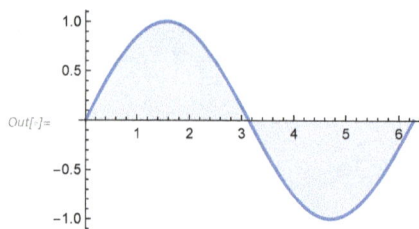

For functions like this, the integral is calculated by adding all the areas above the x axis and subtracting the sum of the areas below the x axis.

This is called the **net area**. For the preceding function, the area above the x axis equals the area below the x axis, so the net area is 0.

Confirm with Integrate:

In[ ]:= **Integrate[f[x], {x, 0, 2 π}]**

Out[ ]= 0

Midpoint Rule

When calculating the integral of a function explicitly, right-aligned or left-aligned rectangles have often been used. In other words, for the ith subinterval, the sample point x_i^* has been chosen to be either the right endpoint or left endpoint of the interval.

However, the **best** approximation to an integral can be found by setting the sample point to be the **midpoint** of the interval.

You can see this in the graph of x^2 by using DiscretePlot and using the option ExtentSize→Full:

In[ ]:= **Show[Plot[x^2, {x, 0, 1}], DiscretePlot[x^2, {x, 0, 1, 1/10}, ExtentSize → Full]]**

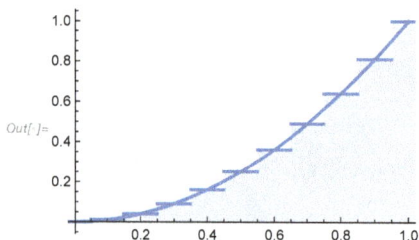

Out[ ]=

You can see that any area that goes above the graph for each rectangle makes up for the missing area when each rectangle is below the graph.

Properties of the Definite Integral 1

It helps to know some properties for the definite integral, as it will make some calculations much easier. For example, every time you calculate the area of a region from a to b, you assume a is less than b. However, you can still calculate the area from b to a.

You would expect it to be the same, but because of the way the integral was defined (which is based on a Riemann sum), you will find the width of each rectangle to be negated (since $\frac{a-b}{n} = -\frac{b-a}{n}$).

In other words, the integral from b to a is the **negative** of the integral from a to b. Here is a formula that is easier to remember:

$$\int_a^b f(x)\,dx = -\int_b^a f(x)\,dx$$

Here is an example with $f[x] = x^2$:

In[ ]:= **Integrate[x^2, {x, 1, 0}]**

Out[ ]= $-\dfrac{1}{3}$

Note that the function was integrated from 1 to 0. If you integrate it from 0 to 1, the answer is just $1/3$:

In[ ]:= **Integrate[x^2, {x, 0, 1}]**

Out[ ]= $\dfrac{1}{3}$

From this, you can deduce that the integral for any function from a to a is 0.

$$\int_a^a f(x)\,dx = 0$$

This makes sense because there is no area covered for the region. It is just one point. Try with the previous function:

In[ ]:= **Integrate[f[x], {x, 1, 1}]**

Out[ ]= 0

Properties 2

Here are some more properties of the definite integral. Assume that $f[x]$ and $g[x]$ are integrable.

For c a constant:

$$\int_a^b c\,dx = c\,(b-a)$$

There is a sum property for functions:

$$\int_a^b (f(x) + g(x))\,dx = \int_a^b f(x)\,dx + \int_a^b g(x)\,dx$$

There is also a difference property for functions:

$$\int_a^b (f(x) - g(x))\,dx = \int_a^b f(x)\,dx - \int_a^b g(x)\,dx$$

There is a scalar multiple property for functions:

$$\int_a^b c\,f(x)\,dx = c\int_a^b f(x)\,dx$$

The first property can be understood as finding the area of a rectangle with length c and running from a to b (i.e. width $b - a$).

The area of the rectangle with height 5 and running from 1 to 5 is $5\,(5-1) = 20$:

In[ ]:= **{Plot[5, {x, 1, 5}, Filling → Axis], Integrate[5, {x, 1, 5}]}**

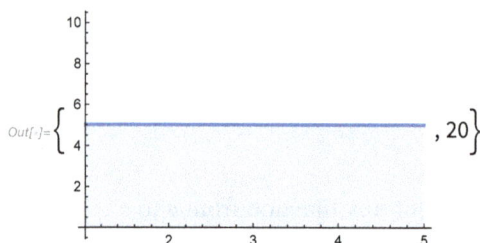

Out[ ]:= $\left\{ \right.$, 20 $\left. \right\}$

Properties 3

If you add the area from a to b to the area from b to c for a function $f[x]$, you get the area of the function from a to c:

$$\int_a^b f(x)\,dx + \int_b^c f(x)\,dx = \int_a^c f(x)\,dx$$

If $f[x]$ is **positive** for $a \le x \le b$, then the area covered by f from a to b is **positive**:

If $f[x] \ge 0$ for $a \le x \le b$, then $\int_a^b f(x)\,dx \ge 0$

If $f[x]$ is always **greater** than $g[x]$ in the region $a \leq x \leq b$, then the area covered by f is **greater** than the area covered by g:

$$\text{If } f[x] \geq g[x] \text{ for } a \leq x \leq b, \text{ then } \int_a^b f(x)\,dx \geq \int_a^b g(x)\,dx$$

If $f[x]$ is always above the value m and below the value M in the region $a \leq x \leq b$, then the area covered by f is **greater** than the area of the rectangle with height m and width $b - a$ and **less** than the area of the rectangle with height M and width $b - a$:

$$\text{If } m \leq f[x] \leq M \text{ for } a \leq x \leq b, \text{ then } m\,(b - a) \leq \int_a^b f(x)\,dx \leq M\,(b - a)$$

Here is a graph depicting the last property:

In[ ]:= `Plot[{3 Sin[x] + 4, 1, 7}, {x, 2, 6},` ⋯ ◆ `]`

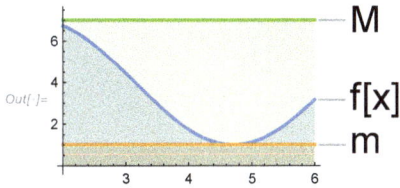

Exercises

Exercise 1—Integrability

Find where the following function is integrable:

In[·]:= **f[x_] := Piecewise[{{2, x < 0}, {(x − 1)^−2, 0 ≤ x ≤ 5}}, 5]**

Solution

Plot the function:

In[·]:= **Plot[f[x], {x, −10, 10}]**

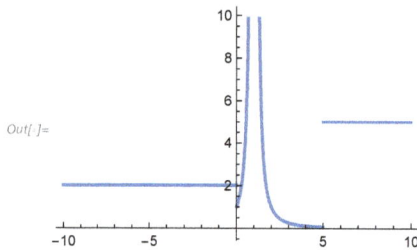

The function approaches ∞ from the right and left at 1, so it is not integrable on any finite interval that contains 1. The function has finitely many jump discontinuities. Therefore, the function is integrable on every finite interval except for those that contain 1.

Exercise 2—Midpoint Rule

The midpoint rules computes the Riemann sum with the midpoints of each subinterval as the sample points.

Use the midpoint rule and four rectangles to find the best approximation to the integral of the following function in the range [1, 4]:

In[·]:= **f[x_] := x^2 + 3 x − Sqrt[x]**

Solution

The width of each rectangle will be $\frac{4-1}{4} = \frac{3}{4}$.

The sample points will be $\frac{x_{i-1}+x_i}{2} = \frac{(1+3\,(i-1)/4+1+3\,i/4)}{2}$.

Now calculate the answer with Sum:

In[ ]:= **Sum[3 / 4 f[(1 + 3 (i − 1) / 4 + 1 + 3 i / 4) / 2], {i, 1, 4}] // N**

Out[ ]= 38.687

Here is the answer with Integrate:

In[ ]:= **Integrate[f[x], {x, 1, 4}] // N**

Out[ ]= 38.8333

Here is a graph of the function with midpoint rectangles:

In[ ]:= **Show[Plot[f[x], {x, 1, 4}], DiscretePlot[f[x], {x, 1, 4, 3/4}, ExtentSize → Full]]**

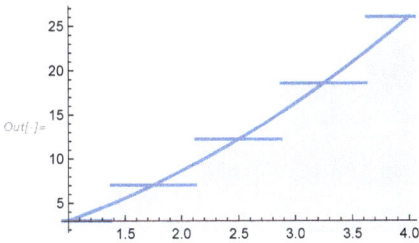

Exercise 3—Common Areas

Find the integral of the following function **without** using Integrate in the interval $[-5, 5]$:

In[ ]:= **f[x_] := RealAbs[x − 2] − 1**

Plot the function:

In[ ]:= **Plot[f[x], {x, −5, 5}, Filling → Axis]**

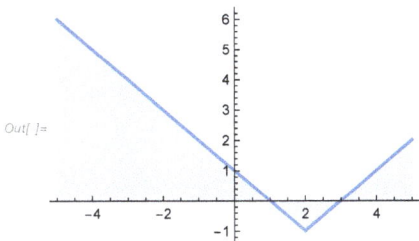

Solution

You can see it is the combination of three triangles, two above the x axis and one below.

The first triangle has base $1 - (-5) = 6$ and height 6, and therefore has area 18.

The second triangle has base $3 - 1 = 2$ and height 1, and therefore has area 1.

The third triangle has base $5 - 3 = 2$ and height 2, and therefore has area 2.

So the integral of the function is $(18 + 2) - 1 = 19$.

Integrate confirms:

In[·]:= **Integrate[f[x], {x, −5, 5}]**

Out[·]= 19

Exercise 4—Integration Properties

Evaluate the following expressions **without** using Integrate:

a: $\int_{\frac{\pi}{12}}^{\frac{\pi}{12}} e^x \sin(x) x^2 \, dx$

b: $\int_{-1}^{1} \sin(x) \, dx$

c: $\int_{2}^{3} f(x) \, dx$, given that $\int_{0}^{3} f(x) \, dx = 5$ and $\int_{0}^{2} f(x) \, dx = 1$.

Solution

a: Since the limits of integration are the same, the answer is 0.

b: Sine is an odd function, so any area left of the y axis will cancel out with the area right of the y axis, as seen in the following graph:

In[·]:= **Plot[Sin[x], {x, −1, 1}, Filling → Axis]**

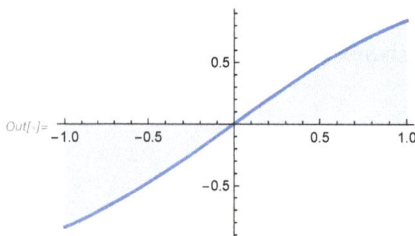

So the answer is 0.

Integrate agrees:

In[·]:= **Integrate[Sin[x], {x, −1, 1}]**

Out[·]= 0

c: Since $\int_{0}^{2} f(x) \, dx + \int_{2}^{3} f(x) \, dx = \int_{0}^{3} f(x) \, dx$, $\int_{2}^{3} f(x) \, dx = \int_{0}^{3} f(x) \, dx - \int_{0}^{2} f(x) \, dx$. So $\int_{2}^{3} f[x] \, dx = 5 - 1 = 4$.

Exercise 5—Integration Estimates

Find upper and lower bounds for the integral of the following function in the range [4, 5]:

In[ ]:= **f[x_] := 4 x^3**

Solution

Based on the function's derivative, you know the function is always increasing:

In[ ]:= **D[f[x], x]**

Out[ ]= $12\,x^2$

Therefore, its minimum in the interval is at 4 while its maximum is at 5.

From the last property for integrals, you know the lower bound for the integral is:

In[ ]:= **f[4] (5 − 4)**

Out[ ]= **256**

And the upper bound for the integral is:

In[ ]:= **f[5] (5 − 4)**

Out[ ]= **500**

Confirm with Integrate:

In[ ]:= **Integrate[f[x], {x, 4, 5}]**

Out[ ]= **369**

25 | The Fundamental Theorem of Calculus

Overview

So far, differential calculus and integral calculus have been covered.

The original goal of differential calculus was finding the tangent line to a function at a point:

In[ ]:= **Plot[{Sin[x], x}, {x, −2, 2}]**

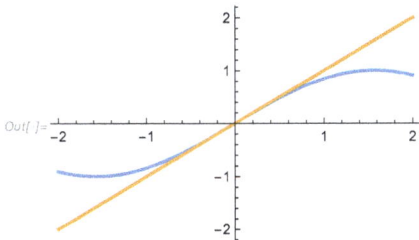

The original goal of integral calculus was finding the area under the curve of a function:

In[ ]:= **Plot[Sin[x], {x, −2, 3}, Filling → Axis]**

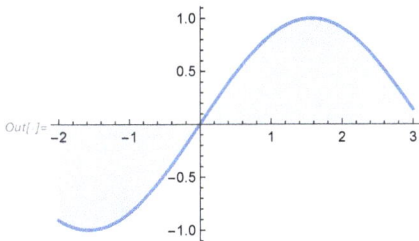

The fundamental theorem of calculus ties these two ideas together. The goal of this lesson is to go over this very important theorem.

Integral Functions

To start, consider the following function, called an **integral function**:

In[ ]:= **f[x_] := Integrate[t, {t, 0, x}]**

It calculates the integral of the function $g[t] = t$ from 0 to x.

So its value at 0 should be 0:

In[·]:= **f[0]**

Out[·]= 0

You can think of it as the function that calculates the area under the curve from 0 to x.

Graph both functions:

In[·]:= **Plot[{f[x], x}, {x, 0, 5}, PlotLegends → {"$\int_0^x t \, dt$", "x"}]**

Out[·]=

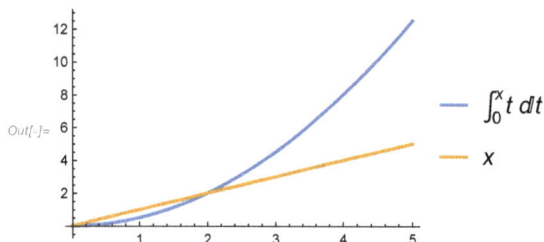

You can see that the function is increasing, because the area increases as x increases.

Explicit

Since integrals are Riemann sums, the next step is to find an explicit, algebraic formula for the integral function.

The width will be $\frac{x}{n}$. Use right-aligned triangles to make the calculations easier. So $x_i^* = 0 + i * \frac{x}{n} = \frac{i*x}{n}$.

Therefore, the sum will be $\sum g[x_i^*] \frac{x}{n} = \sum \frac{i*x}{n} * \frac{x}{n}$ from $i = 1$ to n:

In[·]:= **Limit[Sum[(i * x / n) * x / n, {i, 1, n}], n → Infinity]**

Out[·]= $\dfrac{x^2}{2}$

You can see that the integral is actually $\frac{x^2}{2}$. Recall that the antiderivative of $g[t] = t$ is $\frac{t^2}{2} + C$:

In[·]:= **D[x^2/2, x]**

Out[·]= x

So the integral function for $g[t]$ is actually an **antiderivative of $g[t]$**.

Intuition

To see why this might be true, consider the integral function $f[x] = \int_0^x g[t]\, dt$ and the expression $f[x + h] - f[x]$.

If you make h very small, then visually $f[x + h] - f[x]$ is the area of a tiny strip. If you approximate it to be a rectangle, it has width h and height $g[x]$:

In[ ]:= **Show[Plot[x, {x, 0, 5}], Plot[x, {x, 3, 3.1}, PlotRange → {0, 5}, Filling → Axis]]**

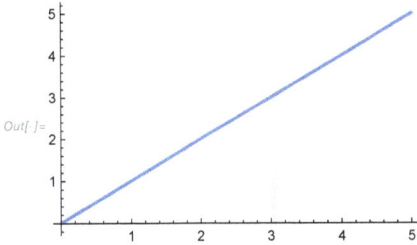

Out[ ]=

So the area of the strip would be $h * g[x]$. Dividing by h gives $g[x]$. In other words, $\frac{f[x+h]-f[x]}{h} = g[x]$.

So if you take the limit as $h \to 0$, the left side becomes the derivative of f at x, so $f'[x] = g[x]$. Another way of saying this is that $f[x]$ is an antiderivative of $g[x]$.

Fundamental Theorem 1

Here is the first part of the **fundamental theorem of calculus**:

Let g be a continuous function on $[a, b]$.

Define $f[x]$ to be the function $\int_a^x g[t]\, dt$ on $[a, b]$.

Then $f[x]$ is continuous on $[a, b]$, differentiable on (a, b) and an antiderivative of g.

In other words, $f'[x] = g[x]$.

Interpretation

The first part of the fundamental theorem of calculus basically says that if you take a function f, integrate it, and then differentiate it with respect to x, you get back the original function f.

For example, the derivative of the function $f[x] = \int_0^x \mathrm{Sin}\left[\sqrt{1 + t^2}\right] dt$ with respect to x is just $\mathrm{Sin}\left[\sqrt{1 + x^2}\right]$.

Confirm with Integrate and D:

In[ ]:= **D[Integrate[Sin[Sqrt[1 + t^2]], {t, 0, x}], x]**

Out[ ]= $\mathsf{Sin}\left[\sqrt{1 + x^2}\right]$

If the upper bound for the integral is a function of x (like x^2), the chain rule must also be used.

For example, the derivative of the function $g[x] = \int_0^{x^2} \mathsf{Sin}\left[\sqrt{1 + t^2}\right] dt$ with respect to x is **not** $\mathsf{Sin}\left[\sqrt{1 + x^4}\right]$.

That is because it has the form $g[x] = f[x^2]$. So its derivative is $g'[x] = f'[x^2] * 2x = 2x \mathsf{Sin}\left[\sqrt{1 + x^4}\right]$ by the chain rule.

Confirm with Integrate and D:

In[ ]:= **D[Integrate[Sin[Sqrt[1 + t^2]], {t, 0, x^2}], x]**

Out[ ]= $2 x \mathsf{Sin}\left[\sqrt{1 + x^4}\right]$

Fresnel and Erf

You may wonder when integral functions are actually used.

Consider the following function, which is used in optics and highway design:

In[ ]:= **f[x_] := Integrate[Sin[π t^2/2], {t, 0, x}]**

The Wolfram Language actually has this function built in. It is called FresnelS.

Here is a plot of the function and its derivative in the range $[-5, 5]$:

In[ ]:= **Plot[{FresnelS[x], FresnelS'[x]}, {x, −5, 5}, PlotLegends → "Expressions"]**

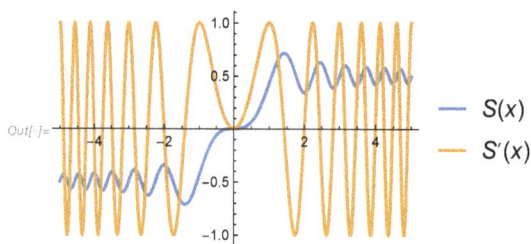

Another integral function is the error function, called Erf in the Wolfram Language. It is used in statistics and probability.

Here is a plot of the function and its derivative in the range $[-5, 5]$:

In[ ]:= **Plot[{Erf[x], Erf'[x]}, {x, −5, 5}, PlotLegends → "Expressions"]**

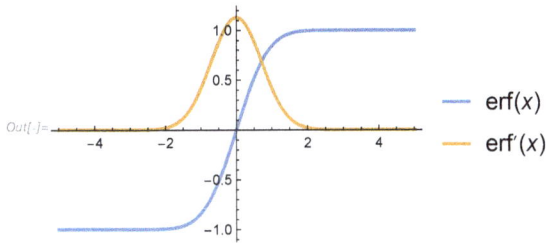

Fundamental Theorem 2

The second part of the fundamental theorem of calculus lets you calculate integrals more easily, without explicitly taking the limit of a Riemann sum.

Let f be continuous on $[a, b]$. Let F be any antiderivative of f (i.e. $F' = f$).

Then $\int_a^b f[x]\, dx = F[b] - F[a]$.

Put another way, $\int_a^b F'[x]\, dx = F[b] - F[a]$.

So if you have a function F, take its derivative, and then integrate it, you get back the original function F but in the form $F[b] - F[a]$.

Taking the first and second parts of the fundamental theorem of calculus together, you can see that differentiation and integration are actually **inverse processes** to each other: What one does, the other undoes, and vice versa.

With the second part in mind, calculate the integral of x^2 from 1 to 4 without using Integrate.

An antiderivative of x^2 is $\frac{x^3}{3}$, so the answer is $\frac{4^3}{3} - \frac{1^3}{3} = \frac{64-1}{3} = 21$.

Confirm with Integrate:

In[ ]:= **Integrate[x^2, {x, 1, 4}]**

Out[ ]= 21

Summary

The fundamental theorem of calculus is one of the most important theorems in calculus.

It establishes differentiation and integration as being inverse processes to one another.

It is especially useful for calculating integrals.

Instead of calculating Riemann sums, which can be rather tedious to do without a computer (and still tedious to input *with a computer*), you can just take antiderivatives and evaluate them.

The next lesson will cover indefinite integrals.

Exercises

Exercise 1—Integral Function

The following integral function is given:

In[ ]:= **f[x_] := Integrate[Sin[t] / t, {t, 0, x}]**

Evaluate it at the points 5, 10 and 20 and plot the corresponding areas they represent.

Solution

Evaluate the function:

In[ ]:= **f /@ {5, 10, 20} // N**

Out[ ]= **{1.54993, 1.65835, 1.54824}**

Here are the plots:

In[ ]:= **Table[Show[Plot[Sin[t] / t, {t, 0, 20}], Plot[Sin[t] / t, {t, 0, i}, Filling → Axis]], {i, {5, 10, 20}}]**

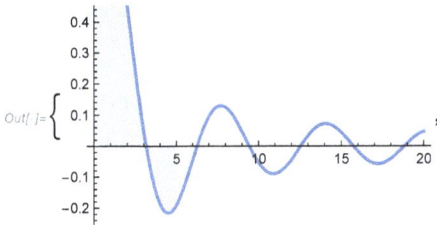

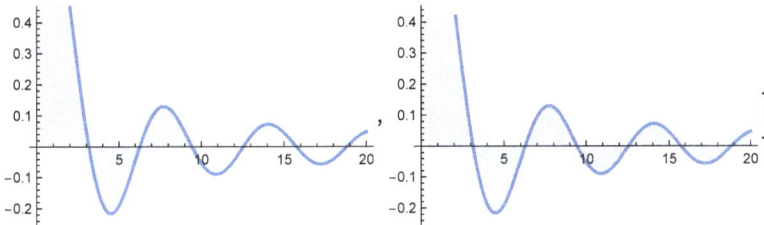

The function is actually special and has the representation SinIntegral in the Wolfram Language:

In[ ]:= **f[x]**

Out[ ]= **SinIntegral[x]**

Here is its plot:

In[·]:= **Plot[SinIntegral[x], {x, 0, 20}]**

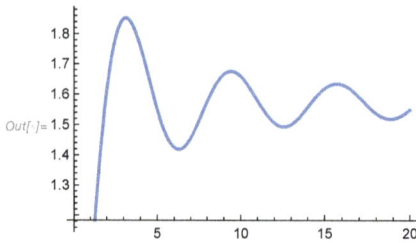

Out[·]=

Exercise 2—Calculating Derivatives

Calculate the derivative of the following function **without** using D:

In[·]:= **f[x_] := Integrate[Cos[CubeRoot[2 t] – 4]^5, {t, 0, x^3/3}]**

Solution

The function has the form $f[x] = g\left[\frac{x^3}{3}\right]$, where both f and g are integral functions. The derivative of $\frac{x^3}{3}$ is x^2.

Therefore, by the first part of the fundamental theorem of calculus and the chain rule, the derivative is the following expression:

In[·]:= **x^2 Cos[CubeRoot[2 x^3/3] – 4]^5**

Out[·]= $x^2 \, Cos\left[4 - \left(\frac{2}{3}\right)^{1/3} \sqrt[3]{x^3}\right]^5$

Confirm with D:

In[·]:= **D[f[x], x]**

Out[·]= $x^2 \, Cos\left[4 - \left(\frac{2}{3}\right)^{1/3} \sqrt[3]{x^3}\right]^5$

Exercise 3—Inflection Points

Find the inflection points of the following function **without** using D:

In[·]:= **f[x_] := Integrate[t^3 – 3 t + 3, {t, 0, x}]**

Solution

The function has the following form:

$$\int_0^x \left(t^3 - 3\,t + 3\right) dt$$

So its first derivative is $x^3 - 3\,x + 3$. Its second derivative is $3\,x^2 - 3$ by the power rule for differentiation.

Its inflection points are the places where the second derivative is 0, so you solve for x when the second derivative is 0:

In[ ]:= **sol = Solve[3 x^2 − 3 == 0, x]**

Out[ ]= **{{x → −1}, {x → 1}}**

So the function has inflection points at 1 and −1. Here is its graph:

In[ ]:= **Plot[f[x], {x, −2, 2}, Epilog → {PointSize[Large], Point[{x, f[x]} /. sol]}]**

Out[ ]=

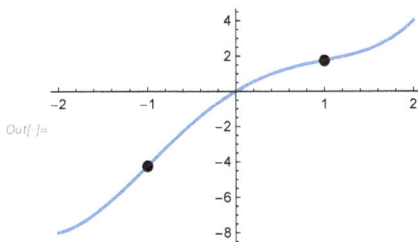

Exercise 4—Integral

Calculate the integral of the following function from $\pi/2$ to $5\,\pi/3$ **without** using Integrate:

In[ ]:= **f[x_] := 5 Cos[x] + 7 Sin[x]**

Solution

An antiderivative of the function is $5\,\mathrm{Sin}[x] - 7\,\mathrm{Cos}[x]$:

In[ ]:= **D[5 Sin[x] − 7 Cos[x], x]**

Out[ ]= **5 Cos[x] + 7 Sin[x]**

So the integral is $5\,\mathrm{Sin}[5\,\pi/3] - 7\,\mathrm{Cos}[5\,\pi/3] - (5\,\mathrm{Sin}[\pi/2] - 7\,\mathrm{Cos}[\pi/2])$:

In[ ]:= **5 Sin[5 π/3] − 7 Cos[5 π/3] − (5 Sin[π/2] − 7 Cos[π/2])**

Out[ ]= $-\dfrac{17}{2} - \dfrac{5\sqrt{3}}{2}$

Confirm with Integrate:

In[·]:= **Integrate[f[x], {x, $\pi/2$, 5 $\pi/3$}]**

Out[·]= $\dfrac{1}{2}\left(-17 - 5\sqrt{3}\right)$

Exercise 5—Physics

A particle moves with the following velocity:

In[·]:= **velocity[$t\_$] := $t\wedge3 - 11\,t\wedge2 + 34\,t - 24$**

Find the particle's displacement and distance traveled in the interval from $t = 0$ to 7.

Solution

Here is a plot of the function in the interval:

In[·]:= **Plot[velocity[t], {t, 0, 7}]**

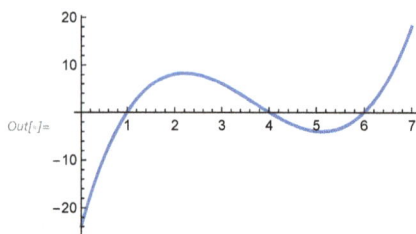

The displacement of the particle can be found by simply taking the integral from 0 to 7:

In[·]:= **Integrate[velocity[t], {t, 0, 7}]**

Out[·]= $\dfrac{91}{12}$

To find the total distance traveled, take the integral of the absolute value of the function:

In[·]:= **Integrate[RealAbs[velocity[t]], {t, 0, 7}]**

Out[·]= $\dfrac{469}{12}$

26 | Indefinite Integrals

Overview

The last lesson covered the very important fundamental theorem of calculus, which connects differential and integral calculus.

For example, the integral of the following function:

In[ ]:= **f[x_] := 4 x^3**

from 2 to 3 is given by the following expression:

In[ ]:= **3^4 − 2^4**

Out[ ]= **65**

This is because an antiderivative of $f[x] = 4 x^3$ is x^4:

In[ ]:= **D[x^4, x]**

Out[ ]= **4 x^3**

And the fundamental theorem of calculus says you can evaluate the integral by evaluating the antiderivative at 3 and 2 and subtracting.

Integrate agrees:

In[ ]:= **Integrate[f[x], {x, 2, 3}]**

Out[ ]= **65**

The general antiderivative is given a special name: the **indefinite integral**.

The goal of this lesson is to review the indefinite integrals of all the functions covered so far.

Indefinite Integrals

From here on, the antiderivative of a function will be referred to as its indefinite integral (sometimes just integral).

Integrate can also calculate integrals of functions, without any range being needed:

In[ ]:= **Integrate[f[x], x]**

Out[ ]= **x^4**

You can see that the indefinite integral of the previous function is just x^4, plus a constant. Integrate does not include the constant when it calculates the integral, so be careful.

If you want to calculate a definite integral, you need the extra parameters for the endpoints a and b:

In[·]:= **Integrate[f[x], {x, 2, 3}]**

Out[·]= **65**

Here is the graph of the function and its integral:

In[·]:= **Plot[{f[x], Evaluate[$\int$f[x] dx]}, {x, −5, 5}, PlotLegends → {"f[x]", "$\int$f[x] dx"}]**

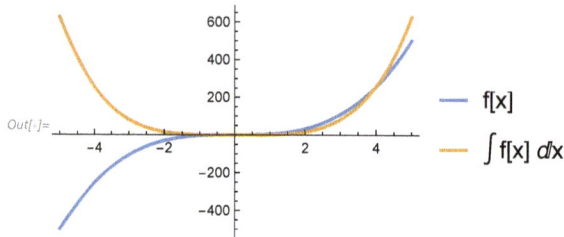

Note that by default, the constant C for the integral is just 0.

Special Note

Recall that the integral is defined for integrable functions. Integrable functions are usually continuous or have finitely many jump discontinuities.

However, you can still calculate integrals for functions with infinite discontinuities. You just stipulate that you are integrating on an interval **without any infinite discontinuities**.

So the following function has a calculable integral on the interval [1, 4]:

In[·]:= **f[x_] := 1 / x ^ 2**

In[·]:= **Integrate[f[x], {x, 1, 4}]**

Out[·]= $\dfrac{3}{4}$

But not on the interval [−1, 1]:

In[·]:= **Integrate[f[x], {x, −1, 1}]**

••• Integrate : Integral of $\dfrac{1}{x^2}$ does not converge on {−1, 1}.

Out[·]= $\displaystyle\int_{-1}^{1} \dfrac{1}{x^2}\, d$x

The indefinite integral is just the antiderivative of the function:

In[ ]:= **Integrate[f[x], x]**

$$Out[\ ]= -\frac{1}{x}$$

Constant

The integral of a constant is given:

In[ ]:= **f[x_] := k**

In[ ]:= **Integrate[f[x], x]**

Out[ ]= **k x**

You can see the indefinite integral is just $k x + C$.

Here are the constant function $g[x] = 3$ and its integral $3 x$ in a graph:

In[ ]:= **Plot[{3, Evaluate[$\int 3\, dx$]}, {x, −5, 5}, PlotLegends → "Expressions"]**

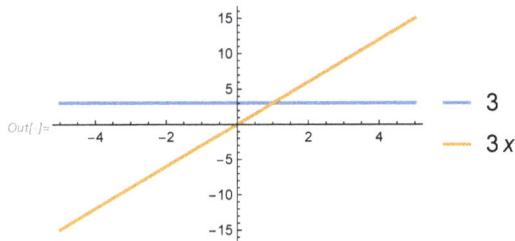

Note that **Evaluate** is needed to plot the graph for **Integrate**.

Linear

The integral of a linear function is given:

In[ ]:= **f[x_] := k x + b**

In[ ]:= **Integrate[f[x], x]**

$$Out[\ ]= b x + \frac{k x^2}{2}$$

You can see the indefinite integral is just $\frac{k x^2}{2} + b x + C$.

Here are the linear function $g[x] = 5\,x + 4$ and its integral $\frac{5\,x^2}{2} + 4\,x$ in a graph:

In[·]:= **Plot[{5 x + 4, Evaluate[$\int (5\,x + 4)\,d\!x$]}, {x, −5, 5}, PlotLegends → "Expressions"]**

Out[·]=

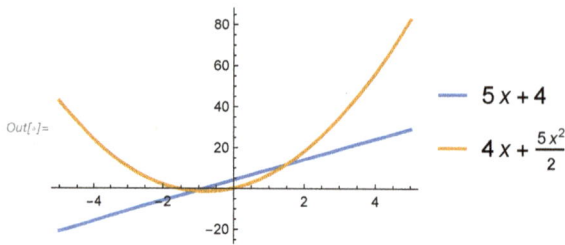

$$— \quad 5\,x + 4$$

$$— \quad 4\,x + \frac{5\,x^2}{2}$$

Note that **Evaluate** is needed to plot the graph for **Integrate**.

Power Rule

The integral of a general power function is given:

In[·]:= **f[x_] := x ^ a**

In[·]:= **Integrate[f[x], x]**

Out[·]= $\dfrac{x^{1+a}}{1+a}$

You can see the indefinite integral is just $x^{1+a}/(1+a) + C$.

Here are the power function $g[x] = x^3$ and its integral $\frac{x^4}{4}$ in a graph:

In[·]:= **Plot[{x ^ 3, Evaluate[$\int x^3\,d\!x$]}, {x, −5, 5}, PlotLegends → "Expressions"]**

Out[·]=

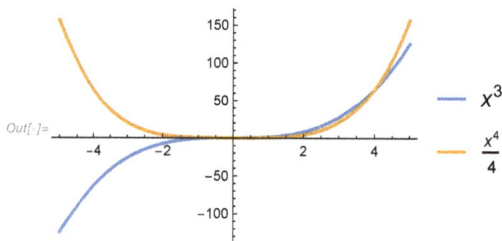

$$— \quad x^3$$

$$— \quad \frac{x^4}{4}$$

Note that **Evaluate** is needed to plot the graph for **Integrate**.

Sine and Cosine

The integrals of sine and cosine are given:

In[ ]:= **f[x_] := Sin[x]**

In[ ]:= **g[x_] := Cos[x]**

In[ ]:= **Integrate[f[x], x]**

Out[ ]= **−Cos[x]**

In[ ]:= **Integrate[g[x], x]**

Out[ ]= **Sin[x]**

You can see the indefinite integral of $\text{Sin}[x]$ is $-\text{Cos}[x] + C$. The indefinite integral of $\text{Cos}[x]$ is $\text{Sin}[x] + C$.

Here are the sine function and its integral in a graph, and the cosine function and its integral in another graph:

In[ ]:= **{Plot[{f[x], Evaluate[$\int$ f[x] d x]}, {x, −5, 5}, ⋯ $\div$],**

Plot[{g[x], Evaluate[$\int$ g[x] d x]}, {x, −5, 5}, ⋯ $\div$]}

Out[ ]= $\Bigg\{$

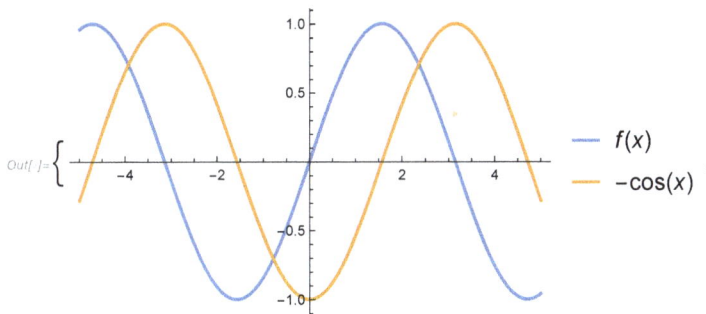

— $f(x)$
— $-\cos(x)$

,

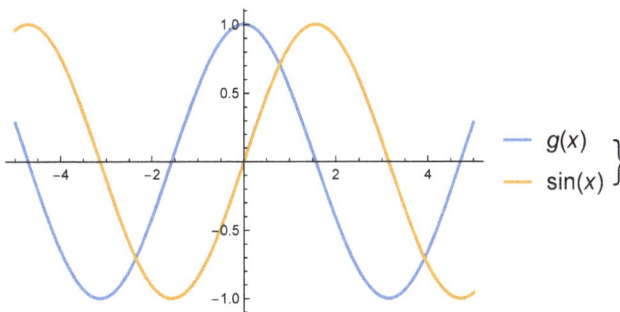

— $g(x)$
— $\sin(x)$

$\Bigg\}$

Note that **Evaluate** is needed to plot the graph for **Integrate**.

Other Trigonometric Functions 1

The integrals of the following trigonometric functions are given:

In[ ]:= **f[x_] := Sec[x] Tan[x]**

In[ ]:= **g[x_] := −Csc[x] Cot[x]**

In[ ]:= **Integrate[f[x], x]**

Out[ ]= **Sec[x]**

In[ ]:= **Integrate[g[x], x]**

Out[ ]= **Csc[x]**

You can see the indefinite integral of Sec[x] Tan[x] is Sec[x] + C. The indefinite integral of −Csc[x] Cot[x] is Csc[x] + C.

Here are the first function and its integral in a graph, and the second function and its integral in another graph:

In[ ]:= **{Plot[{f[x], Evaluate[$\int$ f[x] d x]}, {x, −5, 5}, ⋯ ✦],**

Plot[{g[x], Evaluate[$\int$ g[x] d x]}, {x, −5, 5}, ⋯ ✦]}

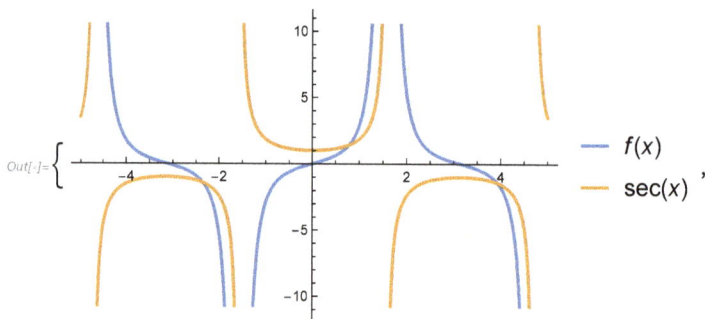

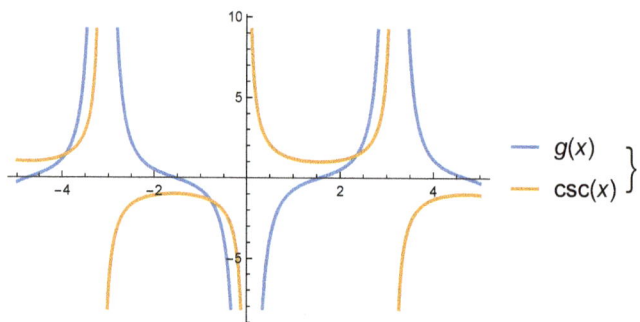

Note that **Evaluate** is needed to plot the graph for **Integrate**.

Other Trig 2

The integrals of $\text{Sec}[x]^2$ and $-\text{Csc}[x]^2$ are given:

In[ ]:= **f[x_] := Sec[x]^2**

In[ ]:= **g[x_] := -Csc[x]^2**

In[ ]:= **Integrate[f[x], x]**

Out[ ]= Tan[x]

In[ ]:= **Integrate[g[x], x]**

Out[ ]= Cot[x]

You can see the indefinite integral of $\text{Sec}[x]^2$ is $\text{Tan}[x] + C$. The indefinite integral of $-\text{Csc}[x]^2$ is $\text{Cot}[x] + C$.

Here are the first function and its integral in a graph, and the second function and its integral in another graph:

In[ ]:= **{Plot[{f[x], Evaluate[$\int$ f[x] d x]}, {x, −5, 5}, ⋯ ▾],**

Plot[{g[x], Evaluate[$\int$ g[x] d x]}, {x, −5, 5}, ⋯ ▾]}

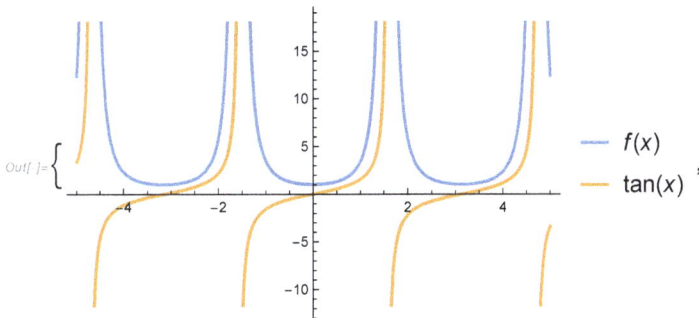

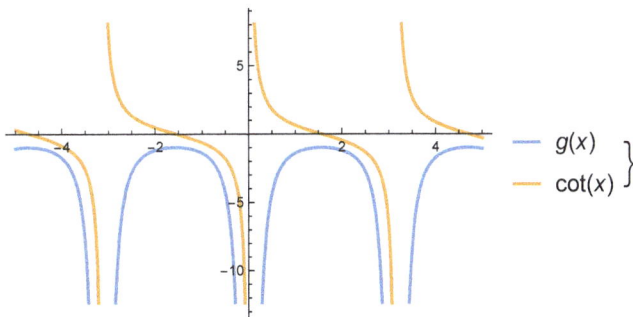

Out[ ]=

Note that **Evaluate** is needed to plot the graph for **Integrate**.

Summary

The general antiderivative of a function is called its indefinite integral.

You can take the indefinite integral of many functions, so long as they are integrable.

You can also take the definite integral of functions with infinite discontinuities, so long as it is not over a range that contains the infinite discontinuities.

The indefinite integrals for many of the functions you study can be found just from Integrate.

The next lesson will cover the substitution rule, which lets you calculate more difficult (and interesting!) integrals.

Exercises

Exercise 1—Calculate a Definite Integral

Find the definite integral of the following function from $[0, 5\pi/4]$:

In[ ]:= **f[x_] := Sin[x] + Cos[x]**

Solution

Use Integrate:

In[ ]:= **Integrate[f[x], {x, 0, 5 π/4}]**

Out[ ]= 1

Here is the area covered in a graph:

In[ ]:= **Plot[f[x], {x, 0, 5 π/4}, Filling → Axis]**

Out[ ]=

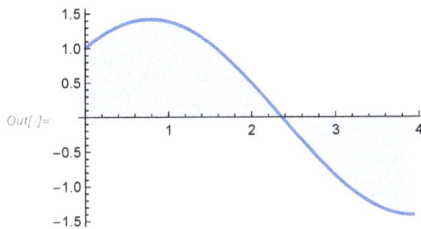

Exercise 2—Error

Calculating the definite integral of the following function from $[-1, 1]$ brings up an error:

In[ ]:= **Integrate[-Csc[x]^2, {x, -1, 1}]**

••• Integrate : Integral of $-Csc[x]^2$ does not converge on $\{-1, 1\}$.

Out[ ]= $\int_{-1}^{1} -Csc[x]^2 \, dx$

Why is this the case?

Solution

Look at the plot of the function in the interval:

In[ ]:= **Plot[−Csc[x]^2, {x, −1, 1}]**

Out[ ]=

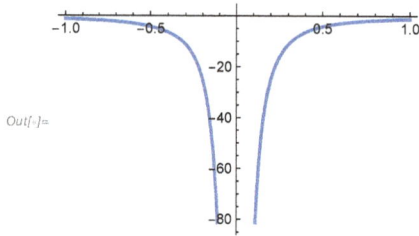

You can see there is an infinite discontinuity at 0, so the integral in the given range is undefined.

Exercise 3—Algebraic

Find the indefinite integral of the following function:

In[ ]:= **f[x_] := 10 x^5 + 5 + Sqrt[x] + 1/x^3**

Solution

The indefinite integral can be calculated with Integrate:

In[ ]:= **Integrate[f[x], x]**

$$Out[\ ]= -\frac{1}{2x^2} + 5x + \frac{2x^{3/2}}{3} + \frac{5x^6}{3}$$

Note that its derivative is the original function:

In[ ]:= **D[−$\frac{1}{2x^2}$ + 5 x + $\frac{2x^{3/2}}{3}$ + $\frac{5x^6}{3}$, x]**

$$Out[\ ]= 5 + \frac{1}{x^3} + \sqrt{x} + 10x^5$$

You also could take the integral of each part, use the power rule, and sum them all together:

In[ ]:= **Integrate[#, x] & /@ {10 x^5, 5, Sqrt[x], 1/x^3}**

$$Out[\ ]= \left\{\frac{5x^6}{3}, 5x, \frac{2x^{3/2}}{3}, -\frac{1}{2x^2}\right\}$$

Here are the function and its integral in a graph:

In[ ]:= **Plot[{f[x], Evaluate@Integrate[f[x], x]}, {x, 0, 5}]**

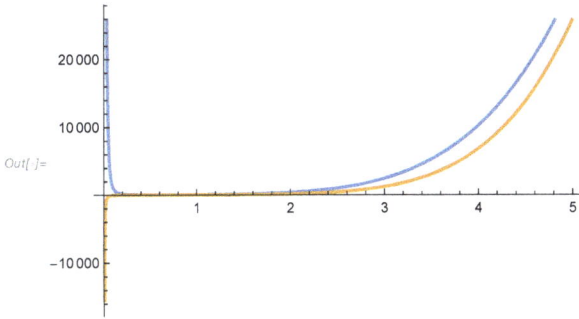

Exercise 4—Mechanics

The acceleration of a particle is given:

In[ ]:= **a[t_] := t^2 + Cos[t]**

Find possible functions for the velocity and position.

Solution

The integral of acceleration is velocity:

In[ ]:= **Integrate[a[t], t]**

Out[ ]= $\dfrac{t^3}{3} + \text{Sin}[t]$

The integral of velocity is position:

In[ ]:= **Integrate[Integrate[a[t], t], t]**

Out[ ]= $\dfrac{t^4}{12} - \text{Cos}[t]$

Here is a graph showing the possible position, velocity and acceleration of the particle:

In[ ]:= **Plot[{a[t], Evaluate@Integrate[a[t], t], Evaluate@Integrate[Integrate[a[t], t], t]},**
　　　{t, 0, 2 π}, PlotLegends → {"acceleration", "velocity", "position"}]

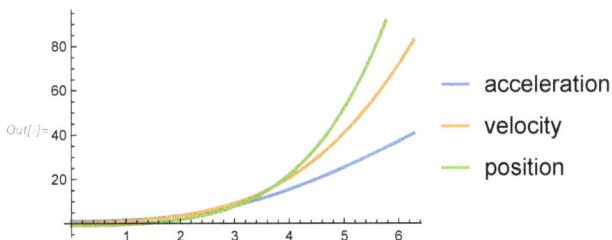

Exercise 5—Current

The current in a wire is given by the following **Heaviside function**:

In[·]:= current[*t*_] := HeavisideTheta[*t* − 1] + HeavisidePi[*t* − 3] − HeavisideLambda[*t* − 7]

Find the total charge that goes through the wire from $t = 0$ to 10 seconds.

Solution

Use Integrate:

In[·]:= Integrate[current[t], {t, 0, 10}]

Out[·]= 9

So 9 coulombs go through the wire in 10 seconds.

Here are the function and its integral in a graph:

In[·]:= Plot[{current[t], Evaluate@Integrate[current[t], t]}, {t, 0, 10}, ··· → ··· +]

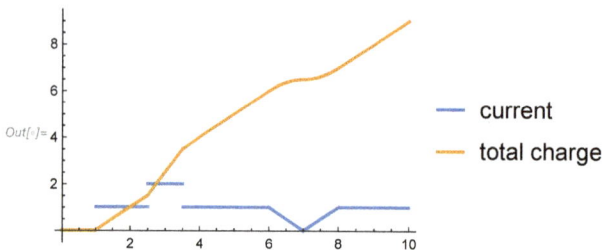

27 | The Substitution Rule

Overview

The rules given for calculating integrals cover many cases.

For example, the integral of x^2 is $\frac{x^3}{3} + C$ by the power rule:

In[ ]:= **Integrate[x^2, x]**

Out[ ]= $\dfrac{x^3}{3}$

Those rules alone cannot find the integral of the following function, however:

In[ ]:= **f[x_] := 10 Sin[3 x]**

Or this function:

In[ ]:= **g[x_] := 2 x^2 Sqrt[x^3 + 3]**

Here are both functions' plots:

In[ ]:= **Plot[{f[x], g[x]}, {x, −5, 5}, PlotLegends → "Expressions"]**

Out[ ]=

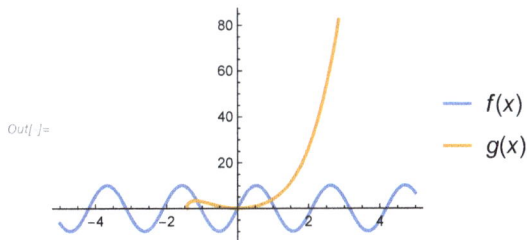

The goal of this lesson is to develop a way for calculating such integrals, with a technique called the **substitution rule**.

Chain Rule

First, recall the chain rule for derivatives.

Given two differentiable functions $f[x]$ and $g[x]$:

In[ ]:= **Clear[f, g]**

The derivative of $f[g[x]]$ is $f\,'[g[x]] * g\,'[x]$:

In[ ]:= **D[f[g[x]], x]**

Out[ ]= $f'[g[x]]\, g'[x]$

The derivative of $g[f[x]]$ is $g'[f[x]] * f'[x]$:

In[·]:= **D[g[f[x]], x]**

Out[·]= **f'[x] g'[f[x]]**

So the integral of $f'[g[x]] * g'[x]$ is $f[g[x]]$:

In[·]:= **Integrate[f'[g[x]] * g'[x], x]**

Out[·]= **f[g[x]]**

And the integral of $g'[f[x]] * f'[x]$ is $g[f[x]]$:

In[·]:= **Integrate[g'[f[x]] * f'[x], x]**

Out[·]= **g[f[x]]**

Substitution Rule

To make this easier, instead of evaluating the expression $\int f'[g[x]]\, g'[x]\, dx$, use the **substitution** $u = g[x]$.

The **substitution rule** says that if f is continuous on an interval, and $u = g[x]$ is a differentiable function with range the same interval, then:

$$\int f(g(x))\, g'(x)\, dx = \int f(u)\, du$$

Note that the new integral $\int f[u]\, du$ has width du instead of dx. You have to calculate that du. Calculating du is easy. Just take $u = g[x]$ and differentiate.

Recall that the derivative of u can be expressed as $\frac{du}{dx}$. Here du and dx can be thought of as **differentials**, which you can multiply.

So $du = g'[x]\, dx$, which you can see in the substitution rule equation.

Here is an example: calculate the integral $\int 3\, \mathrm{Sin}[3\,x]\, dx$.

Since $\mathrm{Sin}[3\,x]$ is the composition of the functions $\mathrm{Sin}[x]$ and $3\,x$, set u to $3\,x$.

$du = 3\, dx$. So the integral is $\int 3\, \mathrm{Sin}[3\,x]\, dx = \int \mathrm{Sin}[u]\, du$:

In[·]:= **Integrate[Sin[u], u]**

Out[·]= **−Cos[u]**

Plugging in $u = 3\,x$, you see that the answer is $-\mathrm{Cos}[3\,x]$.

Integrate agrees:

In[ ]:= **Integrate[3 Sin[3 x], x]**

Out[ ]= **−Cos[3 x]**

Trigonometric Function

Calculate the integral of the first function mentioned in the introduction:

In[ ]:= **f[x_] := 10 Sin[3 x]**

Set $u = 3\,x$. So $du = 3\,dx$.

Here you have to be careful. The previous example already had a 3, so calculating the integral was straightforward. Now you have a $10\,dx$ that you must express in terms of du.

$10\,dx = \frac{10}{3} * 3\,dx = \frac{10}{3}\,du$, so you now calculate the integral
$\int 10\,\text{Sin}[3\,x]\,dx = \int \frac{10}{3}\,\text{Sin}[u]\,du$:

In[ ]:= **Integrate[10/3 Sin[u], u]**

Out[ ]= $-\dfrac{10\,\text{Cos}[u]}{3}$

Plugging in $u = 3\,x$, you see that the answer is $-\frac{10}{3}\,\text{Cos}[3\,x]$:

In[ ]:= **Integrate[10/3 Sin[u], u] /. u → 3 x**

Out[ ]= $-\dfrac{10}{3}\,\text{Cos}[3\,x]$

Integrate agrees:

In[ ]:= **Integrate[f[x], x]**

Out[ ]= $-\dfrac{10}{3}\,\text{Cos}[3\,x]$

Algebraic Function

Calculate the integral of the second function mentioned in the introduction:

In[ ]:= **g[x_] := 2 x^2 Sqrt[x^3 + 3]**

Set $u = x^3 + 3$. So $du = 3\,x^2\,dx$.

The original function has a $2\,x^2\,dx$, which is expressed in terms of du:

$2\,x^2\,dx = \frac{2}{3} * 3\,x^2\,dx = \frac{2}{3}\,du$.

The integral becomes $\int 2\,x^2\ \sqrt{x^3 + 3}\ dx = \int \frac{2}{3}\ \sqrt{u}\ du$:

$In[\cdot]:=$ **Integrate[2 / 3 Sqrt[u], u]**

$Out[\cdot]=$ $\dfrac{4\,u^{3/2}}{9}$

Plugging in $u = x^3 + 3$, you see that the answer is $\frac{4}{9}\left(x^3 + 3\right)^{3/2}$:

$In[\cdot]:=$ **Integrate[2 / 3 Sqrt[u], u] /. u → x^3 + 3**

$Out[\cdot]=$ $\dfrac{4}{9}\left(3 + x^3\right)^{3/2}$

Integrate agrees:

$In[\cdot]:=$ **Integrate[g[x], x]**

$Out[\cdot]=$ $\dfrac{4}{9}\left(3 + x^3\right)^{3/2}$

Rational Algebraic Function

Compute the integral of the following function:

$In[\cdot]:=$ **f[x_] := 2 x / Sqrt[5 − 5 x^2]**

Set $u = 5 - 5\,x^2$. So $du = -10\,x\,dx$.

The original function has a $2\,x\,dx$, which is expressed in terms of du:

$2\,x\,dx = -\frac{2}{10} * -10\,x\,dx = -\frac{1}{5}\,du$.

The integral becomes $\int 2\,x \Big/ \sqrt{5 - 5\,x^2}\ dx = \int -\frac{1}{5} \Big/ \sqrt{u}\ du$:

$In[\cdot]:=$ **Integrate[−1 / (5 Sqrt[u]), u]**

$Out[\cdot]=$ $-\dfrac{2\,\sqrt{u}}{5}$

Plugging in $u = 5 - 5\,x^2$, you see that the answer is $-\frac{2}{5}\,\sqrt{5 - 5\,x^2}$:

$In[\cdot]:=$ **Integrate[−1 / (5 Sqrt[u]), u] /. u → 5 − 5 x^2**

$Out[\cdot]=$ $-\dfrac{2}{5}\,\sqrt{5 - 5\,x^2}$

Integrate agrees:

In[·]:= **Integrate[f[x], x]**

Out[·]= $-\dfrac{2}{5} \sqrt{5 - 5x^2}$

A Tricky Example

Compute the integral of the following function:

In[·]:= **f[x_] := x ^3 Sqrt[3 − x^2]**

Set $u = 3 - x^2$. So $du = -2\,x\,dx$.

This example is a little trickier. The original function has a $x^3\,dx$, but $du = -2\,x\,dx$. So what do you do?

You note that $x^3 = x * x^2$.

$u = 3 - x^2$, so $x^2 = 3 - u$.

$x\,dx = -\dfrac{1}{2} * 2\,x\,dx = -\dfrac{1}{2}\,du.$

So $x^3\,dx = -\dfrac{1}{2}\,(3 - u)\,du.$

The integral becomes $\int x^3 \sqrt{3 - x^2}\,dx = \int -\dfrac{1}{2}\,(3 - u)\,\sqrt{u}\,du$:

In[·]:= **Integrate[−1/2 (3 − u) Sqrt[u], u]**

Out[·]= $\dfrac{1}{5}\,(-5 + u)\,u^{3/2}$

Plugging in $3 - x^2$ for u, the answer is $\dfrac{1}{5}\left(-5 + \left(3 - x^2\right)\right)\left(3 - x^2\right)^{3/2} = \dfrac{1}{5}\left(-2 - x^2\right)\left(3 - x^2\right)^{3/2}$:

In[·]:= **Integrate[−1/2 (3 − u) Sqrt[u], u] /. u → 3 − x^2**

Out[·]= $\dfrac{1}{5}\left(-2 - x^2\right)\left(3 - x^2\right)^{3/2}$

Integrate agrees:

In[·]:= **Integrate[f[x], x]**

Out[·]= $\dfrac{1}{5}\left(-2 - x^2\right)\left(3 - x^2\right)^{3/2}$

Definite Integrals

When calculating definite integrals using the substitution rule, you have two ways of going about it. The first way has you calculate the integral, plug in the value for u, and evaluate the integral at the endpoints of the interval.

For example, the previous example, as an integral from 0 to 1, has the value:

$In[\cdot]:= \left(-\frac{1}{5}(3-1^2)^{3/2}(2+1^2)\right) - \left(-\frac{1}{5}(3-0^2)^{3/2}(2+0^2)\right)$

$Out[\cdot]= -\frac{6\sqrt{2}}{5} + \frac{6\sqrt{3}}{5}$

The second way has you **keep the integral in u form and change the endpoints to accommodate**.

If you keep the previous integral in u form and change the endpoints from 0 to 1 to 3 to 2 by plugging 0 and 1 into u, you get the value:

$In[\cdot]:= \left(\frac{1}{5}(-5+2)2^{3/2}\right) - \left(\frac{1}{5}(-5+3)3^{3/2}\right)$

$Out[\cdot]= -\frac{6\sqrt{2}}{5} + \frac{6\sqrt{3}}{5}$

You can see that the integrals are the same. Sometimes its better to keep the integral in u form and change the endpoints to save space.

In general, if $g'[x]$ is continuous on $[a, b]$ and $f[x]$ is continuous on the range of $u = g[x]$, then:

$$\int_a^b f(g(x))\, g'(x)\, dx = \int_{g(a)}^{g(b)} f(u)\, du$$

As always, confirm the previous calculations with Integrate:

$In[\cdot]:= \{Integrate[-1/2\,(3-x)\,Sqrt[x], \{x, 3, 2\}], Integrate[f[x], \{x, 0, 1\}]\}$

$Out[\cdot]= \left\{-\frac{6}{5}\left(\sqrt{2}-\sqrt{3}\right), -\frac{6}{5}\left(\sqrt{2}-\sqrt{3}\right)\right\}$

Summary

The substitution rule lets you calculate integrals for functions you previously could not calculate with the basic integration rules. It works by basically doing the chain rule in reverse.

For the integral you want to compute, you make a suitable substitution u for some piece of it. Then you calculate du by taking u's derivative and modifying the integral accordingly.

Once you have calculated the integral with u, plug back in what you substituted u for to get the integral.

For definite integrals, you can keep the original endpoints and calculate the integral using the second part of the fundamental theorem of calculus. Or you can change the endpoints and calculate the integral in u form, which can be easier.

The next lesson will apply the techniques you have learned to find the area between two curves.

Exercises

Exercise 1—Algebraic

Find the integral of the following function:

In[·]:= **f[x_] := x CubeRoot[x^2 + 1]**

Solution

Use the substitution $u = x^2 + 1$. $du = 2 x dx$.

The original function has an $x dx$, which equals $\frac{1}{2} * 2 x dx = \frac{1}{2} du$.

So transform the integral $\int x \sqrt[3]{x^2 + 1} \, dx$ to $\int \frac{1}{2} \sqrt[3]{u} \, du$:

In[·]:= **Integrate[CubeRoot[u]/2, u]**

Out[·]= $\dfrac{3 u \sqrt[3]{u}}{8}$

Plugging in $u = x^2 + 1$, you see the answer is $\frac{3}{8} \left(x^2 + 1\right) \sqrt[3]{x^2 + 1}$:

In[·]:= **Integrate[CubeRoot[u]/2, u] /. u → x^2 + 1**

Out[·]= $\dfrac{3}{8} \left(1 + x^2\right) \sqrt[3]{1 + x^2}$

Integrate agrees:

In[·]:= **Integrate[f[x], x]**

Out[·]= $\dfrac{3}{8} \left(1 + x^2\right) \sqrt[3]{1 + x^2}$

Exercise 2—Rational

Find the integral of the following function:

In[·]:= **f[x_] := (18 x^5 − 6)/((x^6 − 2 x)^2)**

Solution

Use the substitution $u = x^6 - 2 x$. $du = \left(6 x^5 - 2\right) dx$.

The original function has the expression $(18 x^5 - 6) dx$, which equals $3 \left(6 x^5 - 2\right) dx = 3 du$.

So transform the integral $\int \dfrac{18 x^5 - 6}{\left(x^6 - 2 x\right)^2} \, dx$ to $\int \dfrac{3}{u^2} \, du$:

In[·]:= **Integrate[3 / u ^ 2, u]**

Out[·]:= $-\dfrac{3}{u}$

Plugging in $u = x^6 - 2\,x$, you see the answer is $-\dfrac{3}{x^6 - 2\,x}$:

In[·]:= **Integrate[3 / u ^ 2, u] /. u → x ^ 6 − 2 x**

Out[·]:= $-\dfrac{3}{-2\,x + x^6}$

Integrate agrees:

In[·]:= **Integrate[f[x], x]**

Out[·]:= $\dfrac{3}{2\,x - x^6}$

Exercise 3—Trig

Find the integral of the following function:

In[·]:= **f[*x*_] := Sin[*x*] Sqrt[Cos[*x*]]**

Solution

Use the substitution $u = \text{Cos}[x]$. $d\,u = -\text{Sin}[x]\,d\,x$.

The original function has a $\text{Sin}[x]\,d\,x$, which equals $-d\,u$.

So transform the integral $\int \text{Sin}[x]\ \sqrt{\text{Cos}[x]}\ d\,x$ to $\int -\sqrt{u}\ d\,u$:

In[·]:= **Integrate[−Sqrt[u], u]**

Out[·]:= $-\dfrac{2\,u^{3/2}}{3}$

Plugging in $u = \text{Cos}[x]$, you see the answer is $-\dfrac{2}{3}\,\text{Cos}[x]^{3/2}$:

In[·]:= **Integrate[−Sqrt[u], u] /. u → Cos[x]**

Out[·]:= $-\dfrac{2}{3}\,\text{Cos}[x]^{3/2}$

Integrate agrees:

In[-]:= **Integrate[f[x], x]**

Out[-]= $-\dfrac{2}{3}\, \text{Cos[x]}^{3/2}$

Exercise 4—Metabolism

The basal metabolism rate for a particular cat is given by the function in kcal/hr:

In[-]:= **cat[$t\_$] := 33 − 0.04 Cos[$\pi\, t$ / 12]**

where t is the time in hours measured from 12 noon. Find the total basal metabolism for the cat over one week.

Solution

To find the total basal metabolism, integrate the function. One week equals 168 hours, so the endpoints are 0 and 168.

The integral is a difference, so integrate the 33 and $-0.04\, \text{Cos}[\pi\, t\, / 12]$ separately:

In[-]:= **Integrate[33, {x, 0, 168}]**

Out[-]= **5544**

You can see that the first integral is easy (it is $33\, x$), and the second uses the substitution rule. Use the substitution $u = \pi\, t\, / 12$. So $du = \pi\, dt\, / 12$.

The function you are integrating has a $0.04\, dt$, which can be expressed as $\dfrac{0.04}{\pi/12}\,(\pi\, dt\, / 12) = \dfrac{0.48}{\pi}\, du.$

So the integral $\int_0^{168} 0.04\, \text{Cos}[\pi\, t\, / 12]\, dt$ becomes $\int_0^{14\pi} \dfrac{0.48}{\pi}\, \text{Cos}[u]\, du$:

In[-]:= **Integrate[0.48 / π Cos[u], {u, 0, 14 π}]**

Out[-]= **0.**

The total basal metabolism is 5544 kcal. Integrate agrees:

In[-]:= **Integrate[cat[t], {t, 0, 168}]**

Out[-]= **5544.**

Exercise 5—Definite Integral

Use the substitution rule to evaluate the following integral in the range $[-2, 5]$. Evaluate it over the endpoints for the new integral:

$In[\cdot]:=$ f[x_] := x^8 $\sqrt[5]{x^3 - 8}$

Solution

Set $u = x^3 - 8$. So $du = 3 x^2 dx$.

There is an $x^8 dx$ expression in the integral: $x^8 = x^2(x^3)^2$.

$x^3 = u + 8$, so $x^8 dx = \frac{1}{3} (x^3)^2 (3 x^2 dx) = \frac{1}{3} (u + 8)^2 du$.

$(-2)^3 - 8 = -16$, and $5^3 - 8 = 117$, so the new integral is $\int_{-16}^{117} \frac{1}{3} (u + 8)^2 \sqrt[5]{u} \, du$, which is evaluated with Integrate:

$In[\cdot]:=$ **Integrate[1/3 $\sqrt[5]{u}$ (u+8)^2, {u, −16, 117}]**

$Out[\cdot]:= -\dfrac{35 \left(4096 \, 2^{4/5} - 9\,146\,475 \, 3^{2/5} \, 13^{1/5}\right)}{1584}$

Confirm by evaluating the original integral:

$In[\cdot]:=$ **Integrate[f[x], {x, −2, 5}]**

$Out[\cdot]:= -\dfrac{35 \left(4096 \, 2^{4/5} - 9\,146\,475 \, 3^{2/5} \, 13^{1/5}\right)}{1584}$

28 | Areas between Curves

Overview

When you calculate integrals, you are finding the area between the curve of the function and the x axis:

In[ ]:= **f[x_] := x^2**

In[ ]:= **Integrate[f[x], {x, −1, 1}]**

Out[ ]= $\dfrac{2}{3}$

In[ ]:= **Plot[f[x], {x, −1, 1}, Filling → Axis]**

Out[ ]=

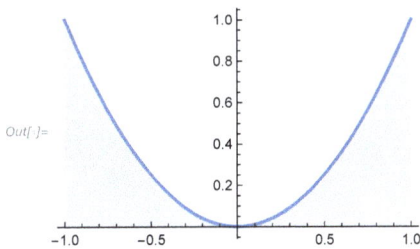

What if you wanted to find the area between two curves?

This lesson will show how to do so and give several examples.

Area between Curves

Consider two continuous functions f and g where $f \geq g$:

In[ ]:= **f[x_] := 5**
g[x_] := −x^2 + 1

To find the area between the two curves in the region $[a, b]$, look again at Riemann sums.

As usual, the width will be $\dfrac{b-a}{n}$ for n rectangles.

The height will be $f[x_i^*] - g[x_i^*]$ for each strip, as seen in the plot:

In[·]:= **Show[Plot[{f[x], g[x]}, {x, −1, 3}], DiscretePlot[{f[x], g[x]}, {x, −1, 3}, $\cdots$ $\div$]]**

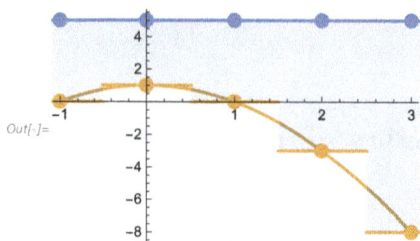

Out[·]=

Therefore, the Riemann sum will be $\sum_{i=0}^{n} (f[x_i^*] - g[x_i^*]) \, \Delta x$. So the integral is $\int_a^b (f[x] - g[x]) \, dx$.

The area between the preceding curves from -1 to 3 is given:

In[·]:= **Integrate[f[x] − g[x], {x, −1, 3}]**

Out[·]= $\dfrac{76}{3}$

Two Polynomials

Find the area of the region bounded above by $f[x] = 3\, x^2 + 4$, bounded below by $g[x] = 4\, x + 1$, bounded on the left by $x = 1$, and bounded on the right by $x = 5$.

First, define the functions:

In[·]:= **f[x_] := 3 x^2 + 4**
g[x_] := 4 x + 1

The endpoints of the interval to be integrated are 1 and 5, so the integral is:

In[·]:= **Integrate[f[x] − g[x], {x, 1, 5}]**

Out[·]= **88**

Here is a plot of the two functions and the area between them:

In[·]:= **Plot[{f[x], g[x]}, {x, 1, 5}, Filling → {1 → {2}}, PlotLegends → "Expressions"]**

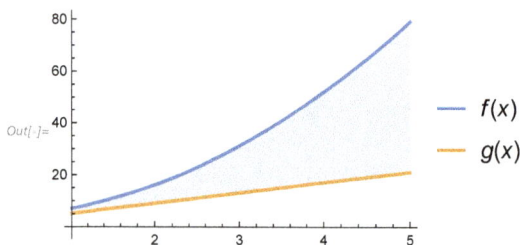

Out[·]=

Enclosed Area

Sometimes you are asked to find the area **enclosed** between two curves, without any endpoints for an integral clearly given.

For example, find the area enclosed between the following two curves:

In[ ]:= **f[x_] := -x^2 + 4**
        g[x_] := x^2 - 2

First, find where the two functions intersect with Solve:

In[ ]:= **sol = Solve[f[x] == g[x], x]**

Out[ ]= $\left\{\left\{x \to -\sqrt{3}\right\}, \left\{x \to \sqrt{3}\right\}\right\}$

You now have the endpoints for the integral. Then plot the functions to see which is above and which is below:

In[ ]:= **Plot[{f[x], g[x]}, {x, sol[[1, 1, 2]], sol[[2, 1, 2]]}, PlotLegends → "Expressions"]**

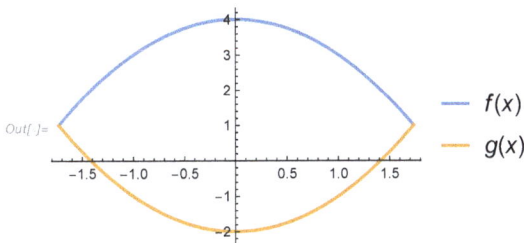

You can see $f[x]$ is always above and $g[x]$ is always below.

So the integral is given by:

In[ ]:= **Integrate[f[x] - g[x], {x, sol[[1, 1, 2]], sol[[2, 1, 2]]}]**

Out[ ]= $8\sqrt{3}$

General Functions

The function $f[x]$ may not always be greater than the function $g[x]$.

For example, for the two functions:

In[ ]:= **f[x_] := Sin[x]**
        g[x_] := Cos[x]

Sine is greater than cosine on regions $2n\pi + \pi/4 \le \theta \le 2n\pi + 5\pi/4$. Here, assume n is an integer.

Cosine is greater than sine on regions $2n\pi - 3\pi/4 \le \theta \le 2n\pi + \pi/4$. Here, assume n is an integer.

To find the area between sine and cosine on the interval $[-3\pi/4, 5\pi/4]$, split the integral in two.

The first integral from $-3\pi/4$ to $\pi/4$ has cosine greater than sine, so it is:

In[ ]:= **Integrate[g[x] – f[x], {x, –3 π/4, π/4}]**

Out[ ]= $2\sqrt{2}$

The second integral from $\pi/4$ to $5\pi/4$ has sine greater than sine, so it is:

In[ ]:= **Integrate[f[x] – g[x], {x, π/4, 5 π/4}]**

Out[ ]= $2\sqrt{2}$

So the total area is $4\sqrt{2}$.

In general, for two functions f and g, the area between on the interval $[a, b]$ is given by $\int_a^b |f[x] - g[x]|\,dx$.

Here is the previous answer with RealAbs and Integrate:

In[ ]:= **Integrate[RealAbs[f[x] – g[x]], {x, –3 π/4, 5 π/4}]**

Out[ ]= $4\sqrt{2}$

Timing

Sometimes it is better to calculate an integral the long way without RealAbs.

For example, the previous calculation with RealAbs took around 0.3 seconds for the Wolfram Language to calculate it:

In[ ]:= **Integrate[RealAbs[f[x] – g[x]], {x, –3 π/4, 5 π/4}] // Timing**

Out[ ]= $\left\{0.259861,\ 4\sqrt{2}\right\}$

Here, use the function Timing to find how long it takes. When you do the integrals separately:

In[ ]:= **(Integrate[g[x] – f[x], {x, –3 π/4, π/4}] + Integrate[f[x] – g[x], {x, π/4, 5 π/4}]) // Timing**

Out[ ]= $\left\{0.024492,\ 4\sqrt{2}\right\}$

it only takes 0.02 seconds to get the answer!

In general it is quicker to find the places on the interval where the two functions intersect:

In[ ]:= **Solve[f[x] == g[x] && −3 π/4 ≤ x ≤ 5 π/4, x] // Simplify**

Out[ ]= $\left\{\left\{x \to \frac{\pi}{4}\right\}, \left\{x \to \frac{5\pi}{4}\right\}, \left\{x \to -\frac{3\pi}{4}\right\}\right\}$

Then plot the functions to see which is above which in each region to calculate the integral:

In[ ]:= **Plot[{f[x], g[x]}, {x, −3 π/4, 5 π/4}, PlotLegends → "Expressions"]**

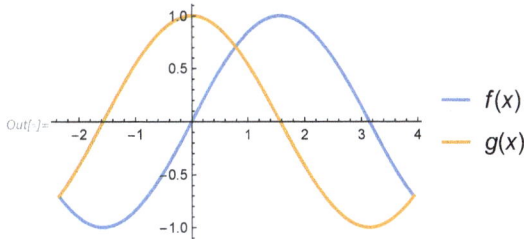

Functions of *y*

Sometimes functions are given in terms of the *y* coordinate and give the value for the *x* coordinate:

In[ ]:= **f[y_] := y − 3**
g[y_] := 2 y + 4

You still want to find the area between the curves, but the Riemann sum changes slightly.

If you want to find the area from $y = c$ to $y = d$, you again split the region into n rectangles and calculate the width to be $\Delta y = \frac{d-c}{n}$.

Now the height will be the value of the **rightmost** function minus the value of the **leftmost** function: $f[y_i^*] - g[y_i^*]$. Therefore, the integral is $\int_{y=c}^{y=d}(f[y] - g[y]) \, d\,y$.

To plot functions as a function of *y* in the *x y* plane, you have to use ParametricPlot:

In[ ]:= **ParametricPlot[{{f[y], y}, {g[y], y}}, {y, −5, 5}, ⋯ → ⋯ ↵]**

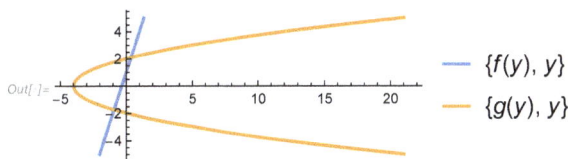

Be sure to plot the functions exactly as shown, with the **function** in the **first** coordinate and y in the **second** coordinate.

Here, f is actually to the left of g, so the integral from -5 to 5 is:

In[·]:= **Integrate[g[y] – f[y], {y, –5, 5}]**

Out[·]= **70**

Area between Two Functions of y

Find the area enclosed by the line $y = 3\,x + 1$ and the parabola $y^2 = x + 4$.

First, solve for x in each case:

In[·]:= **{Solve[y == 3 x + 1, x], Solve[y^2 == x + 4, x]}**

Out[·]= $\left\{\left\{\left\{x \to \frac{1}{3}(-1+y)\right\}\right\}, \left\{\left\{x \to -4 + y^2\right\}\right\}\right\}$

Then define the functions in terms of y:

In[·]:= **f[y_] := 1/3 (–1 + y)**
g[y_] := –4 + y^2

Find where the functions intersect:

In[·]:= **sol = Solve[f[y] == g[y], y]**

Out[·]= $\left\{\left\{y \to \frac{1}{6}\left(1 - \sqrt{133}\right)\right\}, \left\{y \to \frac{1}{6}\left(1 + \sqrt{133}\right)\right\}\right\}$

Then see which function is the rightmost and which is the leftmost in a plot:

In[·]:= **ParametricPlot[{{f[y], y}, {g[y], y}}, {y, sol[[1, 1, 2]], sol[[2, 1, 2]]},** ··· → ··· ◆ **]**

Out[·]=

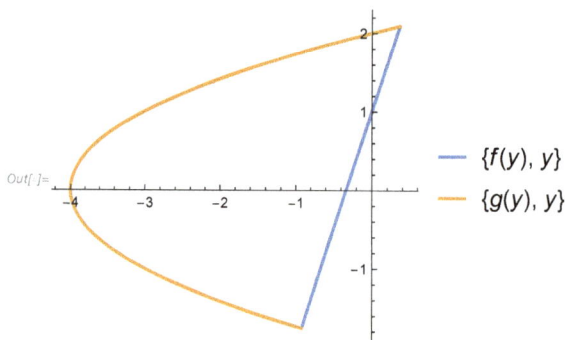

So the integral is:

In[ ]:= **Integrate[f[y] – g[y], {y, sol[[1, 1, 2]], sol[[2, 1, 2]]}]**

Out[ ]= $\dfrac{133 \sqrt{133}}{162}$

Summary

Finding the area between curves is rather easy with integrals. For a function $f[x]$ greater than a function $g[x]$ in the region $a \le x \le b$, the area is simply $\int_a^b (f[x] - g[x]) \, dx$.

When you do not know the region and want to find the area enclosed between two functions:

1. Find where the functions intersect.
2. Find which function is above and which is below.
3. Then calculate the integral.

For general functions $f[x]$ and $g[x]$, the area between the two is given by $\int_a^b |f[x] - g[x]| \, dx$.

Timing shows it is sometimes faster to use Solve, Plot and Integrate to find a general area instead of just using RealAbs and Integrate.

For functions of y in the region $c \le y \le d$, the integral is given by $\int_c^d (f[y] - g[y]) \, dy$, where $f[y]$ is to the right of $g[y]$. Functions of y can be plotted with ParametricPlot.

The next lesson will use integrals to calculate the volume of solids.

Exercises

Exercise 1—Area

Find the area between the curves given by the following functions in the range $[-2, 1]$:

```
In[·]:= f[x_] := x^3 Sin[x^4]
        g[x_] := 3 x - 3
```

Solution

First, plot the two functions:

```
In[·]:= Plot[{f[x], g[x]}, {x, -2, 1}, PlotLegends → "Expressions"]
```

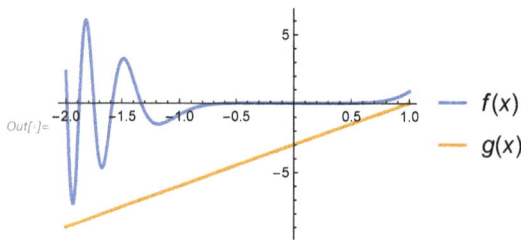

You can see the first function is above the second function in the range.

So the integral is the following:

```
In[·]:= Integrate[f[x] - g[x], {x, -2, 1}]
```

$$Out[·]= \frac{1}{4} (54 - Cos[1] + Cos[16])$$

Alternatively, you could use RealAbs and not have to look at the plots:

```
In[·]:= Integrate[RealAbs[f[x] - g[x]], {x, -2, 1}]
```

$$Out[·]= \frac{1}{4} (54 - Cos[1] + Cos[16])$$

Exercise 2—Enclosed Area

Find the area enclosed between the curves given by the following two functions:

```
In[·]:= f[x_] := 7 + 8 x - 2 x^2
        g[x_] := x + 2
```

Solution

First, find the endpoints of the interval:

In[ ]:= **sol = Solve[f[x] == g[x], x]**

Out[ ]= $\left\{\left\{x \to \frac{1}{4}\left(7 - \sqrt{89}\right)\right\}, \left\{x \to \frac{1}{4}\left(7 + \sqrt{89}\right)\right\}\right\}$

So the area is given by the following integral:

In[ ]:= **Integrate[RealAbs[f[x] – g[x]], {x, sol⟦1, 1, 2⟧, sol⟦2, 1, 2⟧}]**

Out[ ]= $\frac{89 \sqrt{89}}{24}$

Here is a plot of the two functions with the enclosed area:

In[ ]:= **Plot[{f[x], g[x]}, {x, sol⟦1, 1, 2⟧, sol⟦2, 1, 2⟧}, Filling → {1 → {2}}]**

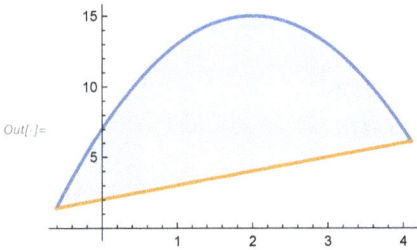

Out[ ]=

Exercise 3—General Area

Find the area between the two curves given by the following two functions in the range [0, 5]:

In[ ]:= **f[x_] := .5 Sin[x / 3 + 4]**
    g[x_] := Cos[2 x – 1]

Solution

The area is given by the following integral:

In[ ]:= **Integrate[RealAbs[f[x] – g[x]], {x, 0, 5}] // Timing**

Out[ ]= **{16.9672, 3.79789}**

The function Timing shows that the integral took several seconds to calculate. Here is a plot of the function:

In[ ]:= **Plot[{f[x], g[x]}, {x, 0, 5}, Filling → {1 → {2}}, PlotLegends → "Expressions"]**

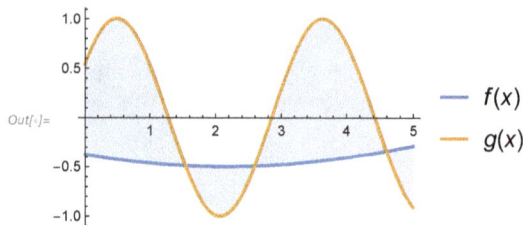

Out[ ]=

Find the places where the functions intersect each other, and use those points and the preceding plot to calculate the area more quickly:

In[ ]:= **sol = Solve[f[x] == g[x] && 0 ≤ x ≤ 5, x] // Quiet**

Out[ ]= **{{x → 1.54154}, {x → 2.59778}, {x → 4.6008}}**

In[ ]:= **Integrate[g[x] – f[x], {x, 0, sol[[1, 1, 2]]}] +**
    Integrate[f[x] – g[x], {x, sol[[1, 1, 2]], sol[[2, 1, 2]]}] +
    Integrate[g[x] – f[x], {x, sol[[2, 1, 2]], sol[[3, 1, 2]]}] +
    Integrate[f[x] – g[x], {x, sol[[3, 1, 2]], 5}] // Timing

Out[ ]= **{0.080868, 3.79789}**

Now the integral took less than a second to calculate.

Exercise 4—Implicit Area

Find the area enclosed between the two curves given as functions of y:

In[ ]:= **f[y_] := y + 2**
    g[y_] := y^2 – y

Solution

First, plot the function with ParametricPlot:

In[ ]:= **ParametricPlot[{{f[y], y}, {g[y], y}}, {y, –3, 4}]**

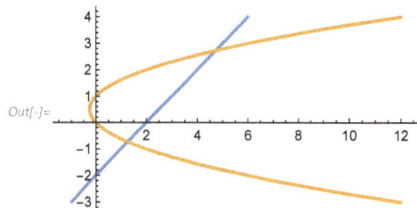

Out[ ]=

f is always to the right of g in the enclosed region, so the integral is $\int (f[y] - g[y]) \, d\, y$.

Now find the places where the curves intersect:

In[ ]:= **sol = Solve[f[y] == g[y], y]**

Out[ ]= $\left\{\left\{y \rightarrow 1 - \sqrt{3}\right\}, \left\{y \rightarrow 1 + \sqrt{3}\right\}\right\}$

So the endpoints are $1 - \sqrt{3}$ and $1 + \sqrt{3}$.

The integral is:

In[ ]:= **Integrate[f[y] – g[y], {y, sol〚1, 1, 2〛, sol〚2, 1, 2〛}] // N**

Out[ ]= 6.9282

Exercise 5—Population

The birth rate of a population is given by the following function:

In[ ]:= **birth[$t\_$] := 4000 + 67.2 t + 0.43 t^2**

The death rate of a population is given by the following function:

In[ ]:= **death[$t\_$] := 2400 + 47 t + 0.32 t^2**

Here t is in years. Find the total change in population between year 0 and year 20.

Solution

Use Integrate:

In[ ]:= **Integrate[RealAbs[birth[t] – death[t]], {t, 0, 20}]**

Out[ ]= 36 333.3

So approximately 36,333 people in total were added to the population between those years.

Here are the plots of the functions:

In[ ]:= **Plot[{birth[t], death[t]}, {t, 0, 20}, Filling → {1 → {2}}]**

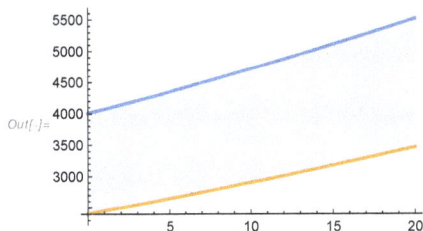

29 | Volumes of Solids

Overview

So far, you have only found the area under curves with the tools of integral calculus:

In[ ]:= **Integrate[x^2, {x, 2, 3}]**

Out[ ]= $\dfrac{19}{3}$

You can also use integral calculus to find the volume of solids!

For example, you can calculate the volume of the following solid:

In[ ]:= **RegionPlot3D[x^2 + y^2 ≤ 1, {x, −2, 2}, {y, −2, 2}, {z, 0, 4}, Boxed → False, Axes → False]**

Out[ ]=

This lesson will show one method for finding the volume of solids and give examples.

Cylinders

The solid in the introduction is a **cylinder**.

Generally, cylinders are constructed by taking a plane region with area B in the xy-plane and sweeping out volume by amount h (called the height) to a parallel plane in xyz-space.

For example, the cylinder in the introduction can be made by taking a disk in the xy-plane:

In[ ]:= **RegionPlot[x^2 + y^2 ≤ 1, {x, −2, 2}, {y, −2, 2}]**

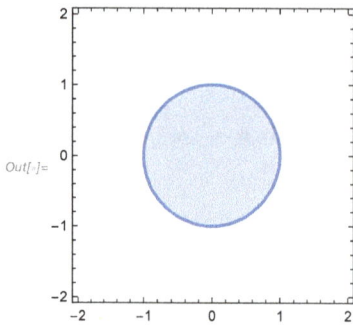

Out[ ]=

and sweeping out volume to the plane $z = 4$:

In[ ]:= **RegionPlot3D[x^2 + y^2 ≤ 1, {x, −2, 2}, {y, −2, 2}, {z, 0, 4}]**

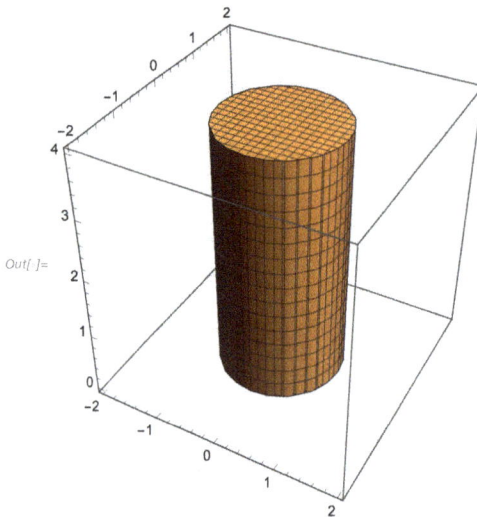

Out[ ]=

Cylinder Volume

To find the volume of a cylinder, you take the base area B and multiply by the height h to get volume Bh.

The cylinder has a disk as a plane region. The area of a disk is just πr^2, where r is the radius. So the disk has area π. The volume of the cylinder is therefore 4π, since the height is 4.

The Wolfram Language lets you find the area and volume of various shapes and solids with **Area** and **Volume**.

To represent the disk, use Disk:

In[ ]:= **disk = Disk[]**

Out[ ]= Disk[{0, 0}]

With Area the area of the disk is:

In[ ]:= **Area[disk]**

Out[ ]= π

To represent the cylinder, use Cylinder:

In[ ]:= **cylinder = Cylinder[{{0, 0, 0}, {0, 0, 4}}, 1]**

Out[ ]= Cylinder[{{0, 0, 0}, {0, 0, 4}}, 1]

With Volume, the volume of the cylinder is:

In[ ]:= **Volume[cylinder]**

Out[ ]= $4\,\pi$

There are many more shapes and solids covered in the Wolfram Language.

General Volume

While there are standard equations for the volume of simple solids like triangular prisms or square pyramids, it is sometimes desired to find the area of more general solids.

For example, consider the following region:

In[ ]:= **RegionPlot3D[−x ≤ y ≤ x && −x ≤ z ≤ x, {x, 1, 3}, {y, −3, 3}, {z, −3, 3}, PlotPoints → 100]**

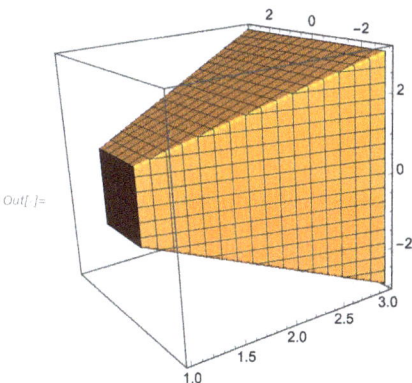

If you slice the region into pieces like a piece of bread, you will find that the area of each piece is $2\,x * 2\,x = 4\,x^2$.

If each piece has width Δx, then each piece has volume $4\,x^2\,\Delta x$. If you add up the volumes of all the pieces, you will get the volume $\int 4\,x^2\,dx$.

In this case, the endpoints are 1 and 3, so the volume for the region is:

In[·]:= **Integrate[4 x^2, {x, 1, 3}]**

Out[·]= $\dfrac{104}{3}$

Volume Formula

For a general solid with cross-sectional area $A[x]$, going from $x = a$ to $x = b$, the volume of the solid is $\int_a^b A[x]\,dx$:

$$\text{volume} = \int_a^b A(x)\,dx$$

Use this to find the volume of a ball.

Here is a picture of a general ball centered at the origin:

For a ball of radius r, each cross section is a disk with radius $\sqrt{r^2 - x^2}$.

So the volume of the ball centered at the origin with radius r can be calculated with the integral $\int_{-r}^{r} \pi\left(r^2 - x^2\right)\,dx$:

In[·]:= **Integrate[π (r^2 − x^2), {x, −r, r}]**

Out[·]= $\dfrac{4\,\pi\,r^3}{3}$

Volume agrees:

In[·]:= **Volume[Ball[{0, 0, 0}, r]]**

Out[·]= $\dfrac{4\,\pi\,r^3}{3}$

Solids of Revolution

Now you can make solids using functions!

Consider the following function:

In[ ]:= **f[x_] := x^2**

If you take the graph of the function and revolve it around the *x* axis, you get a solid.

You can do this in the Wolfram Language with RevolutionPlot3D:

In[ ]:= **RevolutionPlot3D[f[x], {x, −1, 1}, RevolutionAxis → {1, 0, 0}]**

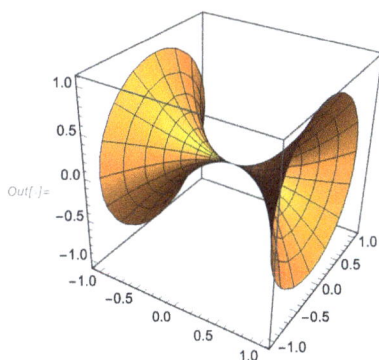

You can also revolve it around the *y* axis:

In[ ]:= **RevolutionPlot3D[f[x], {x, −1, 1}, RevolutionAxis → {0, 0, 1}]**

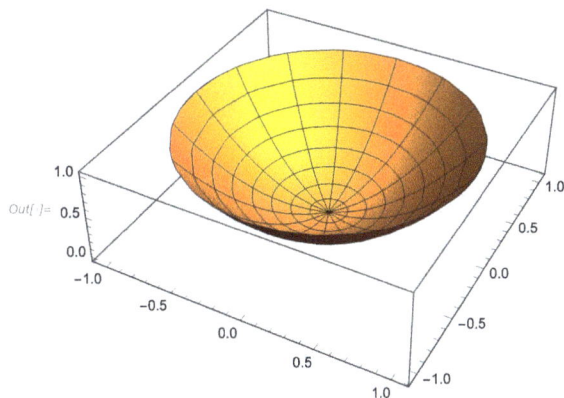

Volume of a Solid of Revolution—Disk Method

When you make a solid of revolution from a function $f[x]$, you are making a solid where each cross section is a disk.

Each disk has radius $f[x]$, so the volume of the solid from $x = a$ to b is $\int_a^b \pi \, f[x]^2 \, dx$.

So for the previous function revolved around the *x* axis, the volume from −1 to 1 is:

In[ ]:= **Integrate[π f[x]^2, {x, −1, 1}]**

Out[ ]= $\dfrac{2\pi}{5}$

Here is the plot of the original function:

In[ ]:= **Plot[f[x], {x, −1, 1}]**

Out[ ]=

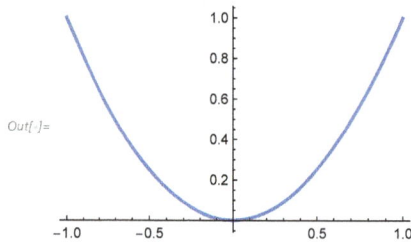

Here is the plot of the function revolved around the *x* axis:

In[ ]:= **RevolutionPlot3D[f[x], {x, −1, 1}, RevolutionAxis → {1, 0, 0}]**

Out[ ]=

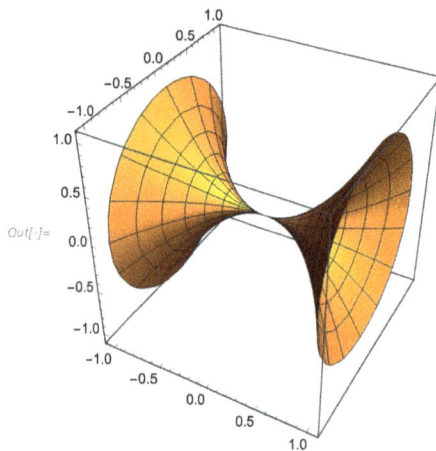

Changing the Axis 1

Suppose you want to revolve the function around the line $y = 1$ instead:

In[ ]:= **RevolutionPlot3D[f[x] – 1, {x, –1, 1}, RevolutionAxis → {1, 0, 0}]**

Out[ ]=

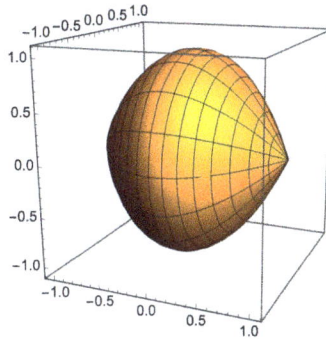

RevolutionAxis always revolves the function around the line $(0, 0, 0)$ and (x, y, z), so you have to shift the function down by one to get the picture you want.

In this case, the radius of each cross-sectional disk is now $1 - f[x]$.

So the volume is $\int_{-1}^{1} \pi (1 - f[x])^2 \, dx$:

In[ ]:= **Integrate[π (1 – f[x])^2, {x, –1, 1}]**

Out[ ]= $\dfrac{16\,\pi}{15}$

Changing the Axis 2

If you revolve around the line $y = -1$, the radius is now $f[x] - (-1) = f[x] + 1$:

In[·]:= **RevolutionPlot3D[f[x] + 1, {x, −1, 1}, RevolutionAxis → {1, 0, 0}]**

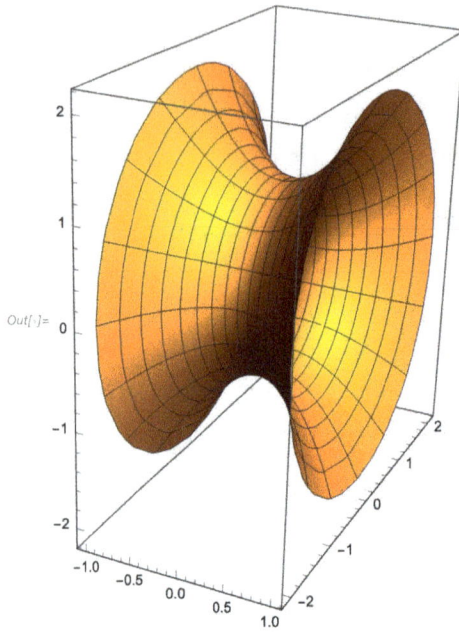

Out[·]=

So the volume is $\int_{-1}^{1} \pi (f[x] + 1)^2 \, dx$:

In[·]:= **Integrate[π (f[x] + 1)^2, {x, −1, 1}]**

Out[·]= $\dfrac{56 \pi}{15}$

Here you shift the function up by one to get the picture you want.

Washer Method Demonstration

Consider the area enclosed by the following two functions:

In[ ]:= **f[x_] := x**
g[x_] := x^3
Plot[{f[x], g[x]}, {x, 0, 1}]

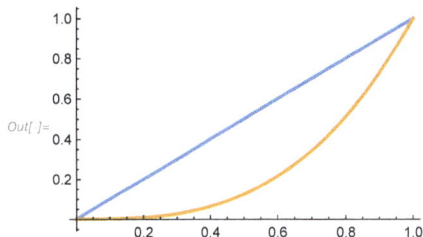

Out[ ]=

The solid made by revolving the enclosed area about the x axis is shown in the following Demonstration:

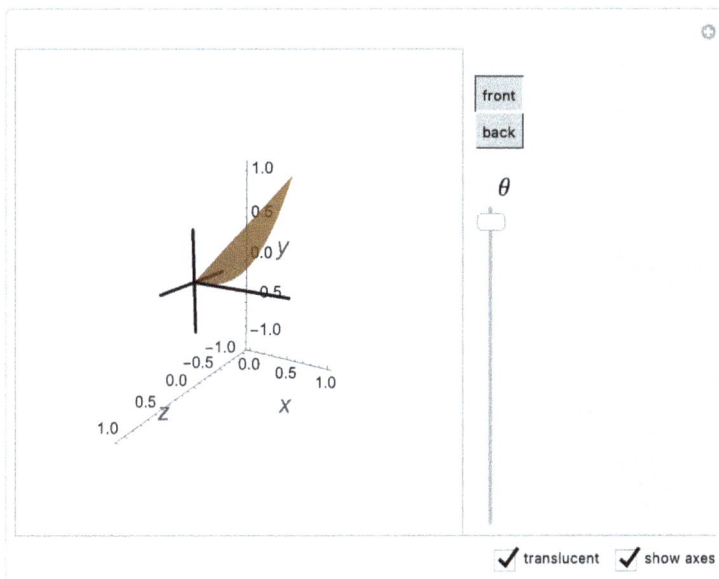

Calculating Volume with the Washer Method

In this case, the cross section is a washer instead of a disk, with outer radius $f[x]$ and inner radius $g[x]$.

The area of the washer is $\pi(f[x]^2 - g[x]^2)$, so the volume of the solid is $\int_0^1 \pi(f[x]^2 - g[x]^2)\,dx$:

In[·]:= **Integrate[π (f[x]^2 – g[x]^2), {x, 0, 1}]**

Out[·]= $\dfrac{4\pi}{21}$

If you revolve the region around the y axis, then the functions become:

In[·]:= **f[y_] := y**

g[y_] := CubeRoot[y]

$g[y]$ is always to the right of $f[y]$, so the radius of each washer is $g[y] - f[y]$ and the volume is $\int_0^1 \pi(g[y]^2 - f[y]^2)\,dy$:

In[·]:= **Integrate[π (g[y]^2 – f[y]^2), {y, 0, 1}]**

Out[·]= $\dfrac{4\pi}{15}$

Summary

Integration can be used to find the volume of solids.

For a general cylinder with base area B and height h, its volume is $B * h$.

For a general solid with cross-sectional area $A[x]$, its volume is $\int_a^b A[x]\,dx$.

The solid of revolution taken by revolving the function $f[x]$ around the line $y = y_0$ has volume $\int_a^b \pi(f[x] - y_0)^2\,dx$.

The solid of revolution taken by revolving the function $g[y]$ around the line $x = x_0$ has volume $\int_c^d \pi(g[y] - x_0)^2\,dy$.

The solid of revolution taken by revolving the area between the functions $f[x]$ and $g[x]$ around the line $y = y_0$ has volume $\int_a^b \pi((f[x] - y_0)^2 - (g[x] - y_0)^2)\,dx$ when $f[x] \geq g[x]$ on the interval $a \leq x \leq b$.

The solid of revolution taken by revolving the area between the functions $f[y]$ and $g[y]$ around the line $x = x_0$ has volume $\int_c^d \pi((f[y] - x_0)^2 - (g[y] - x_0)^2)\,dy$ when $f[y] \geq g[y]$ on the interval $c \leq y \leq d$.

The next lesson will give another method for finding the volume of solids, called the cylindrical shell method.

Exercises

Exercise 1—Cylinder

A cylinder with height 12 has its base to be the plane region given by the area enclosed between the two following functions in the range $[\pi/4,\, 5\,\pi/4]$:

$In[\cdot]:=$ **f[x_] := Sin[x]**
   g[x_] := Cos[x]

Find the volume of the cylinder.

Solution

Plot the functions in the range:

$In[\cdot]:=$ **Plot[{{f[x], g[x]}, {x, π/4, 5 π/4}, Filling → {1 → {2}}, PlotLegends → "Expressions"]**

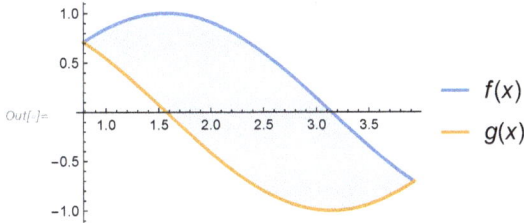

$f[x]$ is greater than $g[x]$ in the range. The area is given by Integrate:

$In[\cdot]:=$ **Integrate[f[x] – g[x], {x, π/4, 5 π/4}]**

$Out[\cdot]=$ $2\ \sqrt{2}$

Therefore, the volume is $2\ \sqrt{2} * 12 = 24\ \sqrt{2}$.

Exercise 2—Triangular Cross-Section

Find the volume of a solid whose cross sections are equilateral triangles with side length x in the range $[0,\, 4]$.

Solution

An equilateral triangle with side length x has area $\frac{1}{2}\left(x * x\ \frac{\sqrt{3}}{2}\right)$:

$In[\cdot]:=$ **equitrianglearea[x_] := x^2 * Sqrt[3] / 4**

So the volume is $\int_0^4 \frac{\sqrt{3}}{4} x^2 \, dx$:

In[·]:= **Integrate[equitrianglearea[x], {x, 0, 4}]**

Out[·]= $\dfrac{16}{\sqrt{3}}$

Here is a plot of the solid:

In[·]:= **RegionPlot3D[−x/2 ≤ y ≤ x/2 && 0 ≤ z ≤ 8 x − 16 y − 8 z && 0 ≤ z ≤ 8 x + 16 y − 8 z,**
{x, 0, 4}, {y, −2, 2}, {z, 0, 4}, PlotPoints → 100]

Out[·]=

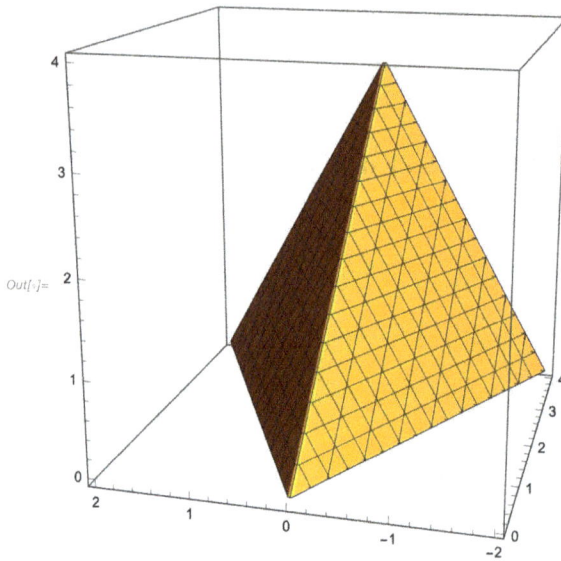

Exercise 3—X Washer

Find the volume of the solid created by rotating the region bounded by
$f[x] = x^2 / 3, \; g[x] = 4 - x^2$ around the x axis.

Solution

Define the functions:

In[·]:= **f[x_] := x^2/3**
g[x_] := 4 − x^2

Find the intersections with Solve:

In[·]:= **Solve[f[x] == g[x], x]**

Out[·]= $\left\{\left\{x \to -\sqrt{3}\right\}, \left\{x \to \sqrt{3}\right\}\right\}$

Then plot the functions with Plot:

In[ ]:= **Plot[{f[x], g[x]}, {x, –Sqrt[3], Sqrt[3]}, PlotLegends → "Expressions"]**

Out[ ]=

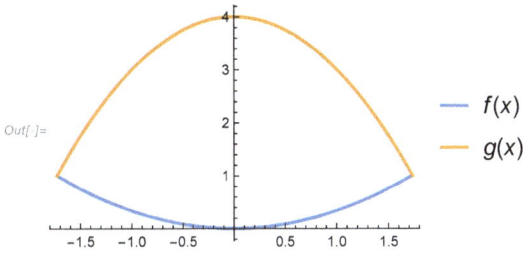

$g[x] \geq f[x]$, so the integral is $\int_{-\sqrt{3}}^{\sqrt{3}} \pi(g[x]^2 - f[x]^2)\,dx$:

In[ ]:= **Integrate[π (g[x]^2 – f[x]^2), {x, –Sqrt[3], Sqrt[3]}]**

Out[ ]= $\dfrac{96\sqrt{3}\,\pi}{5}$

Exercise 4—Y Disk

Find the volume of the solid created by rotating the region bounded by $y^{5/3} = x - 2$, $y = 4$ and $y = 7$ around the y axis.

Solution

Define the function:

In[ ]:= **f[y_] := y^(5/3) + 2**

The endpoints of the integral are 4 and 7. Here is a plot of the function:

In[ ]:= **ParametricPlot[{{f[y], y}, {y, 3, 8}, GridLines → {None, {4, 7}}]**

Out[ ]=

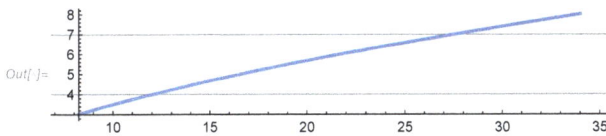

The integral is $\int_4^7 \pi f[y]^2\,d\,y$:

In[ ]:= **Integrate[π f[y]^2, {y, 4, 7}] // N**

Out[ ]= 3727.86

Exercise 5—Ball in a Bowl

A bowl shaped like a hemisphere with diameter 40 cm contains an iron ball with diameter 20 cm. Water is poured to a depth of l cm. Find the volume of water in the bowl.

Solution

Here is the plot to illustrate:

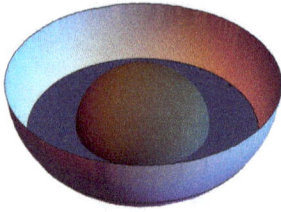

You know the volume of the iron ball is just $\frac{4}{3}\pi(10)^3$:

In[·]:= **vol1 = 4 π / 3 * 10^3**

Out[ ]= $\dfrac{4000\,\pi}{3}$

Now find the volume of the bowl at the given depth. Do it by splitting the volume into cross sections, each being a disk.

The radius of each disk is $\sqrt{20^2 - x^2}$, so the integral is:

In[·]:= **vol2 = Integrate[π (400 − x^2), {x, −20, l}]**

Out[ ]= $\dfrac{16\,000\,\pi}{3} + 400\,l\,\pi - \dfrac{l^3\,\pi}{3}$

So the volume of water in the bowl is the difference of the two:

In[·]:= **vol2 − vol1**

Out[ ]= $4000\,\pi + 400\,l\,\pi - \dfrac{l^3\,\pi}{3}$

30 | Volumes by Cylindrical Shells

Overview

You previously found the volume of solids using the disk and washer method. The cross sections were disks and washers, respectively.

For example, the volume of the solid made by rotating the function $f[x] = x^2$ around the x axis from 1 to 4 is given:

In[ ]:= `Integrate[π (x^2)^2, {x, 1, 4}]`

Out[ ]:= $\dfrac{1023\,\pi}{5}$

And the plot is this:

In[ ]:= `RevolutionPlot3D[x^2, {x, 0, 5}, RevolutionAxis → "X", BoxRatios → {1, 1, 1}]`

Out[ ]:=

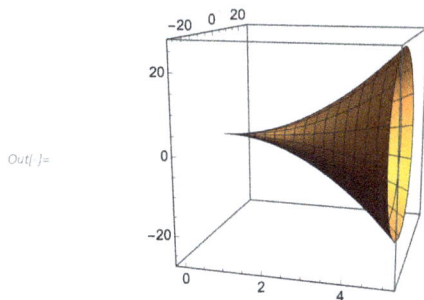

With the washer method, the main idea is to take the volume of one disk and subtract the volume of another disk.

However, consider the following function:

In[ ]:= `Plot[4 x^3 − x^4, {x, 0, 4}]`

Out[ ]:=

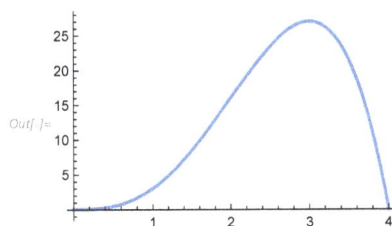

If you rotate it around the y axis, it is not clear what the inner and outer radii are.

The goal of this lesson is to find a way to find the volume in such circumstances with the **cylindrical shell method**.

Intuition

To find the volume, you will still divide the solid into pieces, but the pieces will now be shells instead of cross sections. Each shell will look like a cylinder, which is why the method is called the cylindrical shell method.

Here is a plot to illustrate:

In[·]:= **RevolutionPlot3D[{1, t}, {t, 0, 1}]**

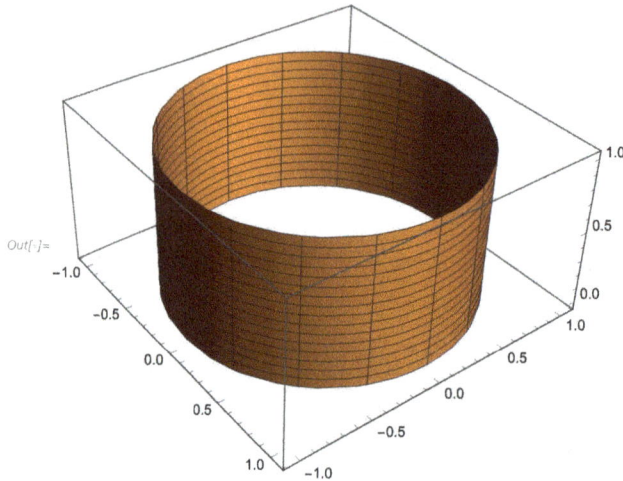

The volume of the given shell is $\pi r_2^2 h - \pi r_1^2 h = 2 \pi h \frac{(r_2 + r_1)}{2} (r_2 - r_1)$ for $r_2 \geq r_1$. If you let $r_2 \to r_1$, then their difference gets small (call it dr) and their average becomes r_1.

Therefore, letting $r_1 = r$, the volume becomes $2 \pi h r\, dr$.

Let f the function be the height, and make it a function of r. If you add up multiple shells, the volume becomes $\int 2 \pi r\, f[r]\, dr$.

Example

Now look back to the introduction.

You wanted to find the volume of the solid made by rotating the following function from 0 to 4 around the y axis:

In[·]:= **f[x_] := 4 x^3 − x^4**

Here is a plot of the solid with RevolutionPlot3D:

In[ ]:= **RevolutionPlot3D[f[x], {x, 0, 4}, RevolutionAxis → "Z", BoxRatios → {1, 1, 1}]**

Out[ ]=

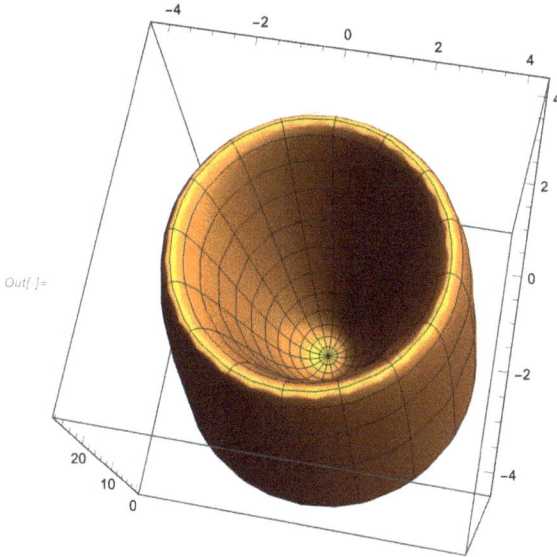

With the cylindrical shell method, the volume is $\int_0^4 2\,\pi\,r\left(4\,r^3 - r^4\right) dr$:

In[ ]:= **Integrate[2 π r (4 r^3 – r^4), {r, 0, 4}]**

Out[ ]= $\dfrac{4096\,\pi}{15}$

Washer versus Cylindrical Shell

Find the volume of the solid made by rotating the region enclosed by x^2 and x^3 about the y axis.

Here is the region in the $x\,y$ plane:

In[ ]:= **Plot[{x^2, x^3}, {x, 0, 1}, Filling → {1 → {2}}, PlotLegends → "Expressions"]**

Out[ ]=

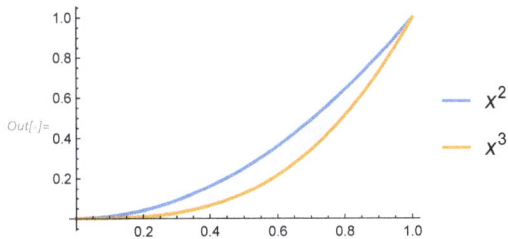

With the washer method, the volume is $\int_0^1 \pi \left(\left(\sqrt[3]{y} \right)^2 - \left(\sqrt{y} \right)^2 \right) d\,y$:

In[·]:= **Integrate[π (y^(2/3) – y), {y, 0, 1}]**

Out[·]= $\dfrac{\pi}{10}$

With the cylindrical shell method, the volume is $\int_0^1 2\,\pi\,x \left(x^2 - x^3 \right) dx$:

In[·]:= **Integrate[2 π x (x^2 – x^3), {x, 0, 1}]**

Out[·]= $\dfrac{\pi}{10}$

The volume is the same with both methods, so it is up to you to determine which to use.

Positive Axis

Suppose you rotated the following function in the range $[0,\,4]$ about the line $x = 6$:

In[·]:= **f[x_] := x^2 – 2 x + 2**

Here it is with RevolutionPlot3D:

In[·]:= **RevolutionPlot3D[$\begin{cases} \text{f[x + 6]} & -6 \le x \le -2 \\ \text{Null} & \text{True} \end{cases}$, {x, –6, 6}, RevolutionAxis → {0, 0, 1}]**

Out[·]=

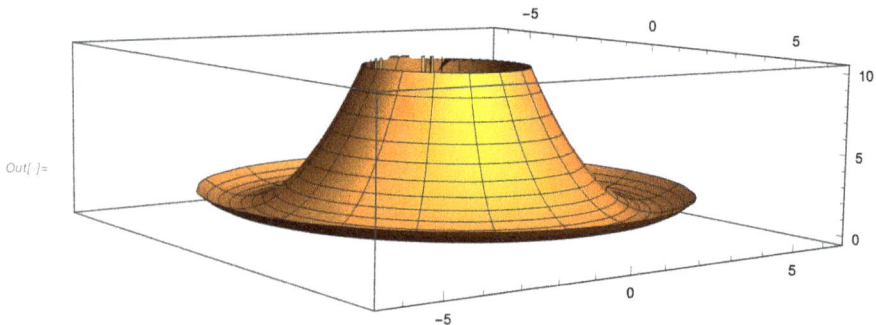

To find volume under the surface, you first need to find the radius. Before, the radii were centered at the origin. Now they are centered at $x = 6$.

Look at the 2D plot of the function:

$In[\,\cdot\,]:=$ **Plot[** $\left\{ \begin{array}{ll} \mathsf{f[x]} & \mathsf{0 \le x \le 4} \\ \mathsf{Null} & \mathsf{True} \end{array} \right.$, **{x, 0, 8},** $\boxed{\cdots \; \oplus}$ **]**

$Out[\,\cdot\,]=$

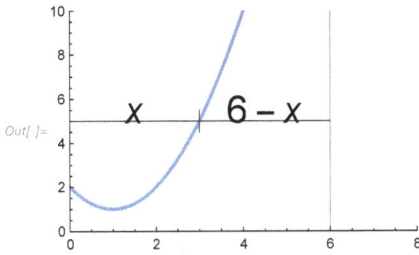

A cylindrical shell centered at 6 would have radius equal to $(6 - x)$, so the volume is $\int_0^4 2\,\pi(6 - x)\,f[x]\,dx$:

$In[\,\cdot\,]:=$ **Integrate[2 π (6 $-$ x) f[x], {x, 0, 4}]**

$Out[\,\cdot\,]=$ $\dfrac{256\,\pi}{3}$

Negative Axis

Suppose you rotated the following function in the range $[0, 3]$ about the line $x = -1$:

$In[\,\cdot\,]:=$ **f[x_] := x^3 $-$ 3 x^2 + 2 x + 1**

Here it is with RevolutionPlot3D:

$In[\,\cdot\,]:=$ **RevolutionPlot3D[** $\left\{ \begin{array}{ll} \mathsf{f[x-1]} & \mathsf{1 \le x \le 4} \\ \mathsf{Null} & \mathsf{True} \end{array} \right.$, **{x, $-$4, 4}, RevolutionAxis $\rightarrow$ {0, 0, 1}]**

$Out[\,\cdot\,]=$

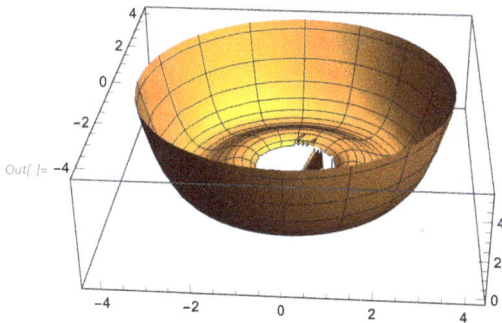

As before, you need to find the new radius since the shells are centered at -1.

Look at the 2D plot of the function:

$In[\cdot]:=$ **Plot[** $\left\{ \begin{array}{ll} f[x] & 0 \le x \le 3 \\ Null & True \end{array} \right.$ **, {x, -2, 3},** $\cdots$ **]**

$Out[\cdot]=$

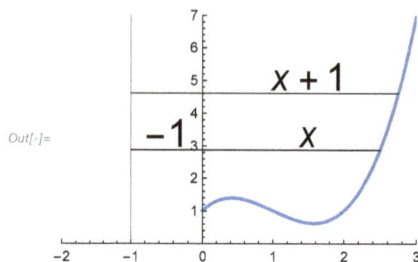

A cylindrical shell centered at -1 would have radius equal to $x - (-1) = x + 1$, so the volume is $\int_0^4 2\pi(x+1) f[x] \, dx$:

$In[\cdot]:=$ **Integrate[2 π (x + 1) f[x], {x, 0, 3}]**

$Out[\cdot]=$ $\dfrac{156\pi}{5}$

Washer over Cylindrical 1

Sometimes the washer/disk method is preferable to the cylindrical method. Consider the following function:

$In[\cdot]:=$ **f[x_] := CubeRoot[x]**

If you revolve around the x axis from 0 to 1, you get the following solid:

$In[\cdot]:=$ **RevolutionPlot3D[{x, f[x]}, {x, 0, 1}, RevolutionAxis → "X"]**

$Out[\cdot]=$

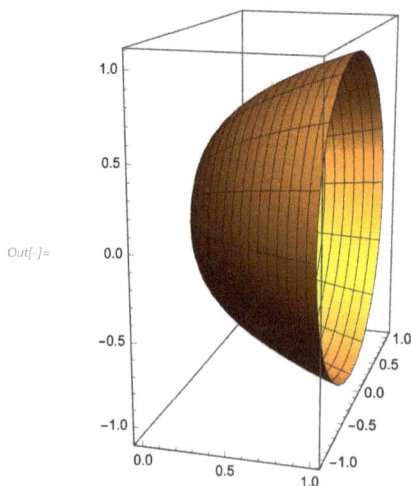

Find the volume using the cylindrical shell method.

The height of each shell will now go left to right instead of top to bottom as shown in this 2D plot:

In[ ]:= **Plot[f[x], {x, 0, 1},** ··· → ··· ✦ **]**

Out[ ]=

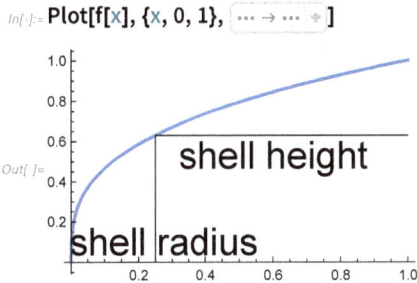

The shell radius goes from top to bottom instead of left to right.

Washer over Cylindrical 2

In this case, you need to express the function in terms of y to set up the integral properly:

In[ ]:= **fy[y_] := y^3**

The shell radius is y and the shell height is $1 - y^3$ (right curve is 1). So the integral is $\int_0^1 2\pi y \left(1 - y^3\right) d\,y$:

In[ ]:= **Integrate[2 π y (1 – fy[y]), {y, 0, 1}]**

Out[ ]= $\dfrac{3\pi}{5}$

If you had used the disk method, you know the radius of each disk is $\sqrt[3]{x}$, so the equivalent integral would be $\int_0^1 \pi \left(\sqrt[3]{x}\right)^2 d\,x$:

In[ ]:= **Integrate[π f[x]^2, {x, 0, 1}]**

Out[ ]= $\dfrac{3\pi}{5}$

Both the cylindrical shell method and washer/disk method have their advantages and disadvantages. It is up to you to determine which is better suited to each problem.

Summary

The cylindrical shell method gives another way to find the volume of solids.

The method works by dividing the solid into many cylindrical shells with infinitesimal width and height a given function.

Sometimes the cylindrical shell method can find volumes that the disk/washer method would make too cumbersome/impossible.

Other times, the disk/washer method is more intuitive and easier to use than the cylindrical shell method.

For a function revolved around the line $x = x_0$ where $x_0 > 0$, the volume is given by $\int_a^b 2\pi(x_0 - x)\, f[x]\, dx$.

For a function revolved around the line $x = x_0$ where $x_0 \leq 0$, the volume is given by $\int_a^b 2\pi(x - x_0)\, f[x]\, dx$.

The next lesson will use the integration techniques to calculate the average value of a function.

Exercises

Exercise 1—Trigonometric Function

Find the volume of the solid created by rotating the following function around the x axis in the range $\left[0,\ \sqrt[3]{\pi/2}\ \right]$:

In[ ]:= **f[x_] := x Sin[x^3] Cos[x^3]**

Solution

Here is the function's plot:

In[ ]:= **Plot[f[x], {x, 0, CubeRoot[(π/2)]}]**

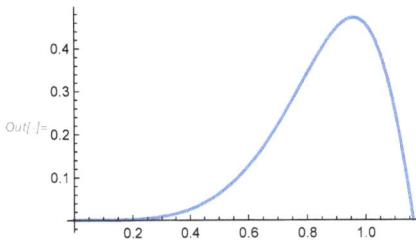

Using the cylindrical shell method, the radius of each shell will be x and the height will be $f[x]$.

Therefore, the volume is $\int_0^{\sqrt[3]{\pi/2}} 2\,\pi\,x\,f[x]$:

In[ ]:= **Integrate[2 π x f[x], {x, 0, CubeRoot[π/2]}] // Simplify**

Out[ ]= $\dfrac{\pi}{3}$

Exercise 2—Positive Axis

The following function is revolved around the line $x = 3$ in the range $[1, 2]$:

In[ ]:= **f[x_] := x**

Find the volume of the solid using the cylindrical shell method.

Solution

The radius of each shell will be $3 - x$. The height will be $f[x]$.

Therefore, the volume is $\int_1^2 2\pi (3 - x) f[x] \, dx$:

In[·]:= **Integrate[2 π (3 – x) f[x], {x, 1, 2}]**

Out[·]= $\dfrac{13\pi}{3}$

Here is a plot of the top of the region:

In[·]:= **RevolutionPlot3D[Piecewise[{{f[x + 3], –2 ≤ x ≤ –1}}, Null], {x, –3, 3}]**

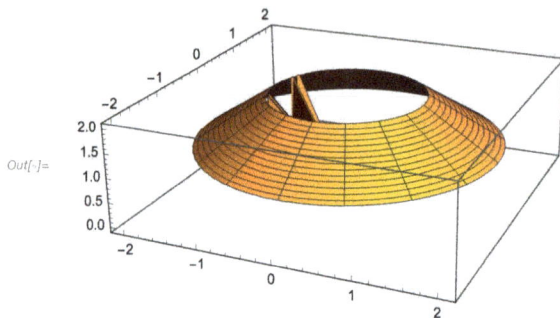

Exercise 3—Revolve around Y

Use the cylindrical method to calculate the volume of the region bounded by the line $y = 2$ and the following function as it is revolved around the line $y = 5$ in the range [0, 4]:

In[·]:= **f[x_] := 1 / (1 + x^2)**

Solution

Plot the region:

In[·]:= **Show[Plot[{f[x], 2, 5}, {x, 0, 4}, Filling → {1 → {2}}], Graphics[Line[{{0, 1}, {4, 1}}]]]**

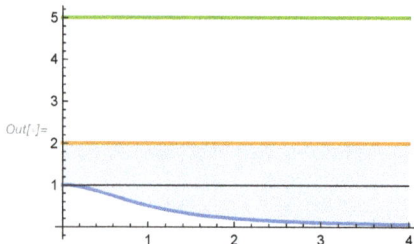

The volume can be calculated by dividing the region into two parts. The volume of the top region is easy since the height of each cylinder is $4 - 0 = 4$, the outer radius is $5 - 1 = 4$, and the inner radius is $5 - 2 = 3$. So the volume is $\pi(4^2 - 3^2) 4 = 28\pi$.

To find the volume of the bottom region, solve the function for x in terms of y:

In[ ]:= **Solve[f[x] == y, x]**

Out[ ]= $\left\{\left\{x \rightarrow -\dfrac{\sqrt{1-y}}{\sqrt{y}}\right\}, \left\{x \rightarrow \dfrac{\sqrt{1-y}}{\sqrt{y}}\right\}\right\}$

The radius will be $5 - y$ and the height will be $4 - \dfrac{\sqrt{1-y}}{\sqrt{y}}$. The endpoints of the region are $y = \dfrac{1}{1+4^2} = \dfrac{1}{17}$ to $y = 1$. So the total volume is:

In[ ]:= **28 π + Integrate[2 π (5 − y) (4 − Sqrt[1 − y] / Sqrt[y]), {y, 1/17, 1}] // N**

Out[ ]= **161.862**

Exercise 4—Volume of a Sphere

Use the cylindrical shell method to find the volume of a sphere of radius r.

Solution

Draw a sphere:

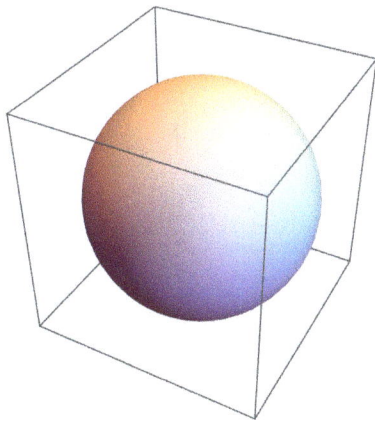

The region you will be integrating over is 0 to r. The radius of each shell is simply x. To find the height, divide the sphere into two hemispheres.

The top hemisphere has the equation $f[x] = \sqrt{r^2 - x^2}$. The bottom hemisphere has the equation $g[x] = -\sqrt{r^2 - x^2}$.

So the volume is $\int_0^r 2\pi x \left(f[x] - g[x] \right) dx$:

In[·]:= **Integrate[2 π x (2 Sqrt[r^2 − x^2]), {x, 0, r}, Assumptions → r > 0]**

Out[·]= $\dfrac{4\pi r^3}{3}$

Exercise 5—Pontoon

Pontoons are used to keep heavy objects afloat. A certain pontoon is designed by rotating the graph of the following function about the x axis in the range $[-5, 5]$:

In[·]:= **pontoon[x_] := 1 − x^2/25**

Here is a graph of the pontoon:

In[·]:= **RevolutionPlot3D[pontoon[x], {x, −5, 5}, RevolutionAxis → "X"]**

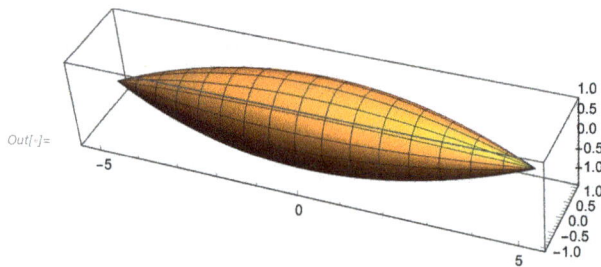

Find the volume of the pontoon using the cylindrical shell method.

Solution

First, convert the function to be in terms of y:

In[·]:= **Solve[pontoon[x] == y, x]**

Out[·]= $\left\{ \left\{ x \to -5\sqrt{1-y} \right\}, \left\{ x \to 5\sqrt{1-y} \right\} \right\}$

The two values give the left and right portions of the pontoon. The radius of each shell of the pontoon is just y. The height of each shell of the pontoon is $5\sqrt{1-y} - \left(-5\sqrt{1-y} \right) = 10\sqrt{1-y}$.

Therefore, the volume of the pontoon is $\int_0^1 2\pi y * 10\sqrt{1-y}\, dy$:

In[·]:= **Integrate[2 π y * 10 Sqrt[1 − y], {y, 0, 1}]**

Out[·]= $\dfrac{16\pi}{3}$

31 | Average Value of a Function

Overview

Suppose you had n values and wanted to find their average value:

```
In[ ]:= list = {1, 3, 4, 3, 5, 7, 10, 2, 11, 13};
```

You would add up all the values and divide by n:

```
In[ ]:= (1 + 3 + 4 + 3 + 5 + 7 + 10 + 2 + 11 + 13) / 10
```

$$Out[ ]= \frac{59}{10}$$

The Wolfram Language has a special function that does this for you, called `Mean`:

```
In[ ]:= Mean[list]
```

$$Out[ ]= \frac{59}{10}$$

Now instead of a list of values, suppose you had a function. Here is an example function with its plot:

```
In[ ]:= f[x_] := x^2 Sin[x] + x
```

```
In[ ]:= Plot[f[x], {x, 0, 5}]
```

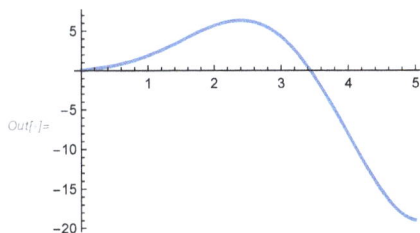

How do you find the "average value" of this function in the range [0, 5]?

The goal of this lesson is to express this problem mathematically and find a way to calculate the average value.

Intuition

To start, consider the function $f[x]$ as running from a to b.

To approximate the average, divide the region into n subintervals, take a sample point from each subinterval, and calculate the average from the function values of the sample points.

Each subinterval will have length $\frac{b-a}{n}$, which will be denoted by Δx. The sample point in the ith subinterval will be x_i^*.

So the average will be $\frac{f[x_1^*]+f[x_2^*]+...+f[x_n^*]}{n} = \Sigma_{i=1}^n \frac{f[x_i^*]}{n}$.

Since $\Delta x = \frac{b-a}{n}$, $n = \frac{b-a}{\Delta x}$, the sum becomes $\Sigma_{i=1}^n \frac{f[x_i^*]}{\frac{b-a}{\Delta x}} = \Sigma_{i=1}^n \frac{f[x_i^*]}{b-a} \Delta x$.

When you let n go to ∞, you see that you just get an integral for f going from a to b that is divided by $b - a$.

In other words:

$$\text{Average Value of f from a to b} = \frac{\int_a^b f(x)\,dx}{b-a}$$

So the average value of the function in the introduction is the following expression:

In[ ]:= **Integrate[f[x], {x, 0, 5}] / 5 // N**

Out[ ]= **−1.12269**

Mean Value Theorem

If the function is continuous, then it will take on its average value at some point in the interval integrated over. This is called the **mean value theorem for integrals**.

Formally, it says that if f is a continuous function on $[a, b]$, then there is at least one number c in the interval such that the following equation holds:

$$f(c) = \frac{\int_a^b f(x)\,dx}{b-a}$$

Multiplying both sides by $b - a$, you get the equation:

$$f(c)\,(b-a) = \int_a^b f(x)\,dx$$

The mean value theorem for integrals is similar to the mean value theorem for derivatives, which is concerned with slopes.

Here is a visual interpretation:

In[ ]:= Plot[{x^2 + 1, Evaluate[1/2 $\int_0^2$ (x^2 + 1) dx]}, {x, 0, 2}, Filling → Axis, PlotRange → {0, 5}]

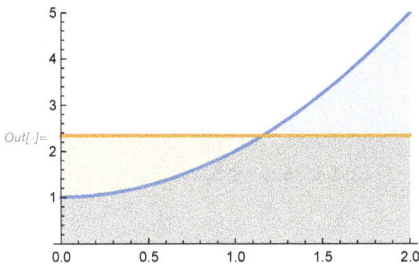

When f is positive, the mean value theorem for integrals shows that there is a point c in the interval $[a, b]$ such that the area of the rectangle with height $f[c]$ and base $b - a$ has the **same area** as the region under the graph of f from a to b.

Mean Value Example

Find the value c in the interval $[-2, 3]$ that has function value equal to the average value of the following function in the interval $[-2, 3]$:

In[ ]:= f[x_] := x^3 + x

The function is continuous, so you can use the mean value theorem for integrals.

Use Solve to find c in the desired interval:

In[ ]:= sol = Solve[f[x] == Integrate[f[x], {x, −2, 3}]/(3 − (−2)) && −2 ≤ x ≤ 3, x] // N

Out[ ]= {{x → 1.34061}}

Here is a plot of the function, the line $x = c$ and the average value for the function in the interval:

In[ ]:= Plot[{f[x], f[sol[1, 1, 2]]}, {x, −2, 3}, PlotRange → All,
    Filling → Axis, Epilog → {Black, PointSize[Large], Point[{x, f[x]} /. sol]}]

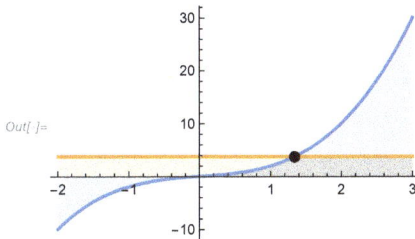

Two Points

The average value of a function can intersect the graph at more than one point.

Consider the following function:

In[ ]:= **f[x_] := x^2 − 3 x + 3**

Find the average value in the interval [−5, 5]:

In[ ]:= **sol = Solve[f[x] == Integrate[f[x], {x, −5, 5}] / (5 − (−5)) && −5 ≤ x ≤ 5, x]**

Out[ ]= $\left\{\left\{x \to \frac{1}{6}\left(9 - \sqrt{381}\right)\right\}, \left\{x \to \frac{1}{6}\left(9 + \sqrt{381}\right)\right\}\right\}$

Here is a plot to illustrate:

In[ ]:= **Plot[{f[x], f[sol[[1, 1, 2]]]}, {x, −5, 5}, PlotRange → All,**
   Filling → Axis, Epilog → {Black, PointSize[Large], Point[{x, f[x]} /. sol]}]

Out[ ]=

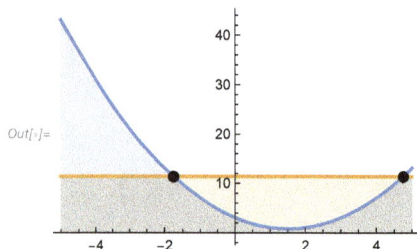

Temperature

If you have a graph that gives the temperature at time t, the mean value theorem for integrals states that over the interval $[t_1, t_2]$, there is some point t^* that has temperature equal to the average temperature over the interval. This assumes the graph is continuous.

Here is a function that gives the temperature for a certain city t hours after noon:

In[ ]:= **temp[t_] := 70 + 24 Cos[π t / 24]**

Here is its graph from 1pm to 3am the next day:

In[ ]:= **Plot[temp[t], {t, 1, 15}]**

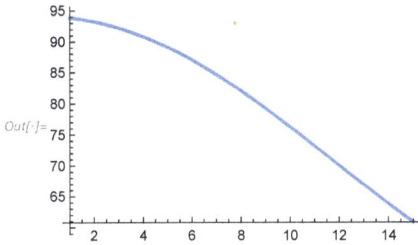

Find the time at which the average temperature equals the temperature:

In[ ]:= **Solve[temp[t] == Integrate[temp[t], {t, 1, 15}]/(15 − 1) && 1 ≤ t ≤ 15, t] // N**

Out[ ]= {{t → 8.57966}}

So at 8:34:46.8pm, the temperature will be equal to the average temperature over the time period from 1pm to 3am the next day.

Velocity

The mean value theorem for integrals also shows something about velocity.

Suppose a particle moves one-dimensionally with position function $s[t]$ and velocity function $v[t]$:

In[ ]:= **Clear[s, v];**

In[ ]:= **v[t] = s'[t];**

The average value of the velocity on the interval $[t_1, t_2]$ is clearly the difference in position divided by the difference in time:

In[ ]:= **avgvelo = (s[t2] − s[t1])/(t2 − t1)**

Out[ ]= $\dfrac{-s[t1] + s[t2]}{-t1 + t2}$

Confirm it by using the mean value theorem for integrals on the derivative of $s[t]$, $v[t]$:

In[ ]:= **SetAttributes[s, {NumericFunction}]; Integrate[v[t], {t, t1, t2}]/(t2 − t1)**

Out[ ]= $\dfrac{-s[t1] + s[t2]}{-t1 + t2}$

Here you explicitly say s is a NumericFunction with SetAttributes in order to evaluate the integral properly.

Car on the Highway

A car moving along the highway has the following velocity function after t minutes:

In[·]:= **carvelocity[t_] := 65 + 5 Cos[π t / 30]**

Use the mean value theorem for integrals to find the average velocity over the time period 15 minutes to 60 minutes:

In[·]:= **avg = Integrate[carvelocity[t], {t, 15, 60}] / (60 − 15) // N**

Out[·]= **63.939**

Here is a graph of the function and average velocity in the given time period:

In[·]:= **Plot[{carvelocity[t], avg}, {t, 15, 60}, PlotLegends → "Expressions"]**

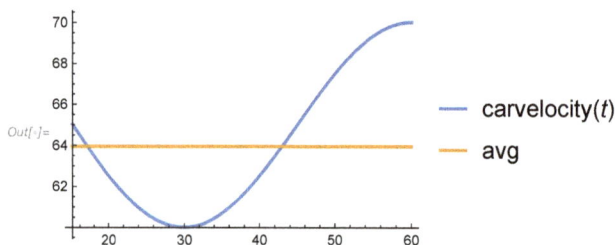

You can see the car's instantaneous velocity is the average velocity twice in the interval.

Summary

With integration, you can find the average value of a function over an interval.

For the interval $[a, b]$, the average value of the function $f[x]$ is $\frac{1}{b-a} \int_a^b f[x]\, dx$.

The mean value theorem for integrals shows that if f is a continuous function, then it takes on its average value in the interval $[a, b]$ at least once in the interval.

Practically speaking, this means that continuous processes like the temperature over the course of a day or the velocity of a moving particle take on their average values at least once in the intervals they are taking place over.

The next lesson will go over ways to approximate the value of an integral.

Exercises

Exercise 1—Average Value

Find the average of the following function in the range [4, 7]:

In[·]:= **f[x_] :=** x**^2 − 0.2** x**^3 + 3** x**^4**

Solution

The average value is $\frac{1}{7-4} \int_4^7 f[x] \, dx = \frac{1}{3} \int_4^7 f[x] \, dx$:

In[·]:= **avg = Integrate[f[x], {x, 4, 7}] / 3**

Out[·]= **3151.85**

Here is a graph of the function and its average value:

In[·]:= **Plot[{f[x], avg}, {x, 4, 7}, PlotRange → All, PlotLegends → "Expressions", Filling → Axis]**

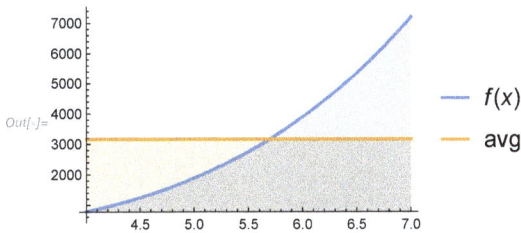

Exercise 2—Mean Value Theorem

Find the point at which the average value over the interval [−5, −2] equals the function value for the following function:

In[·]:= **f[x_] :=** x **/ (** x **^2 − 1)**

Solution

Since the function has no discontinuities in the interval [−5, −2], use the mean value theorem for integrals:

In[·]:= **sol = Solve[f[c] == Integrate[f[x], {x, −5, −2}] / (−2 − (−5)) && −5 ≤ c ≤ −2, c] // N**

Out[·]= **{{c → −3.19808}}**

Here is a plot of the function and its average value:

In[·]:= **Plot[{f[x], f[sol[[1, 1, 2]]]}, {x, −5, −2}, Filling → Axis,**
Epilog → {Black, PointSize[Large], Point[{c, f[c]}] /. sol}]

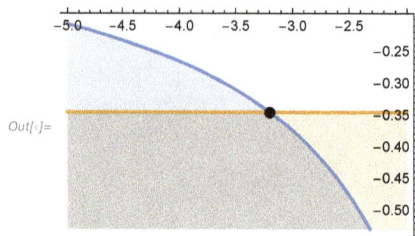

Out[·]=

Exercise 3—Average Current

A switch is turned on, and the current flowing through the switch is modeled by the following function:

In[·]:= **f[t_] := t Cos[t^2 / 15] − Sin[t / 7.5] + t**

If t is in terms of minutes and the switch is left on for 30 minutes, find the average current going through the switch from minute 15 to minute 30.

Solution

The average current through the switch in the given interval is
$\frac{1}{30-15} \int_{15}^{30} f[t]\, dt = \frac{1}{15} \int_{15}^{30} f[t]\, dt$:

In[·]:= **avg = Integrate[f[t], {t, 15, 30}] / 15 // N**

Out[·]= **21.9037**

Here is a plot of the current and its average value:

In[·]:= **Plot[{f[t], avg}, {t, 15, 30}, Filling → Axis]**

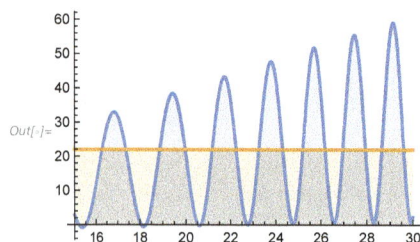

Out[·]=

Exercise 4—Average Cost

A company's cost is modeled by the following function, where x is the number of products sold:

$$\text{In[ ]:= } \textbf{cost[}x\textbf{\_] := 0.0003 } x\textbf{^2 + 304 + 16 Sin[}x\textbf{/25]}$$

If the company produces 500 units, at what point(s) did the average cost per unit equal the cost per unit?

Solution

Since the cost function is continuous, use the mean value theorem for integrals to find the points:

$$\text{In[ ]:= } \textbf{sol = Solve[cost[c] == Integrate[cost[x], \{x, 0, 500\}] / 500 \&\& 0 } \leq \textbf{ c } \leq \textbf{ 500, c] // Quiet}$$

$$\text{Out[ ]= } \{\{c \rightarrow 187.259\}, \{c \rightarrow 215.466\}, \{c \rightarrow 309.133\}\}$$

So when the company made its 188th, 216th and 310th products, the average cost per unit equaled the cost per unit.

Here is a plot of the cost function with its average value, along with the given points:

$$\text{In[ ]:= } \textbf{Plot[\{cost[x], cost[sol[[1, 1, 2]]]\}, \{x, 0, 500\}, PlotRange } \rightarrow \textbf{ \{250, 400\},}$$
$$\textbf{Filling } \rightarrow \textbf{ Bottom, Epilog } \rightarrow \textbf{ \{PointSize[Large], Point[\{c, cost[c]\}] /. sol\}]}$$

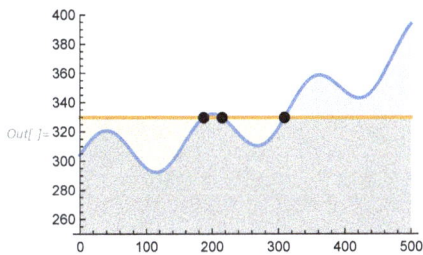

Exercise 5—Hydraulic Press

A hydraulic cylinder pushes down in a press, and the force of a particular one is given by the following function:

$$\text{In[ ]:= } \textbf{hydraulicforce[}x\textbf{\_] := 200 Sec[}x\textbf{]^2}$$

where x is the distance (m) the cylinder is extended in its cycle. Assuming the domain is $[0, \pi/3]$, find the average force of the press over its domain.

Solution

The average force is

$$\frac{1}{\pi/3} \int_0^{\pi/3} 200 \, \text{Sec}[x]^2 \, dx = \frac{3}{\pi} \int_0^{\pi/3} 200 \, \text{Sec}[x]^2 \, dx = \frac{600}{\pi} \left(\text{Tan}\left[\frac{\pi}{3}\right] - \text{Tan}[0] \right) = \frac{600 \sqrt{3}}{\pi}:$$

In[ ]:= **600 Sqrt[3] / π // N**

Out[ ]= **330.797**

Confirm with Integrate:

In[ ]:= **3 / π * Integrate[200 Sec[x]^2, {x, 0, π/3}] // N**

Out[ ]= **330.797**

Here is a 2D interactive representation that gives the force of the cylinder on a certain gas:

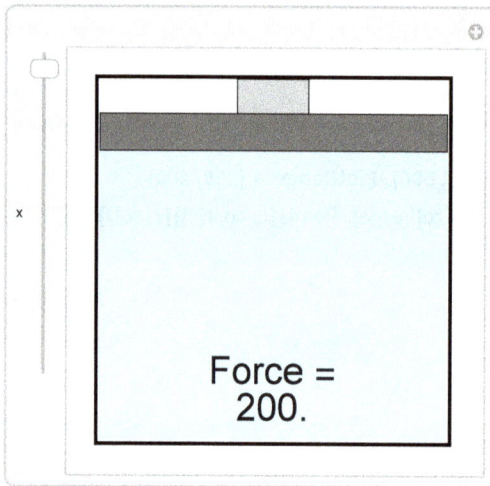

32 | Approximate Integration

Overview

You have been using integration to calculate the area under curves.

For many functions this is possible, and a closed form for the indefinite integral can be calculated:

$In[\ ]:=$ **Integrate[x^2, x]**

$Out[\ ]=$ $\dfrac{x^3}{3}$

The integral of x^2 from 2 to 5 is $\dfrac{5^3-2^3}{3}$:

$In[\ ]:=$ **{(5^3 – 2^3)/3, Integrate[x^2, {x, 2, 5}]}**

$Out[\ ]=$ **{39, 39}**

Some functions have no closed form for their integrals:

$In[\ ]:=$ **Integrate[Exp[Exp[x^2]], {x, 0, 1}]**

$Out[\ ]=$ $\displaystyle\int_0^1 e^{e^{x^2}}\, dx$

The function above $\left(e^{e^{x^2}}\right)$ has no closed form for its integral.

This lesson will go over some ways to evaluate such integrals **numerically**.

Left and Right Approximations

In the beginning lessons about integration, the region being integrated over was often split into n rectangles, then you calculated the area of each rectangle and added up the areas to approximate the area of the region.

The area obtained took the form $\Sigma_{i=1}^{n}\ f[x_i^*]\,\Delta x$, where $\Delta x = \dfrac{b-a}{n}$ for the region from a to b, and x_i^* is some sample point in the ith subinterval $[a + (i – 1)\,\Delta x,\ a + i\,\Delta x]$.

The area obtained took the form $\Sigma_{i=1}^{n}\ f[x_i^*]\,\Delta x$, where $\Delta x = \dfrac{b-a}{n}$ for the region from a to b, and x_i^* is some sample point in the ith subinterval $[a + (i – 1)\,\Delta x,\ a + i\,\Delta x]$.

Visually, when $x_i^* = a + (i – 1)\,\Delta x$, the left endpoint of the ith subinterval, the rectangles had their upper-left corners lying on the curve. When $x_i^* = a + i\,\Delta x$, the right endpoint of the ith subinterval, the rectangles had their upper-right corners lying on the curve.

In both cases, you assumed the function was positive. Here are some example plots to illustrate:

In[ ]:= {Show[Plot[x^2 + 1, {x, 0, 1}], DiscretePlot[x^2 + 1, {x, 0, 1, 1/10}, ExtentSize → Right]],
Show[Plot[x^2 + 1, {x, 0, 1}], DiscretePlot[x^2 + 1, {x, 0, 1, 1/10}, ExtentSize → Left]]}

Out[ ]=
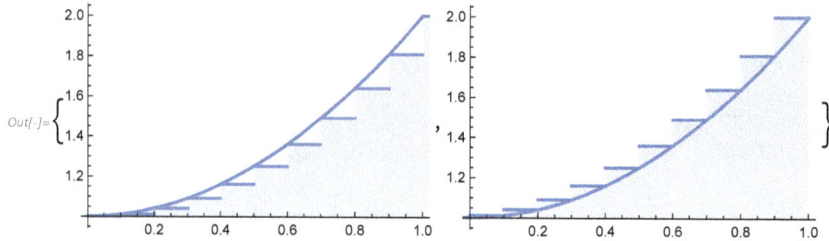

The left plot shows left-aligned plots. The right plot shows right-aligned plots.

The area for each plot is $\Sigma_{i=1}^{10}\left(1 + \left(\frac{i-1}{10}\right)^2\right)\Delta x$ and $\Sigma_{i=1}^{10}\left(1 + \left(\frac{i}{10}\right)^2\right)\Delta x$, respectively:

In[ ]:= {left = Sum[1 + ((i − 1)/10)^2, {i, 1, 10}]/10, right = Sum[1 + ((i)/10)^2, {i, 1, 10}]/10} // N

Out[ ]= {1.285, 1.385}

The actual area is:

In[ ]:= Integrate[x^2 + 1, {x, 0, 1}] // N

Out[ ]= 1.33333

Midpoint Rule

When you let $x_i{}^* = \frac{x_{i-1}+x_i}{2}$, the midpoint of the ith subinterval, the rectangles had the middle of the top go through the curve.

Here is a plot to illustrate:

In[ ]:= Show[Plot[x^2 + 1, {x, 0, 1}], DiscretePlot[x^2 + 1, {x, 0, 1, 1/10}, ExtentSize → Full]]

Out[ ]=
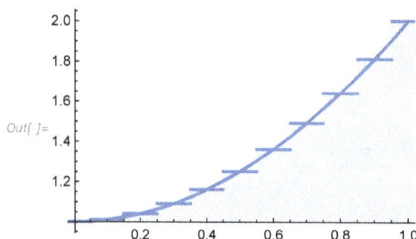

When you used these sample points, you got a better approximation for the area:

In[ ]:= middle = Sum[1 + (((i − 1)/10 + i/10)/2)^2, {i, 1, 10}]/10 // N

Out[ ]= 1.3325

Trapezoidal Rule

In the approximations using left- and right-aligned rectangles, one was above the desired integral, while the other was below:

In[ ]:= **{left, right, Integrate[1 + x^2, {x, 0, 1}]} // N**

Out[ ]= {1.285, 1.385, 1.33333}

If you **averaged** the two results, you got a closer approximation:

In[ ]:= **{(right + left)/2, Integrate[1 + x^2, {x, 0, 1}]} // N**

Out[ ]= {1.335, 1.33333}

Now generalize this. The average of the two approximations symbolically is $(\Sigma_{i=1}^{n} f[x_{i-1}] \Delta x + \Sigma_{i=1}^{n} f[x_i] \Delta x)/2$. These two sums can be put together to make the sum $\frac{1}{2} \Sigma_{i=1}^{n} (f[x_{i-1}] + f[x_i]) \Delta x$.

By taking the average of the left endpoint and right endpoint in each subinterval and multiplying by the width, you get the area of the **trapezoid** with top going from the left endpoint to the right endpoint. As a result, the preceding approximation is called the **trapezoidal rule**.

You can see in the following plot that the rectangles have now become trapezoids:

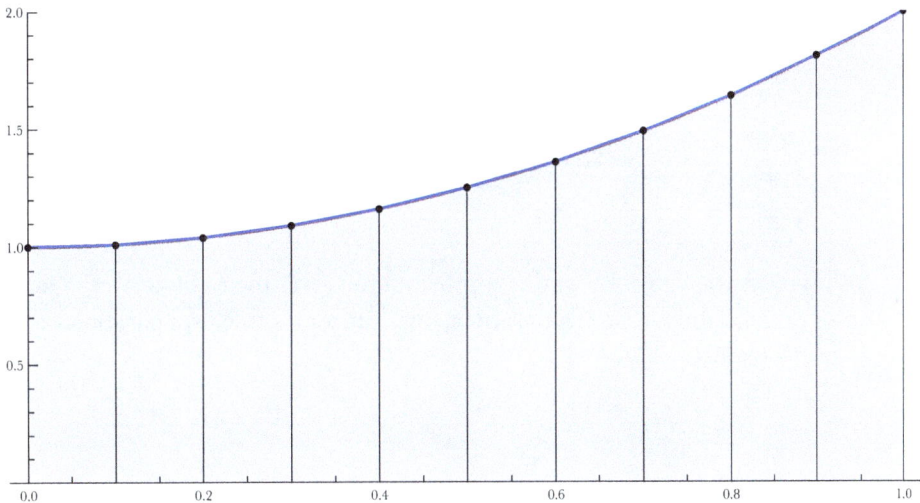

Simpson's Rule

You could also approximate the areas of the regions with polynomials. When the approximations use parabolas, you have what is called **Simpson's rule**, named after the English mathematician Thomas Simpson.

Although there are many ways to derive Simpson's rule, each has a rather involved formulation. Therefore, the derivation will be skipped.

To approximate the area under a curve with Simpson's rule, you use the formula $\frac{1}{3} \sum_{i=1}^{n/2} (f[x_{2\,i-2}] + 4\,f[x_{2\,i-1}] + f[x_{2\,i}])\,\Delta x$. Note that the number of subregions must be **even** in order to use Simpson's rule.

Here is the approximation you get for the previous region using 10 subregions and Simpson's rule:

In[ ]:= **f[x_] := x^2 + 1**

In[ ]:= **{Sum[(f[(2 i − 2)/10] + 4 f[(2 i − 1)/10] + f[2 i/10]), {i, 1, 5}]/30, Integrate[f[x], {x, 0, 1}]}**

Out[ ]= $\left\{\dfrac{4}{3}, \dfrac{4}{3}\right\}$

In this case, you can see that the approximation gives the right answer! This makes sense because the region is a parabola, and Simpson's rule uses parabolas to approximate the region.

Error Bounds

Since you are approximating the integrals in this section, naturally some error will be included in the calculations.

For the midpoint rule, the error is at most $\frac{K(b-a)^3}{24\,n^2}$, where the region goes from a to b, n rectangles are used, and $K \geq |\,f''[x]\,|$, assuming you are integrating the function $f[x]$.

For the trapezoidal rule, the error is at most $\frac{K(b-a)^3}{12\,n^2}$. Notice that the maximum error for the midpoint rule is generally half as much as the maximum error for the trapezoidal rule.

The maximum error for Simpson's rule is $\frac{L(b-a)^5}{180\, n^4}$ if the region goes from a to b, n subregions are used, and $L \geq |f^{(4)}[x]|$, the fourth derivative of the function f.

Since $f[x] = x^2 + 1$ in the previous examples, $f''[x] = 2$, so the maximum error using the midpoint rule and the trapezoidal rule is:

In[ ]:= `{midpointerror = 2 (1 – 0)^3/(24 * 10^2), trapezoiderror = 2 (1 – 0)^3/(12 * 10^2)} // N`

Out[ ]= `{0.000833333, 0.00166667}`

The actual errors were:

In[ ]:= `{RealAbs[Integrate[f[x], {x, 0, 1}] – middle],`
`        RealAbs[Integrate[f[x], {x, 0, 1}] – (left + right)/2]} // N`

Out[ ]= `{0.000833333, 0.00166667}`

Since $f^{(4)}[x] = 0$, the maximum error using Simpson's rule is 0, so there is no error.

NIntegrate

The Wolfram Language also has a way of numerically calculating integrals, using NIntegrate.

For example, the integral in the introduction is calculated:

In[ ]:= `NIntegrate[Exp[Exp[x^2]], {x, 0, 1}]`

Out[ ]= `4.91065`

Here is a plot of the area:

In[ ]:= `Plot[Exp[Exp[x^2]], {x, 0, 1}, Filling → Axis]`

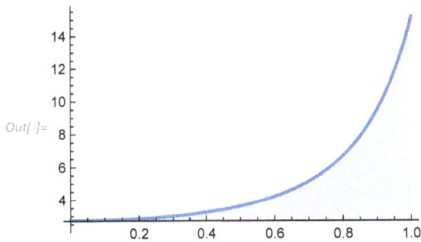

Generally the methods used by NIntegrate are optimized in order to get the best possible answer. As a result, the answers may not be the same answers you would get by hand.

In a later lesson, you will use the Wolfram Language to make your own functions that approximate integrals using the methods you have learned.

Table

Sometimes a function may not have an explicit formula and instead be given via a table.

So long as the function values do not vary too quickly, you can still use the approximate integration techniques to approximate the area for the function.

For example, here is a table that gives the speed of a particle at 11 instances of time:

time (min)	0	1	2	3	4	5	6	7	8	9	10
speed (miles per minute)	45.8	42.3	47.8	50.2	44.2	41.9	38.7	36.8	40.5	39.6	43.7

With the trapezoidal rule, the total distance traveled is:

In[ ]:= $(45.8 + 43.7 + 2 (42.3 + 47.8 + 50.2 + 44.2 + 41.9 + 38.7 + 36.8 + 40.5 + 39.6)) (10 - 0)/(2 * 10)$

Out[ ]:= 426.75

With Simpson's rule, the total distance traveled is:

In[ ]:= $(45.8 + 43.7 + 4 (42.3 + 50.2 + 41.9 + 36.8 + 39.6) + 2 (47.8 + 44.2 + 38.7 + 40.5)) (10 - 0)/(3 * 10)$

Out[ ]:= 425.033

Note that for the trapezoidal rule, the coefficients for the points are 1, 2, 2, ..., 2, 1 and for Simpson's rule they are 1, 4, 2, 4, 2, ..., 2, 4, 1.

Here is a plot of the data:

In[ ]:= **time = {0, 1, 2, 3, 4, 5, 6, 7, 8, 9, 10};**
speed = {45.8`, 42.3`, 47.8`, 50.2`, 44.2`, 41.9`, 38.7`, 36.8`, 40.5`, 39.6`, 43.7`};

In[ ]:= **ListLinePlot[Transpose[{time, speed}], Filling → Axis]**

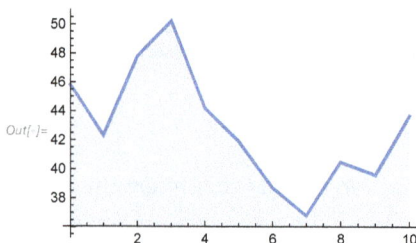

Summary

Even when you cannot find the integral of a function with Integrate, you can still approximate the integral.

Use techniques such as left-aligned rectangles, right-aligned rectangles, the midpoint rule, the trapezoidal rule and Simpson's rule.

NIntegrate is also useful, and gives the best approximation it can come up with when evaluating an integral.

You can also use the approximate integration techniques to approximate the integrals of functions given only a table of their values.

The next lesson will go over exponential functions and some of their properties.

Exercises

Exercise 1—Left, Right and Midpoint Approximations

Use left-aligned rectangles, right-aligned rectangles, and the midpoint rule with 10 rectangles to calculate the integral of the following function from 0 to $\sqrt{2\pi}$:

In[·]:= **f[x_] := Sin[x^2] + x**

Solution

With left-aligned rectangles, the answer is:

In[·]:= **Sum[f[(i − 1) Sqrt[2 π] / 10], {i, 1, 10}] Sqrt[2 π] / 10 // N**

Out[·]= **3.28481**

With right-aligned rectangles, the answer is:

In[·]:= **Sum[f[(i) Sqrt[2 π] / 10], {i, 1, 10}] Sqrt[2 π] / 10 // N**

Out[·]= **3.91312**

With the midpoint rule the answer is:

In[·]:= **Sum[f[((i − 1) + i) / 2 (Sqrt[2 π] / 10)], {i, 1, 10}] Sqrt[2 π] / 10 // N**

Out[·]= **3.55825**

The answer with NIntegrate is:

In[·]:= **NIntegrate[f[x], {x, 0, Sqrt[2 π]}]**

Out[·]= **3.572**

Exercise 2—Trapezoidal Rule

Use the trapezoidal rule with six subintervals to calculate the integral for the following function from -2 to 5:

In[·]:= **f[x_] := Sqrt[49 − x^2]**

Solution

The width of each subinterval is $\frac{5-(-2)}{6} = \frac{7}{6}$.

The answer with the trapezoidal rule is:

In[ ]:= **Sum[(f[−2 + 7 (i − 1)/6] + f[−2 + 7 i/6])/2, {i, 1, 6}] ∗ 7/6 // N**

Out[ ]= 45.3979

The answer with NIntegrate is:

In[ ]:= **NIntegrate[f[x], {x, −2, 5}]**

Out[ ]= 45.5468

Plot the function and trapezoids:

In[ ]:= **Show[Plot[f[x], {x, −2, 5}, ⋯ +],**
ListLinePlot[{Table[{i, f[i]}, {i, −2, 5}], Table[{i, 0}, {i, −2, 5}]}, ⋯ +]]

Out[ ]=

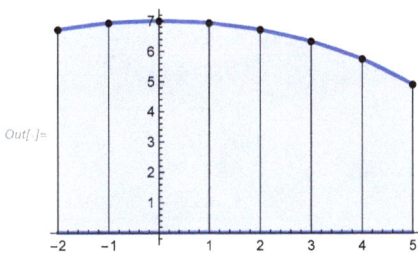

Exercise 3—Simpson's Rule

Use Simpson's rule with 14 subintervals to calculate the integral for the following function from −17 to −10:

In[ ]:= **f[x_] := Tan[(x + 2)/10.5] CubeRoot[x]**

Solution

The width of each subinterval is $\frac{-10-(-17)}{14} = \frac{1}{2}$.

The answer with Simpson's rule is:

In[ ]:= **Sum[f[−17 + (2 i − 2)/2] + 4 f[−17 + (2 i − 1)/2] + f[−17 + 2 i/2], {i, 1, 7}] ∗ 1/(3 ∗ 2) // N**

Out[ ]= 41.7434

The answer with NIntegrate is:

In[·]:= **NIntegrate[f[x], {x, −17, −10}]**

Out[·]:= 41.7339

Plot the function with the parabolas (using Wolfram | Alpha):

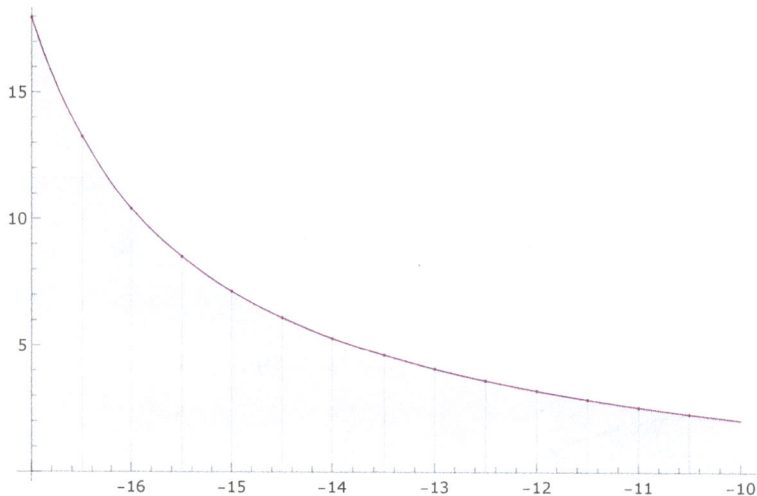

Exercise 4—Table

The following table gives the rate at which water flows out of a hose at nine instances of time:

time (seconds)	0	2	4	6	8	10	12	14	16
flow rate (pints per seconds)	0	0.9	1.42	1.54	1.6	1.67	1.64	1.66	1.69

Use the trapezoidal rule and Simpson's rule to estimate the total amount of water that flows out of the hose in the 16 seconds.

Solution

With the trapezoidal rule, the total water that flows out is:

In[·]:= **(0 + 1.69 + 2 (0.9 + 1.42 + 1.54 + 1.6 + 1.67 + 1.64 + 1.66)) (16 − 0)/(2 ∗ 8)**

Out[·]:= 22.55

With Simpson's rule, the total water that flows out is:

In[·]:= **(0 + 1.69 + 4 (0.9 + 1.54 + 1.67 + 1.66) + 2 (1.42 + 1.6 + 1.64)) (16 − 0)/(3 ∗ 8)**

Out[·]:= 22.7267

Here is a plot of the data:

```
In[ ]:= time = {0, 2, 4, 6, 8, 10, 12, 14, 16};
       hose = {0, 0.9`, 1.42`, 1.54`, 1.6`, 1.67`, 1.64`, 1.66`, 1.69`};
```

```
In[ ]:= ListLinePlot[Transpose[{time, hose}], Filling → Axis]
```

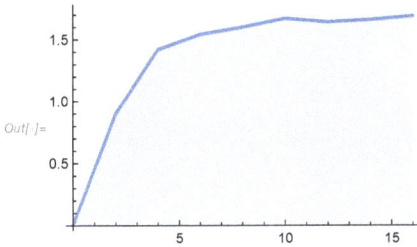

Out[ ]=

Exercise 5—Approximation of π

The following function, when integrated from 0 to 1, gives the value π:

```
In[ ]:= f[x_] := (16 x − 16) / (x^4 − 2 x^3 + 4 x − 4)
```

Approximate the integral using Simpson's rule and the trapezoidal rule with 10 subintervals each, and give the error for each.

Solution

The integral with the trapezoidal rule is:

```
In[ ]:= trapezoid = Sum[(f[(i − 1) / 10] + f[i / 10]) / 2, {i, 1, 10}] / 10 // N
```

Out[ ]= 3.12831

The integral with Simpson's rule is:

```
In[ ]:= simpson = Sum[f[(2 i − 2) / 10] + 4 f[(2 i − 1) / 10] + f[2 i / 10], {i, 1, 5}] / (3 ∗ 10) // N
```

Out[ ]= 3.14141

The integral with NIntegrate and Integrate is:

```
In[ ]:= {NIntegrate[f[x], {x, 0, 1}], Integrate[f[x], {x, 0, 1}]}
```

Out[ ]= {3.14159, π}

Here are the errors for the trapezoidal rule and Simpson's rule:

```
In[ ]:= {RealAbs[π − trapezoid], RealAbs[π − simpson]}
```

Out[ ]= {0.0132832, 0.000180852}

You can see that the error for the trapezoidal rule is two orders of magnitude greater than that for Simpson's rule, as shown in the equations for the maximum error.

33 | Exponential Functions

Overview

Past lessons have mostly covered polynomial functions:

In[ ]:= $f[x\_] := x\wedge3 + 6\,x\wedge2 - 9\,x + 1$

Rational functions:

In[ ]:= $g[x\_] := 2\,x / (x\wedge2 - 4)$

General algebraic functions:

In[ ]:= $h[x\_] := Sqrt[x\wedge3 / (x - 2)]$

And trigonometric functions:

In[ ]:= $p[x\_] := 10\,Sin[x] - Tan[x]$

Here are some graphs to illustrate:

In[ ]:= **Plot[{f[x], g[x], h[x], p[x]}, {x, −5, 5}, PlotLegends → "Expressions"] // Quiet**

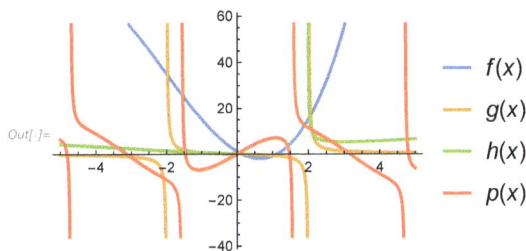

This lesson will focus attention on **exponential functions**.

Definition

In one of the earliest lessons, it was said that exponential functions generally have the form a^x, where $a \geq 0$.

When x is a positive integer, $a^x = a * a * \ldots * a$, where the right-hand side has x a's:

In[ ]:= $\{3\wedge4, 3*3*3*3\}$

Out[ ]= $\{81, 81\}$

When x is a negative integer, $a^x = \frac{1}{a^{-x}} = \frac{1}{a*\ldots*a}$, where the denominator has $-x$ a's:

In[·]:= **{3^−4, 1/(3∗3∗3∗3)}**

Out[·]= $\left\{\dfrac{1}{81}, \dfrac{1}{81}\right\}$

When $x = 0$, $a^x = 1$:

In[·]:= **3^0**

Out[·]= 1

When x is a fraction, i.e. $x = \frac{p}{q}$, $a^x = \sqrt[q]{a^p} = \left(\sqrt[q]{a}\right)^p$:

In[·]:= **{3^(4/5), Surd[3^4, 5], Surd[3, 5]^4} // N**

Out[·]= {2.40822, 2.40822, 2.40822}

In general, when r is rational, the limit of a^x as x approaches r is a^r:

In[·]:= **Limit[a^x, x → r, Assumptions → {a > 0, r ∈ Rationals}]**

Out[·]= a^r

Exponential functions grow faster than any polynomial function.

Graphs

When $a > 1$, the graph of a^x is increasing:

In[·]:= **Plot[3^x, {x, −5, 5}]**

Out[·]=

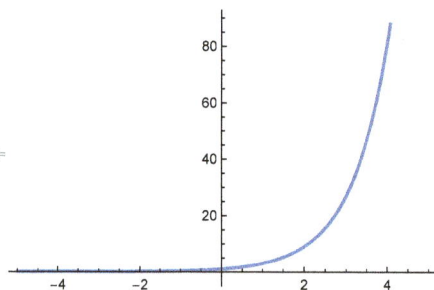

When $0 < a < 1$, the graph of a^x is decreasing:

In[ ]:= **Plot[(1/3)^x, {x, −5, 5}]**

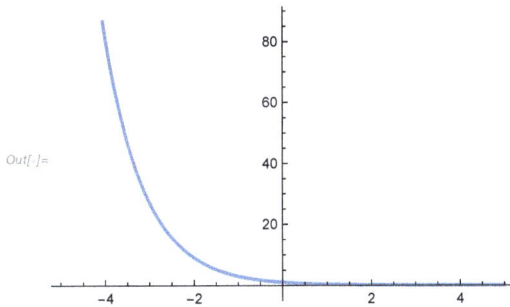

Out[ ]=

When $a = 1$, the graph is constant:

In[ ]:= **Plot[1^x, {x, −5, 5}]**

Out[ ]=

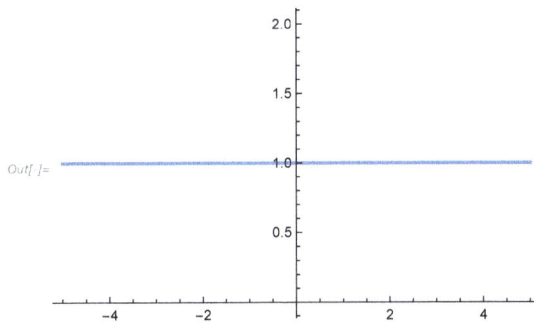

Laws of Exponents

If a, $b > 0$ and x, y are real numbers, then the following **laws of exponents** exist:

$a^{x+y} = a^x \, a^y$:

In[ ]:= **3^(x + y) == 3^x * 3^y**

Out[ ]= **True**

$a^{x-y} = a^x / a^y$:

In[ ]:= **3^(x − y) == 3^x / 3^y**

Out[ ]= **True**

$(a^x)^y = a^{x*y}$:

In[ ]:= **Simplify[(3^x)^y == 3^(x * y), {x, y} ∈ Reals]**

Out[ ]= **True**

$(a * b)^x = a^x * b^x$:

In[·]:= (3 * 5)^x == 3^x * 5^x

Out[·]= True

Function and Limit Properties

If $a > 0$ and $a \neq 1$, then $f[x] = a^x$ is a continuous function.

Its domain is all real numbers:

In[·]:= {FunctionDomain[3^x, x], FunctionDomain[(1/3)^x, x]}

Out[·]= {True, True}

Its range is $(0, \infty)$:

In[·]:= {FunctionRange[3^x, x, y], FunctionRange[(1/3)^x, x, y]}

Out[·]= {y > 0, y > 0}

From the graph of $f[x]$, you can see that if $a > 1$, then the limit of a^x as x approaches ∞ is ∞:

In[·]:= Limit[a^x, x → Infinity, Assumptions → a > 1]

Out[·]= ∞

The limit of a^x as x approaches $-\infty$ is 0:

In[·]:= Limit[a^x, x → −Infinity, Assumptions → a > 1]

Out[·]= 0

If $0 < a < 1$, then the limit of a^x as x approaches ∞ is 0:

In[·]:= Limit[a^x, x → Infinity, Assumptions → 0 < a < 1]

Out[·]= 0

The limit of a^x as x approaches $-\infty$ is ∞:

In[·]:= Limit[a^x, x → −Infinity, Assumptions → 0 < a < 1]

Out[·]= ∞

Derivatives

The derivative of a^x is **NOT** $x * a^{x-1}$. It is $a^x * \text{Log}[a]$, where $\text{Log}[a]$ is the natural logarithm of a:

In[ ]:= **D[a^x, x]**

Out[ ]= $a^x \, \text{Log}[a]$

The value of a that makes $\text{Log}[a] = 1$ has the special value e, in honor of the great mathematician Leonhard Euler:

In[ ]:= **Solve[Log[x] == 1, x]**

Out[ ]= $\{\{x \rightarrow e\}\}$

The Wolfram Language uses E to represent e. Exp can be used to represent e^x.

It is clear that the derivative of the function $f[x] = e^x$ is e^x:

In[ ]:= **D[Exp[x], x]**

Out[ ]= e^x

So at any point on the graph of f, you see that the slope is just e^x, the original function!

The derivative of the function $f[x] = e^{g[x]}$ is $e^{g[x]} * g'[x]$ by the chain rule:

In[ ]:= **Clear[g]**

In[ ]:= **D[Exp[g[x]], x]**

Out[ ]= $e^{g[x]} \, g'[x]$

Integrals

It is clear that the integral of e^x is $e^x + C$:

In[ ]:= **Integrate[Exp[x], x]**

Out[ ]= e^x

The integral of a^x is $a^x / \text{Log}[a]$:

In[ ]:= **Integrate[a^x, x]**

Out[ ]= $\dfrac{a^x}{\text{Log}[a]}$

Here is a graph of e^x with tangent lines going through the curve at various points:

```
In[·]:= f[x_] := Exp[x]
       g[x_] := f[E] + f'[E] (x − E)
       h[x_] := f[1/2] + f'[1/2] (x − 1/2)
       p[x_] := f[4] + f'[4] (x − 4)
       Plot[{f[x], g[x], h[x], p[x]}, {x, −5, 5}, PlotLegends → "Expressions"]
```

Out[·]=

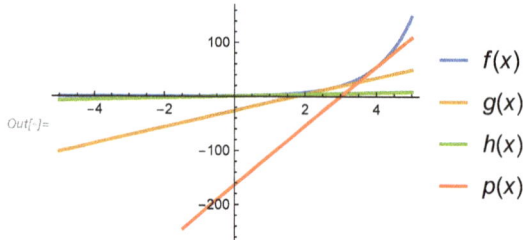

— $f(x)$
— $g(x)$
— $h(x)$
— $p(x)$

The Natural Number e

There are many ways to define e.

You already saw it is the number x that makes $\text{Log}[x] = 1$:

```
In[·]:= Solve[Log[x] == 1, x]
```

Out[·]= $\{\{x \to e\}\}$

It can also be defined with limits.

For example, using the difference quotient definition for derivatives, e^x is the exponential function with slope 1 at 0:

```
In[·]:= Solve[Limit[(x^h − 1)/h, h → 0] == 1, x]
```

Out[·]= $\{\{x \to e\}\}$

The limit of $(1 + 1/n)^n$ as n approaches ∞ is also e:

```
In[·]:= Limit[(1 + 1/n)^n, n → Infinity]
```

Out[·]= e

The area under the curve $f[x] = \frac{1}{x}$ from 1 to a is 1 when $a = e$:

```
In[·]:= Solve[Integrate[1/x, {x, 1, a}] == 1, a]
```

Out[·]= $\{\{a \to e\}\}$

Its value is approximately 2.71828:

In[ ]:= **E // N**

Out[ ]= **2.71828**

The number *e* shows up in a vast number of fields in mathematics, science and engineering.

Growth and Decay

Exponential functions are useful when modeling quantities that grow or decay at a constant rate.

For example, a bacteria colony that starts out with 10 members and doubles every hour can be modeled by the following function:

In[ ]:= **bacterialgrowth[*t*_] := 10 * 2 ^ *t***

In[ ]:= **Plot[bacterialgrowth[t], {t, 0, 5}]**

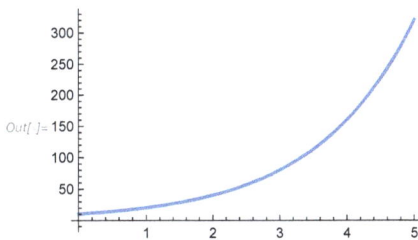

A radioactive material that starts out with 5000 atoms and has half of its members decay every hour can be modeled by the following function:

In[ ]:= **radioactivedecay[*t*_] := 5000 * (1 / 2) ^ *t***

In[ ]:= **Plot[radioactivedecay[t], {t, 0, 5}]**

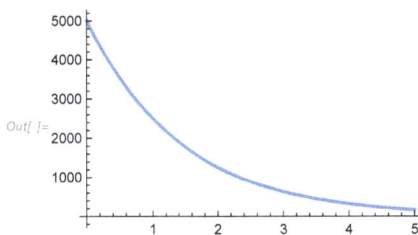

Summary

Exponential functions generally take the form a^x where $a > 0$.

When $a > 1$, the graph of a^x is increasing.

When $0 < a < 1$, the graph of a^x is decreasing.

Exponential functions have many unique properties, and follow the laws of exponents.

The derivative and integral of an exponential function are constant multiples of the exponential function.

The derivative and integral of the exponential function e^x are both e^x (plus C for the integral).

The number e is useful in many different fields.

Exponential functions are good for modeling growth and decay.

The next lesson will cover logarithmic functions.

Exercises

Exercise 1—Derivative and Integral

Find the derivative of the first function and find the integral of the second function from 0 to 3:

In[ ]:= **f[x_] := Exp[Cos[x]]**
   g[x_] := x * 3^(x^2)

Solution

By the chain rule, the derivative of the first function is $e^{\text{Cos}[x]} * \text{Cos}'[x] = e^{\text{Cos}[x]} * -\text{Sin}[x]$:

In[ ]:= **D[f[x], x]**

Out[ ]= $-e^{\text{Cos}[x]} \text{Sin}[x]$

By the substitution rule and letting $u = x^2$, the integral of the second function is
$\frac{1}{2} \int_0^9 3^u \, du = \frac{3^9 - 3^0}{2 \, \text{Log}[3]} = \frac{3^9 - 1}{2 \, \text{Log}[3]}$:

In[ ]:= **Integrate[g[x], {x, 0, 3}] == (3^9 − 1)/(2 Log[3])**

Out[ ]= **True**

Here are their graphs:

In[ ]:= **{Plot[f[x], {x, −10, 10}, ImageSize → Medium], Plot[g[x], {x, −3, 3}, ImageSize → Medium]}**

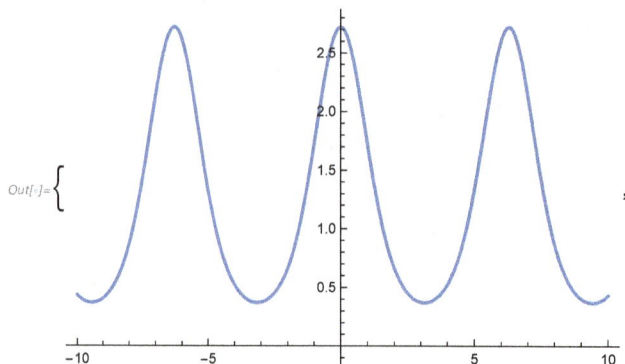

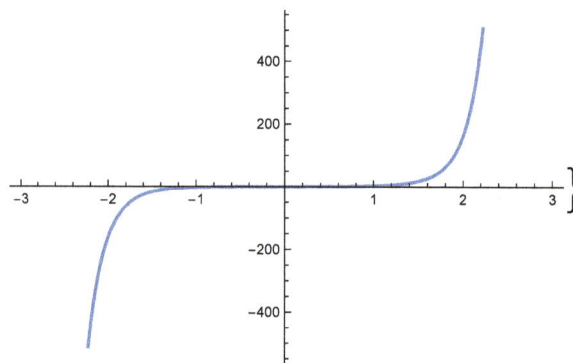

Out[ ]=

Exercise 2—Growth

An artificial lake is stocked with 1000 fish, and the population obeys the following model after t months:

In[ ]:= **population[t_] := 2000 / (1 + Exp[−t / 10])**

Find the limiting size of the lake.

Solution

Plot the function:

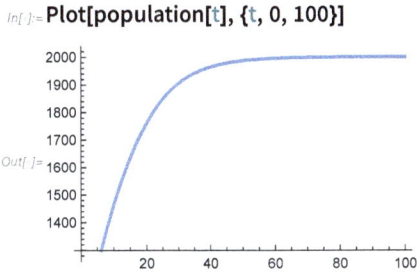

In[ ]:= **Plot[population[t], {t, 0, 100}]**

It appears the limiting size of the lake is somewhere around 2000 fish.

Confirm with Limit:

In[ ]:= **Limit[population[t], t → Infinity]**

Out[ ]= 2000

Exercise 3—Continuously Compounded Interest

When a principal amount of money P is deposited into a bank account and has its interest compounded n times a year with annual interest rate r, the amount of money in the account after t years is given by the following equation:

In[ ]:= **eqn = principalprime == principal (1 + r / n)^(n * t);**

Suppose a man decides to deposit 1000 dollars into his bank account with 3% interest and lets the interest compound *continuously*.

Find the amount of money in the account after 20 years, and compare it to the amount of money when the interest is compounded *daily*.

Solution

Define a function that gives the amount of money in the account based on the length of time and number of times the interest is compounded:

In[ ]:= **money[n_, t_] := 1000 (1 + 0.03 / n)^(n * t)**

To find the amount of money when the interest is compounded continuously after 20 years, let $t = 20$ and take the limit as n goes to ∞:

In[ ]:= **Limit[money[n, 20], n → Infinity]**

Out[ ]= 1822.12

If the money is compounded daily, then the interest is compounded 365 times a year:

In[·]:= **money[365, 20]**

Out[·]= **1822.07**

You can see that the man makes more money when he compounds his interest continuously than when he compounds it daily.

In general, the equation for continuously compounded interest has the following form:

$$P' = P \, e^{rt}$$

Here P is the original amount deposited and P' is the amount of money in the account after t years. This is true for the function **money**:

In[·]:= **Limit[money[n, t], n → Infinity]**

Out[·]= $1000 \, e^{0.03\, t}$

Exercise 4—Hyperbolic Functions

Find the first and second derivatives of the following function:

In[·]:= **f[x_] := (Exp[x] – Exp[–x])/2**

Solution

The first derivative is calculated with **D**:

In[·]:= **D[f[x], x]**

Out[·]= $\dfrac{1}{2} \left(e^{-x} + e^{x} \right)$

The second derivative is calculated with **D**:

In[·]:= **D[f[x], {x, 2}]**

Out[·]= $\dfrac{1}{2} \left(-e^{-x} + e^{x} \right)$

You can see that the function equals its own second derivative!

This function and its derivative are known as **hyperbolic sine** and **hyperbolic cosine**, respectively.

They are given by Sinh and Cosh in the Wolfram Language. Here are their plots:

In[ ]:= `Plot[{Sinh[x], Cosh[x]}, {x, −2, 2}, PlotLegends → "Expressions"]`

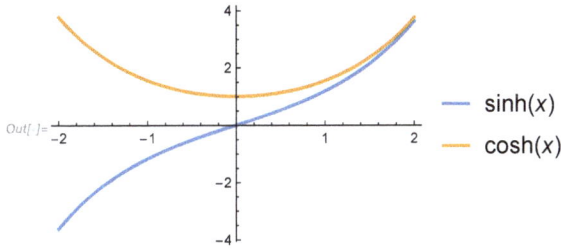

Exercise 5—Gaussian

The following function is known as a **Gaussian**, and is used in many fields, like statistics and quantum mechanics:

In[ ]:= `gaussian[x_] := Exp[−x^2/2]/Sqrt[2 π]`

For the preceding function, it represents a normal distribution with mean 0 and standard deviation 1.

Use Simpson's rule with 20 subintervals to estimate the integral of the function from −5 to 5.

Solution

The width of each interval is $\frac{5-(-5)}{20} = \frac{1}{2}$:

In[ ]:= `simpson =`
`Sum[gaussian[−5 + (2 i − 2)/2] + 4 gaussian[−5 + (2 i − 1)/2] + gaussian[−5 + (2 i)/2],`
`{i, 1, 10}]/(2 ∗ 3) // N`

Out[ ]= `0.999999`

Confirm with Integrate:

In[ ]:= `Integrate[gaussian[x], {x, −5, 5}] // N`

Out[ ]= `0.999999`

In this case, the error is on the order of -8!

In[·]:= **RealAbs[simpson – Integrate[gaussian[x], {x, –5, 5}]]**

Out[·]= 7.20633×10^{-8}

Here is a plot of the Gaussian:

In[·]:= **Plot[gaussian[x], {x, –5, 5}]**

Out[·]=

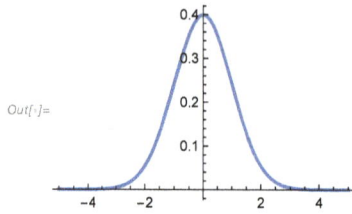

34 | Logarithmic Functions

Overview

The last lesson covered exponential functions, like this one:

In[ ]:= **f[x_] := 2^x**

In[ ]:= **Plot[f[x], {x, −1, 1}]**

Out[ ]=

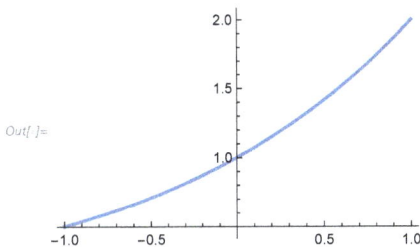

Suppose you wanted to solve the following equation:

In[ ]:= **eqn = 2^x == 5**

Out[ ]= $2^x == 5$

Since 5 is not a power of 2, the answer is not that obvious:

In[ ]:= **Solve[eqn, x, Reals]**

Out[ ]= $\left\{\left\{x \rightarrow \dfrac{Log[5]}{Log[2]}\right\}\right\}$

Solve gives this answer, which is equivalent to $Log_2 5$. $Log_2 5$ is an example of a **logarithm**.

This lesson will cover logarithmic functions and some of their properties.

Inverses

The logarithm is known as an inverse function, because it is the inverse of the function $f[x] = a^x$.

If $f[x]$ is a function, then the **inverse of f**, denoted $f^{-1}[x]$, is the function that when composed with f gives the identity function $i[x] = x$. In other words, $f^{-1}[f[x]] = f[f^{-1}[x]] = x$.

Not all functions have inverses, but the logarithmic function $f[x] = Log_a x$ can be defined as the function such that $a^{Log_a x} = Log_a a^x = x$.

The subscript a for $\text{Log}_a\ x$ is known as the **base** of the logarithm. Log lets you calculate the logarithm of any positive number with any positive base.

For example, the $\text{Log}_3\ 27 = 3$, since $3^3 = 27$:

In[ ]:= **{Log[3, 27], 3^3}**

Out[ ]:= {3, 27}

$\text{Log}_2\ \frac{1}{4} = -2$, since $2^{-2} = \frac{1}{4}$:

In[ ]:= **{Log[2, 1/4], 2^−2}**

Out[ ]:= $\left\{-2, \dfrac{1}{4}\right\}$

By default, the base of Log is e:

In[ ]:= **Log[E^2]**

Out[ ]:= 2

The logarithm with base e is known as the **natural logarithm**.

Graphs of Logarithmic Functions 1

Here is a graph of the natural logarithmic function:

In[ ]:= **f[x_] := Log[x]**

In[ ]:= **Plot[f[x], {x, −1, 10}]**

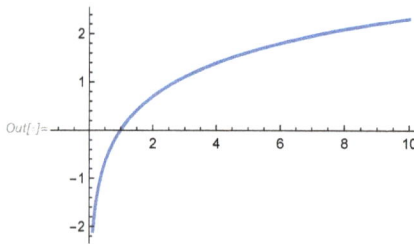

From this you can see that the domain of the logarithm is $(0, \infty)$. The range is all real numbers:

In[ ]:= **{FunctionDomain[f[x], x], FunctionRange[f[x], x, y]}**

Out[ ]:= {x > 0, True}

The limit as x approaches ∞ is ∞. The limit as x approaches 0 from the right is $-\infty$:

In[ ]:= **{Limit[f[x], x → Infinity], Limit[f[x], x → 0, Direction → "FromAbove"]}**

Out[ ]= **{∞, −∞}**

The logarithm has an x-intercept at $x = 1$:

In[ ]:= **Log[1]**

Out[ ]= **0**

Graphs of Logarithmic Functions 2

Here is a graph of the logarithmic function with base $\frac{1}{e}$:

In[ ]:= **g[x_] := Log[1/E, x]**

In[ ]:= **Plot[g[x], {x, −1, 10}]**

Out[ ]=

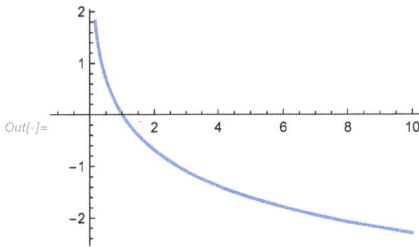

When the base is less than 1, the domain of the logarithm is still $(0, \infty)$. The range is still all real numbers:

In[ ]:= **{FunctionDomain[g[x], x], FunctionRange[g[x], x, y]}**

Out[ ]= **{x > 0, True}**

The limit as x approaches ∞ is $-\infty$. The limit as x approaches 0 from the right is ∞:

In[ ]:= **{Limit[g[x], x → Infinity], Limit[g[x], x → 0, Direction → "FromAbove"]}**

Out[ ]= **{−∞, ∞}**

The logarithm still has an x-intercept at $x = 1$:

In[ ]:= **Log[1/E, 1]**

Out[ ]= **0**

Inverse Graphs

Inverse functions have their graphs as the reflection of the original function with respect to the line $y = x$.

Here is the graph of the natural logarithmic function e^x and the line $y = x$:

$In[\cdot]:=$ **Plot[{E^x, Log[x], x}, {x, −10, 10}, ⋯ ⊕]**

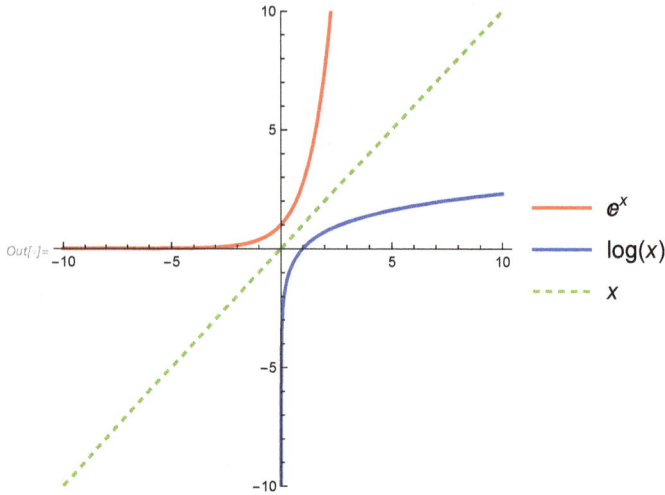

Here is the graph of the logarithmic function with base $\frac{1}{e}$, $\left(\frac{1}{e}\right)^x = e^{-x}$ and the line $y = x$:

$In[\cdot]:=$ **Plot[{E^−x, Log[1/E, x], x}, {x, −10, 10}, ⋯ ⊕]**

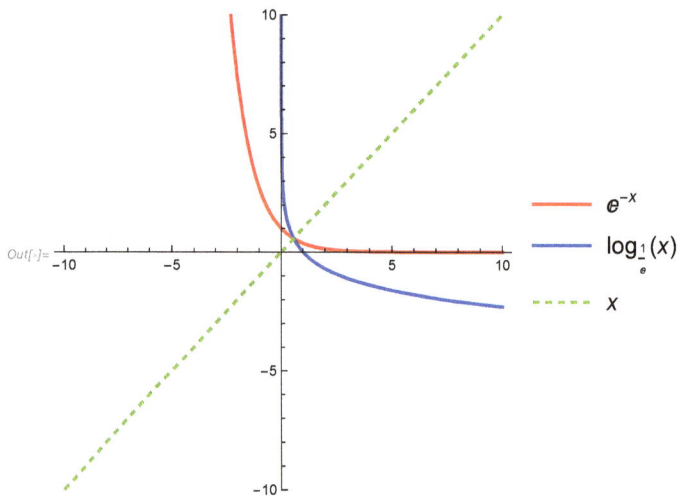

Properties

From the laws of exponents, you can get similar laws for logarithms.

$\text{Log}_a[x * y] = \text{Log}_a[x] + \text{Log}_a[y]$:

In[ ]:= **Log[3, 1/4 * 5] == Log[3, 1/4] + Log[3, 5]**

Out[ ]= True

$\text{Log}_a[x / y] = \text{Log}_a[x] - \text{Log}_a[y]$:

In[ ]:= **Log[1/10, 3/7] == Log[1/10, 3] – Log[1/10, 7]**

Out[ ]= True

$\text{Log}_a[x^b] = b * \text{Log}_a[x]$:

In[ ]:= **Log[4, 5^6.1] == 6.1 Log[4, 5]**

Out[ ]= True

Note that the logarithm with base 1 is generally ignored:

In[ ]:= **Log[1, x] // Quiet**

Out[ ]= ComplexInfinity

Change of Base Formula

The logarithm of an expression can be given as the quotient of two logarithms with the same base.

In other words:

$$\log_x(y) = \frac{\log_a(y)}{\log_a(x)}$$

Note that the logarithm of the base is taken to be the denominator, and the logarithm of y goes in the numerator. a can be any positive number not equal to 1.

This explains why you got the quotient $\text{Log}[5] / \text{Log}[2]$ in the beginning:

In[ ]:= **Solve[2^x == 5, x, Reals]**

Out[ ]= $\left\{\left\{x \to \dfrac{\text{Log}[5]}{\text{Log}[2]}\right\}\right\}$

As before, this is equal to $\text{Log}_2[5]$:

In[·]:= **Log[2, 5] == Log[5] / Log[2]**

Out[·]= **True**

Since you can choose a to be any positive number not equal to 1, a calculator that only has logarithmic functions with base e and 10 can calculate the logarithm for any base.

Derivative of Log

The derivative of the natural logarithmic function is surprisingly $\frac{1}{x}$:

In[·]:= **D[Log[x], x]**

Out[·]= $\dfrac{1}{x}$

Logarithm functions grow more slowly than any polynomial function.

From the change of base formula, you can find the derivative of a logarithm with any base a:

In[·]:= **D[Log[a, x], x]**

Out[·]= $\dfrac{1}{x\,\text{Log}[a]}$

From the chain rule, it is clear that the derivative of $f[x] = \text{Log}_a[g[x]] = \frac{g'[x]}{g[x]\,\text{Log}[a]}$:

In[·]:= **Clear[g]**

In[·]:= **D[Log[a, g[x]], x]**

Out[·]= $\dfrac{g'[x]}{g[x]\,\text{Log}[a]}$

Here is a plot of the natural logarithmic function and its derivative:

In[·]:= **Plot[{Log[x], Log'[x]}, {x, −5, 5}, PlotLegends → "Expressions"]**

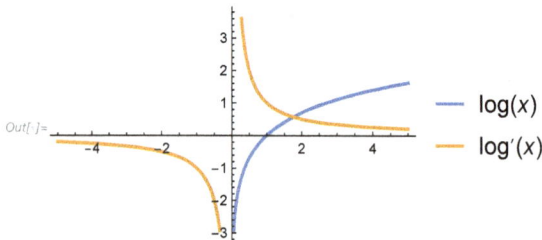

Integral of Log

The integral of the natural logarithm is not obvious:

In[ ]:= **Integrate[Log[x], x]**

Out[ ]= −x + x Log[x]

Here is a graph of the integral of the natural logarithm (with constant $C = 0$):

In[ ]:= **Plot[x Log[x] − x, {x, −1, 10}]**

Out[ ]=

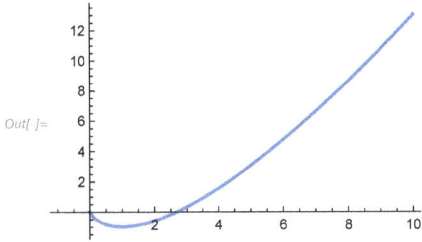

Integral of x^{-1}

To find the integral of $\frac{1}{x}$, you must alter the logarithmic function a little bit so it can have the same domain as $\frac{1}{x}$.

Here is a plot of the function $f[x] = \text{Log}[\,|\,x\,|\,]$:

In[ ]:= **Plot[Log[RealAbs[x]], {x, −5, 5}]**

Out[ ]=

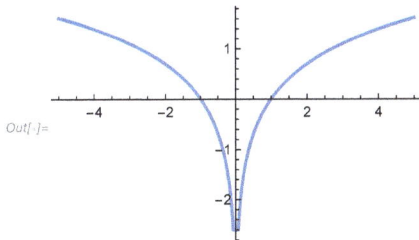

You can see that it now takes on values less than 0. The integral of $\frac{1}{x}$ is $\text{Log}[\,|\,x\,|\,] + C$:

In[ ]:= **Integrate[1 / x, x]**

Out[ ]= Log[x]

Log is specifically designed to handle positive and negative numbers. For negative numbers, the logarithm is imaginary, which is why the real-valued logarithm only has the domain all positive numbers:

In[·]:= **Log[−2]**

Out[·]= $\boldsymbol{i}\,\pi + \mathsf{Log[2]}$

Either way, the integral of $\frac{1}{x}$ is still Log[| x |] + C.

Half-Life

Consider the following exponential decay function:

In[·]:= **decay[*t*_] := n * E^(−λ *t*)**

At time $t = 0$, it equals n. Suppose you wanted to find the time it takes for the function to reach $\frac{n}{2}$, half its initial value:

In[·]:= **Solve[decay[t] == n / 2, t, Reals]**

Out[·]= $\left\{\left\{\mathsf{t} \to \dfrac{\mathsf{Log[2]}}{\lambda}\right\}\right\}$

This value (call it $t_{1/2}$), is known as the **half-life** of the function.

λ is usually called the **decay constant** and its inverse, $\frac{1}{\lambda} = \tau$, is called the **mean lifetime** of the function.

It may be worth remembering the value of Log[2]:

In[·]:= **Log[2] // N**

Out[·]= **0.693147**

Summary

Logarithms are very useful for solving equations of the form $a^x = b$.

Many of the properties of logarithms can be found by looking at the analogous properties of exponentials.

The derivative of the natural logarithmic function is $\frac{1}{x}$.

The integral of the natural logarithmic function is $x * \text{Log}[x] - x + C$.

The integral of $\frac{1}{x}$ is $\text{Log}[|x|] + C$.

Logarithms are useful when finding the half-life of a function.

The next lesson will cover l'Hôpital's rule, which is useful for calculating difficult limits.

Exercises

Exercise 1—Derivative and Integral

Find the derivative of the first function and then evaluate the integral of the second function from $\pi/6$ to $\pi/3$:

```
In[·]:= f[x_] := Log[Sqrt[x]]
        g[x_] := Tan[x]
```

Solution

By the chain rule, the derivative of the first function is $\frac{1}{\sqrt{x}} * \left(\sqrt{x}\right)' = \frac{1}{\sqrt{x}} * \frac{1}{2\sqrt{x}} = \frac{1}{2x}$.

Confirm with D:

```
In[·]:= D[f[x], x]
```

$$Out[·]= \frac{1}{2x}$$

The tangent can be represented as a quotient of sine and cosine $\left(\mathrm{Tan}[x] = \frac{\mathrm{Sin}[x]}{\mathrm{Cos}[x]}\right)$:

```
In[·]:= Tan[x] == Sin[x]/Cos[x]
```

```
Out[·]= True
```

By the substitution rule, and letting $u = \mathrm{Cos}[x]$, you get the integral $\int_{\frac{\sqrt{3}}{2}}^{\frac{1}{2}} -\frac{du}{u} = -\mathrm{Log}\left[\frac{1}{2}\right] - \left(-\mathrm{Log}\left[\frac{\sqrt{3}}{2}\right]\right) = \mathrm{Log}\left[\sqrt{3}\right]$.

Confirm with Integrate:

```
In[·]:= Integrate[g[x], {x, π/6, π/3}] == Log[Sqrt[3]]
```

```
Out[·]= True
```

Exercise 2—Bits and Bytes

A bit can take the values 0 or 1. A byte contains $2^3 = 8$ bits. For a sorted array, it takes at most $\mathrm{Log}_2 n$ comparisons for binary search to find a specific value in an array of size n.

Find the maximum number of comparisons it takes binary search to find a specific bit in a sorted array that takes up 10 GiB of space (GiB means a gibibyte, which has 2^{30} bytes):

```
In[ ]:= UnitConvert[Quantity[1, "Gibibytes"], "Bytes"]
```

```
Out[ ]= 1 073 741 824 B
```

```
In[ ]:= 2^30
```

```
Out[ ]= 1 073 741 824
```

Solution

First, find the number of bits there are in 10 GiB. Since 1 GiB = 2^{30} bytes, there are $2^{30} * 2^3 = 2^{33}$ bits in one GiB. So there are $10 * 2^{33}$ bits in 10 GiB.

Confirm with UnitConvert:

```
In[ ]:= UnitConvert[Quantity[10, "Gibibytes"], "Bits"]
```

```
Out[ ]= 85 899 345 920 b
```

```
In[ ]:= 10 * 2^33
```

```
Out[ ]= 85 899 345 920
```

Therefore, it takes at most $\text{Log}_2[10 * 2^{33}] = \text{Log}_2[10] + 33$ comparisons:

```
In[ ]:= {Log[2, 10 * 2^33], Log[2, 10] + 33} // N
```

```
Out[ ]= {36.3219, 36.3219}
```

So if each comparison takes 0.01 seconds, it will take at most 0.36 seconds to find the bit.

Exercise 3—Half-Life

The decay equation for a certain radioactive isotope is given by the following function, where t is in minutes:

```
In[ ]:= isotope[t_] := Exp[-0.00016045 t]
```

Find the half-life and mean lifetime of the isotope.

Solution

Find the half-life with Solve:

```
In[ ]:= Solve[1/2 == isotope[t], t] // Quiet
```

```
Out[ ]= {{t → 4320.02}}
```

You also could have found the half-life by dividing Log[2] by the decay constant:

In[·]:= **Log[2]/0.00016045**

Out[·]= **4320.02**

The mean lifetime of the isotope is 1 divided by the decay constant:

In[·]:= **1/0.00016045**

Out[·]= **6232.47**

So the half-life of the isotope is ~ 4320 minutes or 3 days, and the mean lifetime is ~ 6232 minutes or ~ 4.33 days.

Here is a plot of the function:

In[·]:= **Plot[isotope[t], {t, 0, 10 000}]**

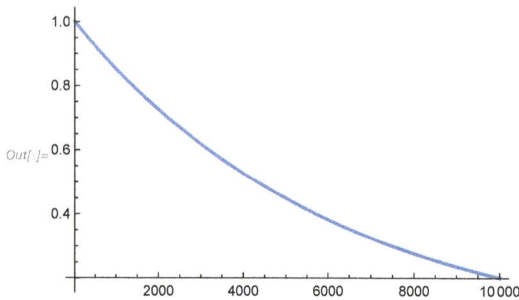

Exercise 4—Richter Scale

The Richter scale is used to find the magnitude of earthquakes.

For an earthquake of intensity I, $\text{Log}_{10}[I/S]$ can be used to find the magnitude of the earthquake:

In[·]:= **magnitude[*intensity_*] := Log[10, *intensity*/s]**

Here S is the intensity of some standard earthquake.

If the magnitude of the earthquake that hit Tohoku, Japan in 2011 was 9.1, and the magnitude of the earthquake that hit Nepal in 2015 was 7.8, how many times more intense was the Tohoku earthquake than the Nepal earthquake?

Solution

Since $\text{Log}_a[x/y] = \text{Log}_a[x] - \text{Log}_a[y]$, 9.1–7.8 is the logarithm with base 10 of the ratio of the intensities.

Solve for the ratio to get the answer:

In[ ]:= **Solve[Log[10, x] == 9.1 – 7.8, x]**

Out[ ]= **{{x → 19.9526}}**

So the Tohoku earthquake was about 20 times more intense than the Nepal earthquake.

Here is a plot of the magnitude function, where s equals 1:

In[ ]:= **Plot[magnitude[intensity] /. s → 1, {intensity, 0, 10}]**

Out[ ]=

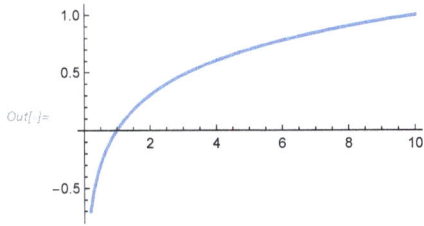

Exercise 5—Euler–Mascheroni Constant

The **Euler–Mascheroni Constant** γ is a number that shows up often in analysis and number theory. It can be calculated by taking the integral of the following function from 0 to 1:

In[ ]:= **em[x_] := –Log[Log[1 / x]]**

Use the midpoint rule with 10 rectangles to approximate γ.

Solution

The rectangles will have width 1 / 10:

In[ ]:= **Sum[em[(i + i – 1) / (2 * 10)], {i, 1, 10}] / 10 // N**

Out[ ]= **0.550187**

γ has the special value EulerGamma in the Wolfram Language:

In[ ]:= **EulerGamma // N**

Out[ ]= **0.577216**

Confirm with Integrate:

In[·]:= **Integrate[em[x], {x, 0, 1}]**

Out[·]= EulerGamma

Here is a plot of the function with the area equal to γ:

In[·]:= **Plot[em[x], {x, 0, 1}, Filling → Axis]**

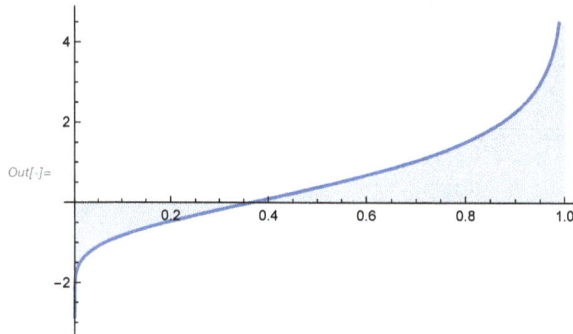

35 | L'Hôpital's Rule

Overview

When calculating the limit, you could just evaluate the function for continuous functions:

In[ ]:= **f[x_] := x^2**

In[ ]:= **f[5] == Limit[f[x], x → 5]**

Out[ ]= True

When the function has a point discontinuity, this would not work:

In[ ]:= **g[x_] := (x^2 − 1)/(x − 1)**

In[ ]:= **g[1] // Quiet**

Out[ ]= Indeterminate

So you try to simplify the function algebraically and find the limit of the simplified version:

In[ ]:= **Simplify[g[x]]**

Out[ ]= 1 + x

In[ ]:= **Limit[Simplify[g[x]], x → 1] == Limit[g[x], x → 1]**

Out[ ]= True

Sometimes the functions are even more complicated, so trying to algebraically simplify the function will **not** work:

In[ ]:= **h[x_] := Log[x − 1]/(x − 2)**

In[ ]:= **Simplify[h[x]]**

Out[ ]= $\dfrac{\text{Log}[-1 + x]}{-2 + x}$

This lesson will show a way to find the limit of the preceding function at 2 and even more complicated expressions using l'Hôpital's rule.

Indeterminate Forms

Consider the following two functions:

In[ ]:= **f[*x*_] := Sin[*x*]**
    g[*x*_] := *x*

If you want to find the limit of the expression $f[x]/g[x]$ as x approaches 0, plugging in 0 gives an **indeterminate form**:

In[ ]:= **f[0]/g[0] // Quiet**

Out[ ]= Indeterminate

In this case, it has the **type 0/0**.

Finding the limit of the following function as x approaches ∞ by plugging in ∞ gives the indeterminate form with **type ∞/∞**:

In[ ]:= **h[*x*_] := (2 *x* − 1)/(3 *x* + 2)**

In[ ]:= **h[∞] // Quiet**

Out[ ]= Indeterminate

Finding the limit of the following function as x approaches 0 from the right by plugging in 0 gives the indeterminate form with **type 0 ∗ ∞**:

In[ ]:= **y[*x*_] :=** *x* **Exp[1/*x*]**

In[ ]:= **y[0] // Quiet**

Out[ ]= Indeterminate

Finding the limit of the following function as x approaches 0 by plugging in 0 gives the indeterminate form with **type $\infty - \infty$**:

In[ ]:= **z[*x*_] := Csc[*x*] − Cot[*x*]**

In[ ]:= **z[0] // Quiet**

Out[ ]= Indeterminate

L'Hôpital's Rule

With some manipulation, all of the previous examples can have limits calculated with what is called **l'Hôpital's rule**.

It is named after the French mathematician Guillaume de l'Hôpital (l'Hospital is also used).

L'Hôpital's rule states that if f and g are two differentiable functions on an open interval I that contains a (except maybe at a) and $g'[x] \neq 0$ on I, then

Limit$_{x\to a}$ $f[x]\,/\,g[x]$ = **Limit**$_{x\to a}$ $f\,'[x]\,/\,g\,'[x]$ if the limits at a for f and g are **both** 0 or $\pm\infty$, **and Limit**$_{x\to a}$ $f\,'[x]\,/\,g\,'[x]$ **exists (or equals $\pm\infty$).**

Find the limit for the function in the introduction: $Log[x-1]\,/\,(x-2)$.

Since $Log[x-1]$ and $x-2$ are both differentiable on $(1,\,3)$, $(x-2)' = 1 \neq 0$ on $(1,\,3)$, and the limits of $Log[x-1]$ and $x-2$ are both 0 at 2, use l'Hôpital's rule to find the limit at 2 for $Log[x-1]\,/\,(x-2)$:

In[ ]:= **D[Log[x – 1], x]/D[x – 2, x]**

Out[ ]= $\dfrac{1}{-1+x}$

L'Hôpital's rule says that finding the limit at 2 for $1\,/\,(x-1)$ will give the same limit at 2 for $Log[x-1]\,/\,(x-2)$.

Since the limit at 2 exists for $1\,/\,(x-1)$, the answer is $1\,/\,(2-1) = 1$:

In[ ]:= **D[Log[x – 1], x]/D[x – 2, x] /. x → 2**

Out[ ]= **1**

Confirm with Limit:

In[ ]:= **Limit[Log[x – 1]/(x – 2), x → 2]**

Out[ ]= **1**

Indeterminate Form 0/0

For the expression $f[x]\,/\,g[x]$:

In[ ]:= **f[x]/g[x]**

Out[ ]= $\dfrac{Sin[x]}{x}$

$Sin[x]$ and x are both differentiable on $(-1,\,1)$, $x' = 1 \neq 0$ on $(-1,\,1)$, and the limits of $Sin[x]$ and x are both 0 at 0:

In[ ]:= **{Limit[f[x], x → 0], Limit[g[x], x → 0]}**

Out[ ]= **{0, 0}**

Use l'Hôpital's rule to find the limit at 0 for $\text{Sin}[x]/x$:

In[·]:= **D[Sin[x], x] / D[x, x]**

Out[·]= **Cos[x]**

L'Hôpital's rule says that finding the limit at 0 for cosine will give the same limit at 0 for $\text{Sin}[x]/x$.

Since the limit at 0 exists for cosine, the answer is $\text{Cos}[0] = 1$:

In[·]:= **Cos[0]**

Out[·]= **1**

Confirm with Limit:

In[·]:= **Limit[Sin[x] / x, x → 0]**

Out[·]= **1**

Indeterminate Form ∞ / ∞

For the function $h[x]$:

In[·]:= **h[x]**

$$Out[·]= \frac{-1+2x}{2+3x}$$

You already know the limit at ∞ should be $2/3$ from an earlier lesson. Confirm with l'Hôpital's rule.

$-1 + 2x$ and $2 + 3x$ are both differentiable on (a, ∞) for any a and the limit at ∞ for each is ∞:

In[·]:= **{Limit[−1 + 2 x, x → Infinity], Limit[2 + 3 x, x → Infinity]}**

Out[·]= **{∞, ∞}**

$(2 + 3x)' = 3 \neq 0$ on $(-\infty, \infty)$, so with l'Hôpital's rule:

In[·]:= **D[−1 + 2 x, x] / D[2 + 3 x, x]**

$$Out[·]= \frac{2}{3}$$

You can see that the limit at ∞ is $2/3$, as it should be:

In[·]:= **Limit[h[x], x → Infinity]**

$$Out[·]= \frac{2}{3}$$

Indeterminate Form 0 ∗ ∞

The function $y[x]$ was not expressed as a quotient:

In[ ]:= **y[x]**

Out[ ]= $e^{\frac{1}{x}}$ x

However, rewriting it as $\left(e^{1/x}\right)/(1/x)$ will let you use l'Hôpital's rule to calculate the right-hand limit at 0, since all the conditions are fulfilled:

In[ ]:= **D[Exp[1/x], x]/D[1/x, x]**

Out[ ]= $e^{\frac{1}{x}}$

The right-hand limit at 0 for $e^{1/x}$ is ∞:

In[ ]:= **Limit[Exp[1/x], x → 0, Direction → "FromAbove"]**

Out[ ]= ∞

Therefore, the right-hand limit at 0 for $y[x]$ is also ∞. Confirm with Limit:

In[ ]:= **Limit[y[x], x → 0, Direction → "FromAbove"]**

Out[ ]= ∞

Here is a plot of the function:

In[ ]:= **Plot[y[x], {x, −1, 1}]**

Out[ ]=

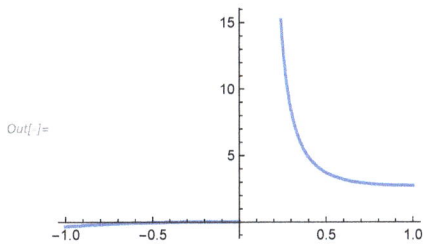

Indeterminate Form ∞ − ∞

The function $z[x]$ was also not expressed as a quotient:

In[ ]:= **z[x]**

Out[ ]= −Cot[x] + Csc[x]

To use l'Hôpital's rule, rewrite the function as a quotient.

$$\text{Csc}[x] - \text{Cot}[x] = \frac{1}{\text{Sin}[x]} - \frac{\text{Cos}[x]}{\text{Sin}[x]} = \frac{(1-\text{Cos}[x])}{\text{Sin}[x]}:$$

In[·]:= **Simplify[z[x] == (1 – Cos[x]) / Sin[x]]**

Out[·]= **True**

At 0, $\frac{(1 - \text{Cos}[x])}{\text{Sin}[x]}$ has the form 0/0:

In[·]:= **{Limit[1 – Cos[x], x → 0], Limit[Sin[x], x → 0]}**

Out[·]= **{0, 0}**

All the conditions are fulfilled for l'Hôpital's rule:

In[·]:= **D[1 – Cos[x], x] / D[Sin[x], x]**

Out[·]= **Tan[x]**

The limit exists at 0 for tangent, so the limit at 0 for $z[x]$ is Tan[0] = 0.

Confirm with Limit:

In[·]:= **Limit[z[x], x → 0]**

Out[·]= **0**

Repeated l'Hôpital's Rule

Sometimes, you have to use l'Hôpital's rule more than once to get the answer.

Consider the following function:

In[·]:= **f[x_] := Exp[x] / (3 x^2)**

To find the limit at ∞, note that the function has the form ∞/∞. The conditions for l'Hôpital's rule are met, so you can use it:

In[·]:= **D[Exp[x], x] / D[3 x^2, x]**

Out[·]= $\dfrac{e^x}{6x}$

At ∞, the form is still ∞/∞, so you have to use l'Hôpital's rule again:

In[·]:= **D[Exp[x], x] / D[6 x, x]**

Out[·]= $\dfrac{e^x}{6}$

At ∞, you now have ∞, so the limit at ∞ for the function is ∞.

Confirm with Limit:

In[ ]:= **Limit[f[x], x → Infinity]**

Out[ ]= ∞

Other Forms

Other indeterminate forms you can encounter when evaluating limits are 0^0, ∞^0 and 1^∞.

Consider the following function, for example:

In[ ]:= **f[x_] := x^(1/x)**

The limit as x approaches ∞ will have the form ∞^0:

In[ ]:= **f[∞] // Quiet**

Out[ ]= Indeterminate

To find the limit in this case, **apply a logarithm to** $y = f[x]$.

Here is the equation:

In[ ]:= **eqn = y == x^(1/x)**

Out[ ]= $y == x^{\frac{1}{x}}$

Take the logarithm of both sides:

In[ ]:= **logeqn = Log[y] == Log[x^(1/x)]**

Out[ ]= $\text{Log}[y] == \text{Log}\left[x^{\frac{1}{x}}\right]$

Using properties of logarithms, then **rewrite the expression** so that you can use l'Hôpital's rule:

In[ ]:= **logeqn = Log[y] == Log[x]/x**

Out[ ]= $\text{Log}[y] == \dfrac{\text{Log}[x]}{x}$

Logarithmic l'Hôpital—Indeterminate Form ∞^0

Log[x] / x evaluates to ∞ / ∞ at ∞, so you can now use l'Hôpital's rule:

In[·]:= **D[Log[x], x] / D[x, x]**

Out[·]= $\dfrac{1}{x}$

You know the limit as x approaches ∞ is 0 for $1/x$. So you now have the equation:

In[·]:= **logeqn = Log[y] == 0**

Out[·]= **Log[y] == 0**

Now solve for y by **exponentiating** to find the limit for the original function:

In[·]:= **eqn = Exp[Log[y]] == Exp[0]**

Out[·]= **y == 1**

Therefore the limit is 1. Confirm with Limit:

In[·]:= **Limit[f[x], x → Infinity]**

Out[·]= **1**

Here is a plot of the function:

In[·]:= **Plot[f[x], {x, 0, 1000}, PlotRange → {0.5, 1.5}]**

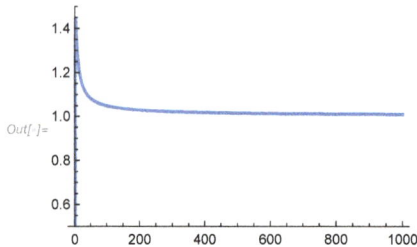

Indeterminate Form 0^0

Find the right-hand limit of the following function at 0:

In[·]:= **f[x_] := x ^ x**

First, note that at 0 the function has the form 0^0. Therefore, you apply a logarithm to $y = x^x$:

In[·]:= **eqn = y == f[x]**

Out[·]= $y == x^x$

Use properties of logarithms:

In[·]:= **logeqn = Log[y] == x Log[x]**

Out[·]= $\text{Log}[y] == x \text{Log}[x]$

Then change it to a quotient:

In[·]:= **logeqn = Log[y] == Log[x] / (1 / x)**

Out[·]= $\text{Log}[y] == x \text{Log}[x]$

Then use l'Hôpital's rule:

In[·]:= **D[Log[x], x] / D[1 / x, x]**

Out[·]= $-x$

From the right the limit is 0, so the limit of the original function is $e^0 = 1$.

Confirm with Limit:

In[·]:= **Limit[f[x], x → 0, Direction → "FromAbove"]**

Out[·]= 1

Indeterminate Form 1^∞

Find the limit of the following function at $\pi / 2$:

In[·]:= **f[x_] := Sin[x] ^ Tan[x]**

Note that plugging in 0 gives the expression 1^∞, so apply a logarithm to $y = \text{Sin}[x]^{\text{Tan}[x]}$:

In[·]:= **eqn = y == f[x]**

Out[·]= $y == \text{Sin}[x]^{\text{Tan}[x]}$

Using properties of logarithms, you get:

In[·]:= **logeqn = Log[y] == Tan[x] * Log[Sin[x]]**

Out[·]= $\text{Log}[y] == \text{Log}[\text{Sin}[x]] \text{Tan}[x]$

Use l'Hôpital's rule:

In[·]:= **D[Log[Sin[x]], x] / D[1 / Tan[x], x]**

Out[·]:= **−Cos[x] Sin[x]**

This gives $-\mathrm{Cos}[\pi/2]\,\mathrm{Sin}[\pi/2] = -0 * 1 = 0$:

In[·]:= **D[Log[Sin[x]], x] / D[1 / Tan[x], x] /. x → π / 2**

Out[·]:= **0**

Therefore, the limit of the original function at $\pi/2$ is $e^0 = 1$.

Confirm with Limit:

In[·]:= **Limit[f[x], x → π / 2]**

Out[·]:= **1**

Summary

L'Hôpital's rule is very useful for calculating limits with indeterminate forms $0/0$ and ∞/∞.

With some algebraic manipulation, you can also calculate limits with indeterminate forms $0 * \infty$ and $\infty - \infty$.

By taking logarithms, you can calculate limits with indeterminate forms 0^0, ∞^0 and 1^∞.

Sometimes l'Hôpital's rule must be used multiple times to arrive at a suitable answer.

The next lesson will cover slope fields and Euler's method.

Exercises

Exercise 1—Trigonometric Functions

Find the limit at π for the following function:

In[ ]:= **f[x_] := Sin[x]/(π^2 − x^2)**

Solution

At π, the function has the form $0/0$. The conditions for l'Hôpital's rule are met, so you can use it:

In[ ]:= **D[Sin[x], x]/D[π^2 − x^2, x]**

Out[ ]= $-\dfrac{\text{Cos}[x]}{2\,x}$

At π you now have $-\text{Cos}[\pi]/2\,\pi = 1/2\,\pi$, so that is the answer.

Confirm with Limit:

In[ ]:= **Limit[f[x], x → π]**

Out[ ]= $\dfrac{1}{2\,\pi}$

Here is a graph of the function around π:

In[ ]:= **Plot[f[x], {x, π − 1, π + 1}]**

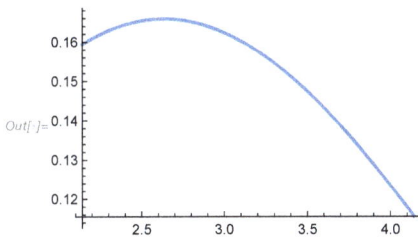

Exercise 2—Difference

Find the right-hand limit of the following function at $\pi/2$:

In[ ]:= **f[x_] := Sec[x] + 1/(x − π/2)**

Solution

At $\pi/2$, the function has the form $\infty - \infty$. Therefore, you can express the function as a quotient and use l'Hôpital's rule.

The function can be written as
$1/\text{Cos}[x] + 1/(x - \pi/2) = ((x - \pi/2) + \text{Cos}[x])/((x - \pi/2)\,\text{Cos}[x])$:

In[·]:= **Simplify[f[x] == ((x − 𝜋/2) + Cos[x])/((x − 𝜋/2) Cos[x])]**

Out[·]= **True**

The limit now has the form $0/0$ at $\pi/2$. Use l'Hôpital's rule:

In[·]:= **D[(x − 𝜋/2) + Cos[x], x]/D[(x − 𝜋/2) Cos[x], x]**

Out[·]= $\dfrac{1 - \text{Sin}[x]}{\text{Cos}[x] - \left(-\frac{\pi}{2} + x\right)\text{Sin}[x]}$

At $\pi/2$, the limit still has the form $0/0$, so use l'Hôpital's rule again:

In[·]:= **D[1 − Sin[x], x]/D[Cos[x] − (−𝜋/2 + x) Sin[x], x]**

Out[·]= $-\dfrac{\text{Cos}[x]}{-\left(\left(-\frac{\pi}{2} + x\right)\text{Cos}[x]\right) - 2\,\text{Sin}[x]}$

At $\pi/2$, you can see that the limit is $0/(0 - 2*1) = 0$:

In[·]:= **D[1 − Sin[x], x]/D[Cos[x] − (−𝜋/2 + x) Sin[x], x] /. x → 𝜋/2**

Out[·]= **0**

Confirm with Limit:

In[·]:= **Limit[f[x], x → 𝜋/2]**

Out[·]= **0**

Exercise 3—Indeterminate Form 1^{∞}

Find the limit of the following function at 0:

In[ ]:= **f[*x*_] := Cos[*x*]^(1/*x*^2)**

Solution

Plugging in 0, you get 1^{∞}, so apply a logarithm to $y = \text{Cos}[x]^{\frac{1}{x^2}}$:

In[·]:= **eqn = y == f[x]**

Out[·]= $y == \text{Cos}[x]^{\frac{1}{x^2}}$

Using properties of logarithms, you get:

In[ ]:= **logeqn = Log[y] == Log[Cos[x]]/(x^2)**

Out[ ]= $\text{Log[y]} == \dfrac{\text{Log[Cos[x]]}}{x^2}$

Use l'Hôpital's rule:

In[ ]:= **D[Log[Cos[x]], x]/D[x^2, x]**

Out[ ]= $-\dfrac{\text{Tan[x]}}{2x}$

This gives $0/0$ at 0, so use l'Hôpital's rule again:

In[ ]:= **D[−Tan[x], x]/D[2 x, x]**

Out[ ]= $-\dfrac{1}{2}\text{Sec[x]}^2$

This gives $-1/2$ at 0, so the limit of the original function at 0 is $e^{-1/2}$. Confirm with Limit:

In[ ]:= **Limit[f[x], x → 0]**

Out[ ]= $\dfrac{1}{\sqrt{e}}$

Exercise 4—Euler's Constant e

Use l'Hôpital's rule to confirm the limit of the following function at ∞ is e:

In[ ]:= **euler[x_] := (1 + 1/x)^x**

Solution

The function has the form $(1 + 1/\infty)^{\infty} = 1^{\infty}$ at ∞, so use a logarithm:

In[ ]:= **logeqn = Log[euler] == Log[(1 + 1/x)^x]**

Out[ ]= $\text{Log[euler]} == \text{Log}\left[\left(1 + \dfrac{1}{x}\right)^x\right]$

Then use properties of logarithms:

In[ ]:= **logeqn = Log[euler] == x Log[(1 + 1/x)]**

Out[ ]= $\text{Log[euler]} == x\,\text{Log}\left[1 + \dfrac{1}{x}\right]$

Rewrite the right-hand side as $\text{Log}[1 + 1/x]/(1/x)$ and then use l'Hôpital's rule:

```
In[·]:= D[Log[1 + 1/x], x] / D[1/x, x]
```

$$Out[·]= \frac{1}{1 + \frac{1}{x}}$$

Evaluating this limit at ∞ gives the value 1, so by exponentiating the equation, you get the limit of the original function at ∞ to be e:

```
In[·]:= Exp[Log[euler]] == Exp[1]
```

```
Out[·]= euler == e
```

Confirm with Limit:

```
In[·]:= Limit[euler[x], x → Infinity]
```

```
Out[·]= e
```

Exercise 5—The Original Application of l'Hôpital's Rule

L'Hôpital wrote the first book on differential calculus. In his book you can see l'Hôpital's rule used on the following function as x approaches a:

```
In[·]:= lhopital[x_] := (Surd[2 a^3 x − x^4, 2] − a CubeRoot[a^2 x]) / (a − Surd[a * x^3, 4])
```

Use l'Hôpital's rule on the function to find the limit. Assume $a > 0$.

Solution

The function has the form $0/0$ at a:

```
In[·]:= Simplify[lhopital[a], a > 0] // Quiet
```

```
Out[·]= ComplexInfinity
```

The conditions for l'Hôpital's rule apply, so you can use it here:

```
In[·]:= D[Surd[2 a^3 x − x^4, 2] − a CubeRoot[a^2 x], x] / D[a − Surd[a * x^3, 4], x]
```

$$Out[·]= -\frac{4\sqrt[4]{a x^3}^{13}\left(-\frac{a^3}{3\sqrt[3]{a^2 x}^2} + \frac{2 a^3 - 4 x^3}{2\sqrt[2]{2 a^3 x - x^4}}\right)}{3 a x^2}$$

Plugging in a, you get:

In[ ]:= **Simplify[**
     D[Surd[2 a^3 x − x^4, 2] − a CubeRoot[a^2 x], x] / D[a − Surd[a∗x^3, 4], x] /. x → a, a > 0]

Out[ ]= $\dfrac{16\,a}{9}$

Confirm with Limit:

In[ ]:= **Limit[lhopital[x], x → a, Assumptions → a > 0]**

Out[ ]= $\dfrac{16\,a}{9}$

36 | Slope Fields and Euler's Method

Overview

It was previously mentioned that some differential equations are very difficult to solve by hand.

For example, the following simple-looking differential equation has a nasty solution when $y[1] = 0$:

In[ ]:= **eqn = y '[x] == y[x]^2 − x**

Out[ ]= $y'[x] == -x + y[x]^2$

Here is the solution with DSolve:

In[ ]:= **DSolve[{eqn, y[1] == 0}, y[x], x]**

Out[ ]= $\left\{\left\{y[x] \rightarrow -\left(\left(i\,x^{3/2}\,\text{BesselJ}\!\left[-\frac{4}{3}, \frac{2}{3}\,i\,x^{3/2}\right]\text{BesselJ}\!\left[-\frac{2}{3}, \frac{2i}{3}\right] - i\,x^{3/2}\,\text{BesselJ}\!\left[-\frac{4}{3}, \frac{2i}{3}\right]\right.\right.\right.$

$\text{BesselJ}\!\left[-\frac{2}{3}, \frac{2}{3}\,i\,x^{3/2}\right] - x^{3/2}\,\text{BesselJ}\!\left[-\frac{2}{3}, \frac{2}{3}\,i\,x^{3/2}\right]\text{BesselJ}\!\left[-\frac{1}{3}, \frac{2i}{3}\right] +$

$\text{BesselJ}\!\left[-\frac{2}{3}, \frac{2i}{3}\right]\text{BesselJ}\!\left[-\frac{1}{3}, \frac{2}{3}\,i\,x^{3/2}\right] + i\,x^{3/2}\,\text{BesselJ}\!\left[-\frac{2}{3}, \frac{2}{3}\,i\,x^{3/2}\right]$

$\text{BesselJ}\!\left[\frac{2}{3}, \frac{2i}{3}\right] - i\,x^{3/2}\,\text{BesselJ}\!\left[-\frac{2}{3}, \frac{2i}{3}\right]\text{BesselJ}\!\left[\frac{2}{3}, \frac{2}{3}\,i\,x^{3/2}\right]\right) \Big/$

$\left(x\left(2\,\text{BesselJ}\!\left[-\frac{2}{3}, \frac{2i}{3}\right]\text{BesselJ}\!\left[-\frac{1}{3}, \frac{2}{3}\,i\,x^{3/2}\right] - \text{BesselJ}\!\left[-\frac{4}{3}, \frac{2i}{3}\right]\right.\right.$

$\text{BesselJ}\!\left[\frac{1}{3}, \frac{2}{3}\,i\,x^{3/2}\right] + i\,\text{BesselJ}\!\left[-\frac{1}{3}, \frac{2i}{3}\right]\text{BesselJ}\!\left[\frac{1}{3}, \frac{2}{3}\,i\,x^{3/2}\right] +$

$\left.\left.\left.\left.\text{BesselJ}\!\left[\frac{1}{3}, \frac{2}{3}\,i\,x^{3/2}\right]\text{BesselJ}\!\left[\frac{2}{3}, \frac{2i}{3}\right]\right)\right)\right)\right\}\right\}$

Even though the solution is rather hard to swallow, it is much easier to visualize what the curve looks like, for any initial values.

This lesson will show how to use slope fields to see the behavior of solutions to differential equations.

Slope Fields

Slope fields, or direction fields, are not hard to create, in principle. For each point (x, y) in the xy-plane and a differential equation $y'[x] = f[x, y]$, you plug in the values for x and y and plot the slope at that point.

"Plotting the slope" means drawing a small line segment, with slope close to the value you get from $f[x, y]$. When you do this for enough points in the xy-plane, you get a slope field that gives a general picture of the solutions to the differential equation.

In the Wolfram Language, StreamPlot or VectorPlot can be used.

Here is an example with the differential equation in the introduction:

In[ ]:= `f[x_, y_] := y^2 - x`

Here are slope fields with VectorPlot and StreamPlot:

In[ ]:= `{VectorPlot[{1, f[x, y]}, {x, -2, 2}, {y, -2, 2}], StreamPlot[{1, f[x, y]}, {x, -2, 2}, {y, -2, 2}]}`

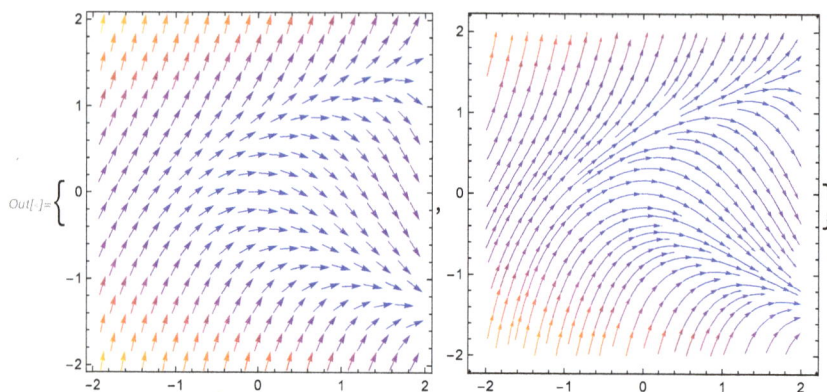

Particular Paths

To isolate particular curves, use the option StreamPoints.

Here is the particular curve for the solution curve in the introduction with VectorPlot:

In[ ]:= `VectorPlot[{1, f[x, y]}, {x, -2, 2}, {y, -2, 2}, StreamPoints → {{{{1, 0}, Red}}}]`

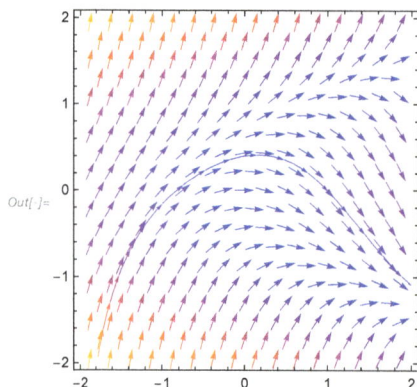

Here is the same curve with StreamPlot:

In[ ]:= **StreamPlot[{1, f[x, y]}, {x, −2, 2}, {y, −2, 2}, StreamPoints → {{{{1, 0}, Blue}, Automatic}}]**

Out[ ]=

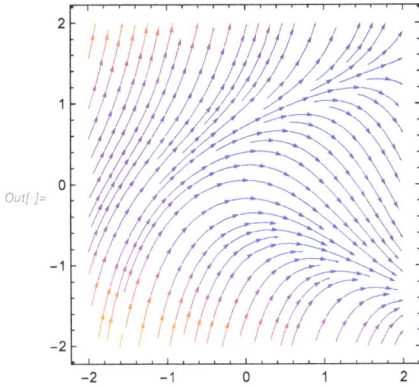

Slope Field for $y' = x - y$

Here is the slope field for the differential equation $y' = x - y$ with $y[0] = 0$:

In[ ]:= **f[x_, y_] := x − y**

In[ ]:= **StreamPlot[{1, f[x, y]}, {x, −2, 2}, {y, −2, 2}]**

Out[ ]=

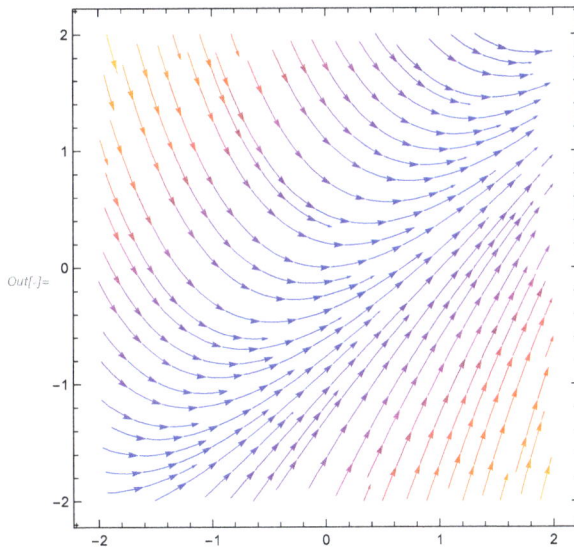

Here is a plot with the particular curve:

In[•]:= **VectorPlot[{1, f[x, y]}, {x, −2, 2}, {y, −2, 2}, StreamPoints → {{{{0, 0}, Red}}}]**

Out[•]=

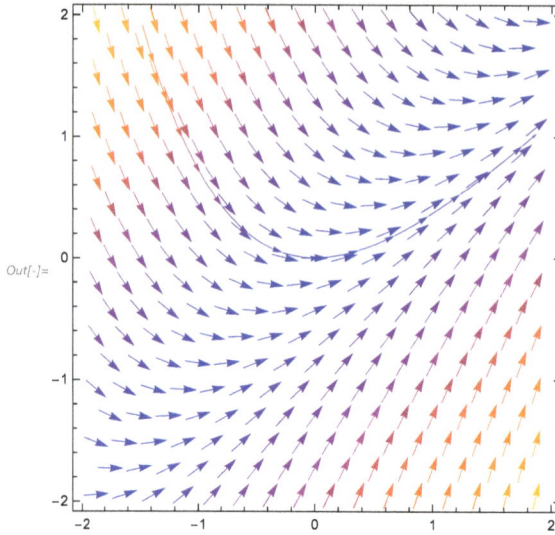

Slope Field for $y' = x * y$

Here is the slope field for the differential equation $y' = x * y$ with $y[1] = -1$:

In[•]:= **f[x_, y_] := x * y**

In[•]:= **StreamPlot[{1, f[x, y]}, {x, −2, 2}, {y, −2, 2}]**

Out[•]=

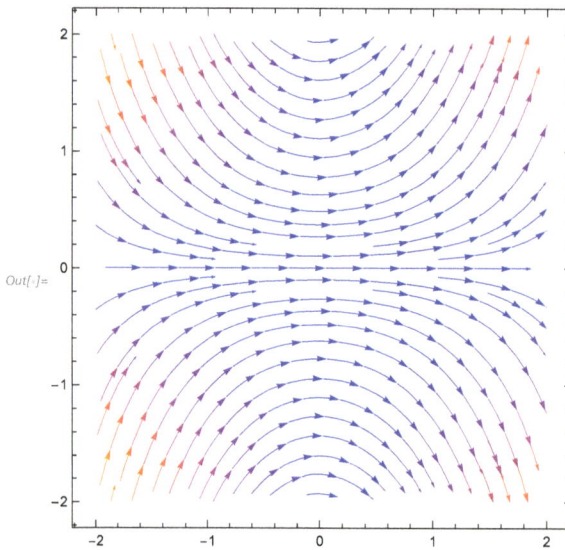

Here is a plot with the particular curve:

In[ ]:= **VectorPlot[{1, f[x, y]}, {x, −2, 2}, {y, −2, 2}, StreamPoints → {{{{1, −1}, Red}}}]**

Out[ ]=

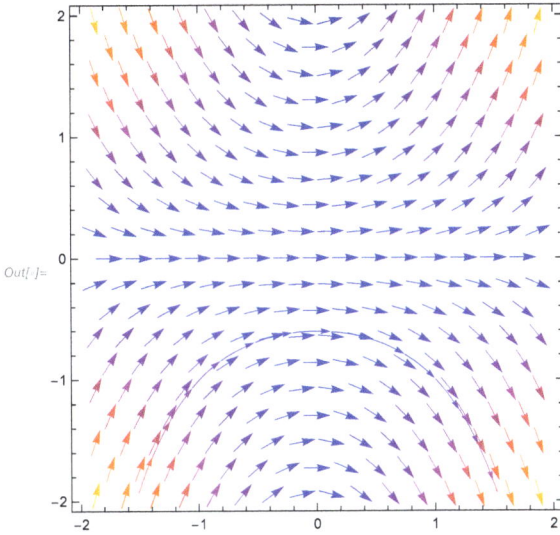

Slope Field for $y' = -x/y$

Here is the slope field for the differential equation $y' = x/y$ with $y[0] = 0$:

In[ ]:= **f[x_, y_] := −x/y**

In[ ]:= **StreamPlot[{1, f[x, y]}, {x, −2, 2}, {y, −2, 2}]**

Out[ ]=

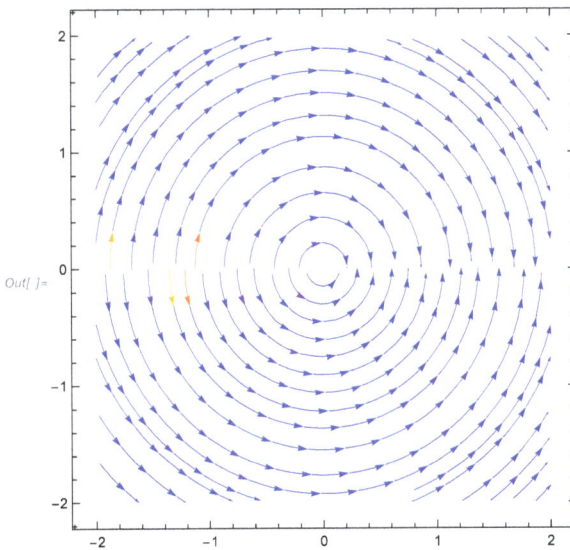

Here is a plot with the particular curve:

In[·]:= **VectorPlot[{1, f[x, y]}, {x, −2, 2}, {y, −2, 2}, StreamPoints → {{{{1, 1}, Red}}}]**

Out[·]=

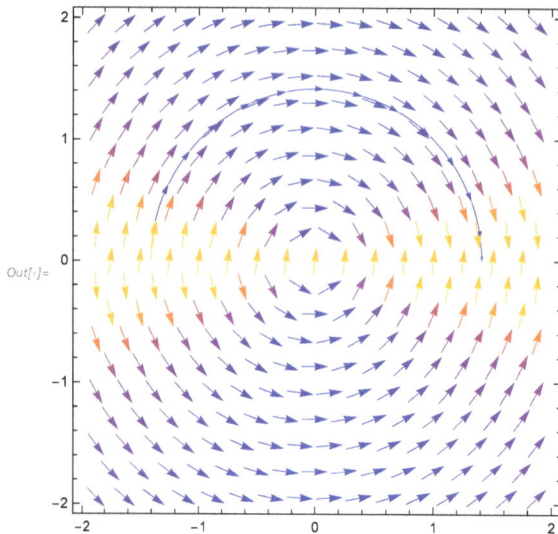

Euler's Method

You can use the differential equation to estimate the value of the solution at different points, given some initial value.

The method to do so is known as **Euler's method**, after Leonhard Euler. The method is rather simple to understand. Point-slope form and the slopes from the differential equation are used to estimate the value at different points.

For example, use the differential equation used in the introduction:

In[·]:= **f[x_, y_] := y ^ 2 − x**

You know that $y[1] = 0$ and would like to estimate the value of y at 2.

Since the slope at $(1, 0)$ is -1, you can estimate the value of y with the linear approximation $y = -(x - 1)$. So $y \sim -1$:

In[·]:= **−(x − 1) /. x → 2**

Out[·]= **−1**

Go a little bit on the tangent line, and then change direction according to the slope field. This will give a better approximation.

For example, if you only went to (1.5, 0.5) on the tangent line, and used the new slope:

In[ ]:= **f[1.5, 0.5]**

Out[ ]= **−1.25**

Then use the equation $y = 0.5 - 1.25 (x - 1.5)$ and follow that tangent line. Plugging in $x = 2$, you get the better estimate $y \sim -0.125$:

In[ ]:= **0.5 − 1.25 (x − 1.5) /. x → 2**

Out[ ]= **−0.125**

Euler's Method in General

It is clear that taking short steps along the tangent lines calculated with the help of the differential equation will make better estimates than longer ones. It should also be clear that if you want to do this by hand, then it is best to choose points you want to estimate that are near the initial value point, to save time and space.

To estimate the value at x_n for the solution of a differential equation $y'[x, y]$, use the equation $y_n = y_{n-1} + h * y'[x_{n-1}, y_{n-1}]$ for $n = 1, 2, 3, \ldots$.

Here is a good visualization with the differential equation $y'[x, y] = y$ with $y[0] = 1$. The solution is $y[x] = e^x$, and here are points you get with Euler's method to approximate $y[1]$ with various step sizes:

In[ ]:= **Show[ListLinePlot[{{{0, 1}, {1, 2}}, {{0.`, 1}, {0.5`, 1.5`}, {1.`, 2.25`}},**
    {{0.`, 1}, {0.1`, 1.1`}, {0.2`, 1.2100000000000002`},
      {0.30000000000000004`, 1.3310000000000002`}, {0.4`, 1.4641000000000002`},
      {0.5`, 1.61051`}, {0.6000000000000001`, 1.7715610000000002`},
      {0.7000000000000001`, 1.9487171`}, {0.8`, 2.1435888100000002`},
      {0.9`, 2.357947691`}, {1.`, 2.5937424601`}}}, PlotStyle → {Red, Blue, Green},
    PlotLegends → {"One Point", "Two Points", "Ten Points"}],
  Plot[Exp[x], {x, 0, 1}]]

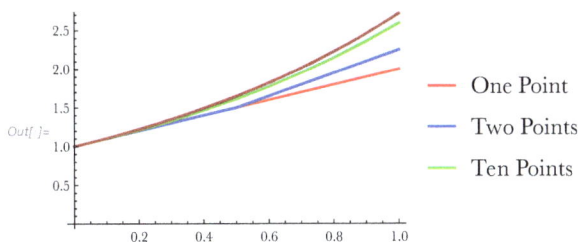

You can see that using one point just gives the tangent to the solution at 0, two points improves the solution, and 10 points gives a much better approximation to the solution.

The plot of e^x is the top-most maroon graph.

Recurrence Relations

The equation used in Euler's method is known as a **recurrence relation**, because calculating the values for the equation depends on values for previous calculations.

Here $y_n = y_{n-1} + h * y'[x_{n-1}, y_{n-1}]$ for $n = 1, 2, 3, \ldots$ means, for example, that to calculate the fourth value of y, you must use the third value of x and the third value of y, which depend on the second value of x and the second value of y, which depend on the first value of x and the first value of y, which were the initial conditions.

Because of this recursive quality, the method is tiresome to use by hand with smaller step sizes, but incredibly easy to program on a computer.

A later lesson will show how to make a program that calculates the points from Euler's method.

Alternatively, you can use NDSolveValue, which gives a way to approximate values with Euler's method, but only with certain step sizes:

```
In[·]:= NDSolveValue[{y'[x] == y[x], y[0] == 1}, y[1],
            {x, 0, 1}, Method → "ExplicitEuler", StartingStepSize → 1/2]

Out[·]= 2.59374
```

Here, NDSolveValue gives the value 2.59374 when the step size is $1/2$, when it should be 2.25 (it is actually using step size $1/10$):

```
In[·]:= NDSolveValue[{y'[x] == y[x], y[0] == 1}, y[1], {x, 0, 1},
            Method → "ExplicitEuler", StartingStepSize → 1/10]

Out[·]= 2.59374
```

Summary

Slope fields are very useful when trying to visualize the behavior of solutions to differential equations.

VectorPlot or StreamPlot can be used to plot them in the Wolfram Language.

Euler's method uses the values of differential equations to estimate the solution at various points.

Euler's method can be used via NDSolve and NDSolveValue or be programmed with a computer.

The next lesson will cover separable differential equations.

Exercises

Exercise 1—Hyperbola

Make a slope field from the differential equation $y' = x / y$.

Solution

First, make a function to model the equation:

```
In[ ]:= f[x_, y_] := x / y
```

Then use StreamPlot:

```
In[ ]:= StreamPlot[{1, f[x, y]}, {x, -2, 2}, {y, -2, 2}, ImageSize -> Large]
```

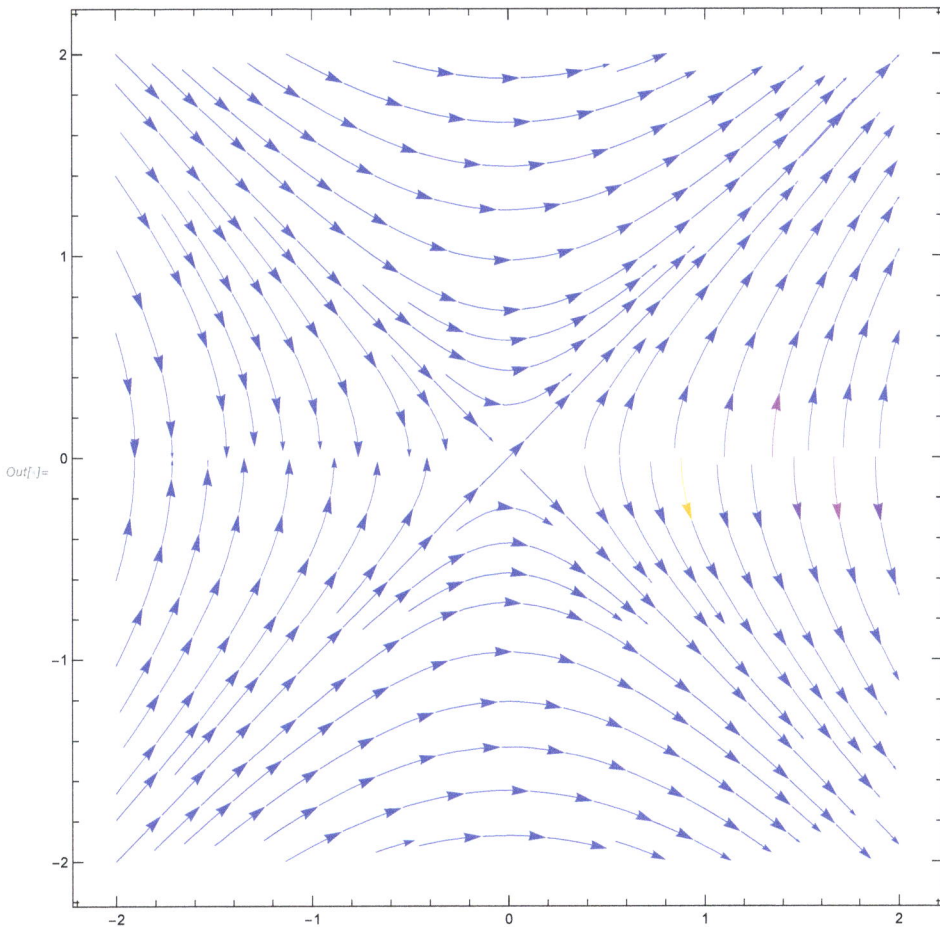

Exercise 2—Trigonometric Slope Field

Make a slope field from the differential equation $y' = y \, \text{Cos}[x] + x \, \text{Tan}[y]$.

Solution

First, make a function to model the equation:

In[-]:= **f[x_, y_] := y Cos[x] + x Tan[y]**

Then use StreamPlot:

In[-]:= **StreamPlot[{1, f[x, y]}, {x, −2, 2}, {y, −2, 2}, ImageSize → Large]**

Out[-]=

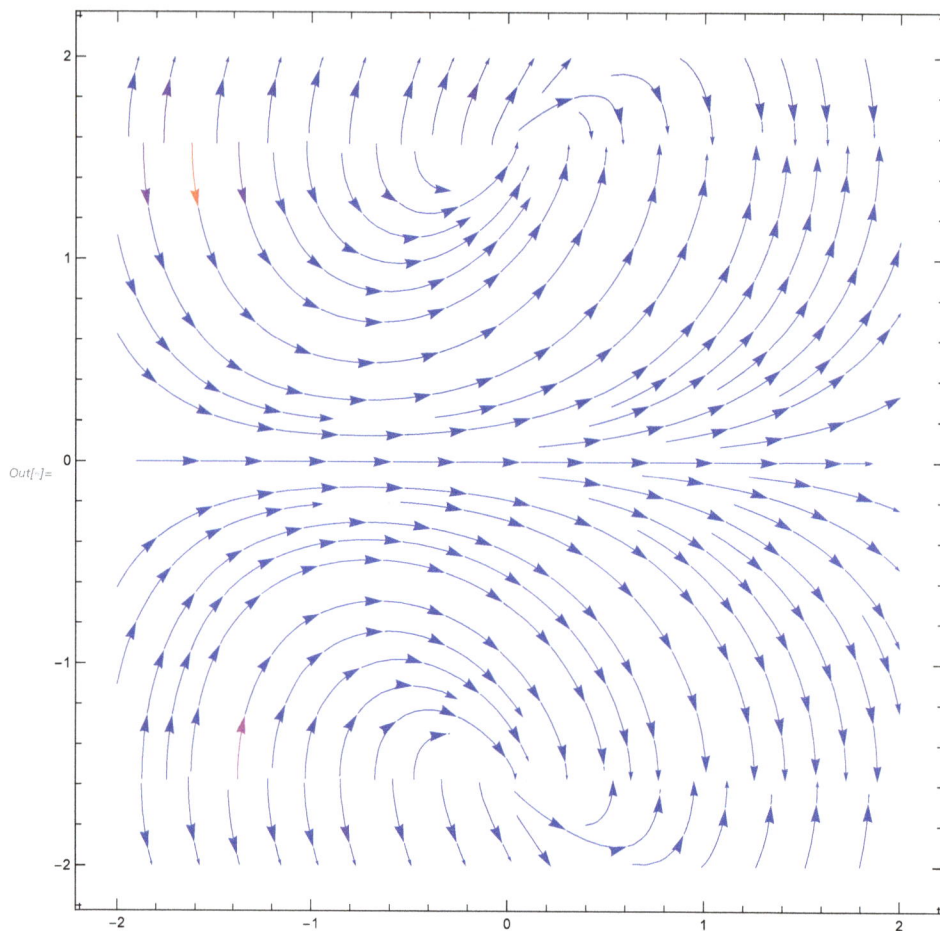

Exercise 3—Euler's Method

Use Euler's method to approximate the solution to the equation $y' = y + x * y$ with initial condition $y[0] = 1$ at $x = 0.15$ with step size 0.05.

Solution

First, model the equation with a function:

In[ ]:= **f[*x_*, *y_*] := *y* + *x* * *y***

At $x = 0.05$, you get the estimate:

In[ ]:= **1 + 0.05 f[0, 1]**

Out[ ]= **1.05**

At $x = 0.1$, you get the estimate:

In[ ]:= **val = 1.05 + 0.05 f[0.05, 1.05]**

Out[ ]= **1.10513**

At $x = 0.15$, you get the estimate:

In[ ]:= **val + 0.05 f[0.1, val]**

Out[ ]= **1.16591**

NDSolveValue gets the estimate:

In[ ]:= **NDSolveValue[{y '[*x*] == f[*x*, y[*x*]], y[0] == 1}, y[0.15],**
**   {*x*, 0, 0.15}, Method → "ExplicitEuler", StartingStepSize → 0.1]**

Out[ ]= **1.17217**

You can see that using NDSolve improves the estimate by roughly 0.007.

Exercise 4—Newton's Law of Cooling

A boiling pot of soup at temperature 115°C is placed in a room with temperature 22°C. After an hour, the soup has temperature 66°C.

Newton's law of cooling lets you model the temperature of the soup using the following differential equation:

In[ ]:= **pot[*time_*, *temp_*] := −Log[93 / 44] (*temp* − 22)**

Here, it is assumed the temperature is a function of the time. Use a slope field to predict the temperature after 10 hours.

Solution

Here is a plot with StreamPlot:

```
In[·]:= StreamPlot[{1, pot[time, temp]}, {time, 0, 10},
        {temp, 20, 120}, StreamPoints → {{{{0, 115}, Blue}}}]
```

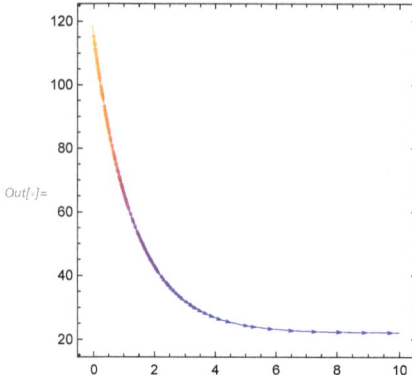

```
Out[·]=
```

Based on the plot, you can predict the temperature will be somewhere near 22°C.

NDSolveValue agrees:

```
In[·]:= NDSolveValue[{temp '[time] == pot[time, temp[time]], temp[0] == 115},
        temp[10], {time, 0, 10}]

Out[·]= 22.0523
```

Exercise 5—Circuit

Kirchhoff's law gives the following relationship between the capacitance, resistance, voltage, current and charge in a circuit:

```
In[·]:= resistance * D[charge[time], time] + charge / capacitance == voltage[time]
```

$$Out[·]= \frac{charge}{capacitance} + resistance\ charge'[time] == voltage[time]$$

Here, assume the voltage and charge are functions of time.

If you have a circuit with resistance 10 Ω, capacitance 0.1 F and a battery with constant voltage 120 V, plot the slope field for the differential equation.

Solution

The function to model the equation will be:

```
In[·]:= current[time_, charge_] := (−charge + 0.1 * 120) / (10 * 0.1)
```

Here is a slope field:

In[ ]:= **StreamPlot[{1, current[time, charge]}, {time, 0, 10}, {charge, 0, 20}]**

Out[ ]=

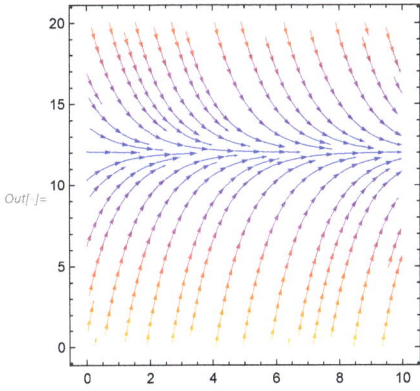

You can see that as time goes on, the charge approaches the value 12 coulombs.

37 | Separable Differential Equations

Overview

Most differential equations cannot be solved easily.

For example, the solution to the following differential equation with DSolve is:

In[ ]:= **DSolve[y'[x] == y[x]^2 − x, y[x], x]**

$$\text{Out[ ]=}\left\{\left\{y[x]\to\left(-\text{BesselJ}\left[-\frac{1}{3},\frac{2}{3}\,i\,x^{3/2}\right]c_1+\right.\right.\right.$$
$$i\,x^{3/2}\left(-2\,\text{BesselJ}\left[-\frac{2}{3},\frac{2}{3}\,i\,x^{3/2}\right]-\text{BesselJ}\left[-\frac{4}{3},\frac{2}{3}\,i\,x^{3/2}\right]c_1+\text{BesselJ}\left[\frac{2}{3},\frac{2}{3}\,i\,x^{3/2}\right]c_1\right)\right)\Big/$$
$$\left.\left.\left(2x\left(\text{BesselJ}\left[\frac{1}{3},\frac{2}{3}\,i\,x^{3/2}\right]+\text{BesselJ}\left[-\frac{1}{3},\frac{2}{3}\,i\,x^{3/2}\right]c_1\right)\right)\right\}\right\}$$

However, the solution to the following differential equation looks rather simple comparatively:

In[ ]:= **DSolve[y'[x] == x * y[x], y[x], x]**

$$\text{Out[ ]=}\left\{\left\{y[x]\to e^{\frac{x^2}{2}}\,c_1\right\}\right\}$$

This is because the equation is **separable**.

This lesson will cover how to solve separable differential equations.

Definition

A **separable differential equation** is an equation of the form $y' = f[x] * g[y]$. It is called separable because x and y can be **separated** into two functions, as shown. To solve separable differential equations, recall that $y' = \frac{dy}{dx}$.

Using the differential notation, you can put all the y's on one side of the equation and all the x's on the other. Integrating both sides of the equation leaves an equivalent equation relating y and x, which gives the general solution.

Here is an example with the differential equation in the introduction, $y' = x * y$:

In[ ]:= **y' == x * y;**

In this case, $f[x] = x$ and $g[y] = y$.

Rearranging the equation, you get the equivalent equation:

In[·]:= $(1/y)\, dy == x\, dx;$

Integrating both sides gives:

In[·]:= $\int (1/y)\, dy == \int x\, dx + c$

Out[·]= $Log[y] == c + \dfrac{x^2}{2}$

So $y = e^{x^2/2+C} = C * e^{x/2}$ is a general solution to the differential equation.

$y' = y$

Solve the differential equation $y' = y$:

In[·]:= $y' == y;$

In this case, $f[x] = 1$ and $g[y] = y$:

In[·]:= $y' == (1)*y;$

Rearranging, it is equivalent to:

In[·]:= $dy/y == 1\, dx;$

Now integrate:

In[·]:= $\int (1/y)\, dy == \int 1\, dx$

Out[·]= $Log[y] == x$

Solving for y, you get $y = e^{x+C} = e^x * e^C = K * e^x$, for some constant K.

Confirm with DSolveValue:

In[·]:= $DSolveValue[y'[x] == y[x], y[x], x]$

Out[·]= $e^x\, c_1$

A Harder Example

Solve the differential equation $y' = 3\, x^3 / (4\, y + Sin[y])$:

In[·]:= $y' == 3\, x\wedge 3/(4\, y + Sin[y]);$

In this case, $f[x] = 3\, x^3$ and $g[y] = 1/(4\, y + Sin[y])$:

In[·]:= $y' == (3\, x\wedge 3)*1/(4\, y + Sin[y]);$

Rearranging, it is equivalent to:

In[ ]:= **(4 y + Sin[y]) *d* y == 3 x^3 *d* x;**

Now integrate:

In[ ]:= $\int$ **(4 y + Sin[y]) *d* y ==** $\int$ **3 x^3 *d* x**

Out[ ]= **2 y^2 − Cos[y] == $\dfrac{3 x^4}{4}$**

The solutions obey the equation $2\ y^2 - \text{Cos}[y] == 3\ x^4 / 4 + C.$

Here is a plot of several of them with ContourPlot:

In[ ]:= **ContourPlot[Evaluate[Table[2 y^2 − Cos[y] == 3 x^4 / 4 + i, {i, 0, 5}]],**
{x, −5, 5}, {y, −5, 5}, Axes → True]

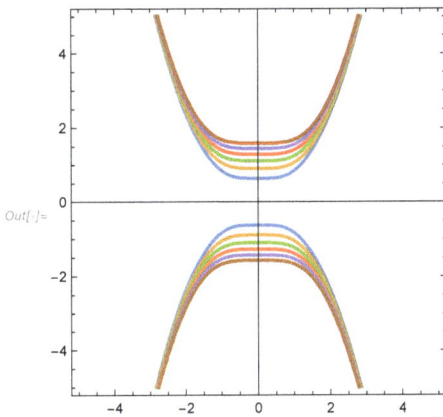

Initial Values

Solve the equation $y' = x * y$, with $y[0] = 1$.

The preceding problem is called an **initial value problem** and lets you find a particular solution to the differential equation.

$f[x] = x$ and $g[y] = y$, so rearrange and integrate:

In[ ]:= $\int$ **(1 / y) *d* y ==** $\int$ **x *d* x**

Out[ ]= **Log[y] == $\dfrac{x^2}{2}$**

The general solution is $y = K * e^{x^2/2}$.

Plugging in the initial conditions, you can see that $1 = K * e^0 = K$, so the solution to the initial value problem is $y = e^{x^2/2}$.

Confirm with DSolve:

In[·]:= DSolve[{y'[x] == x * y[x], y[0] == 1}, y[x], x]

Out[·]= $\left\{\left\{y[x] \rightarrow e^{\frac{x^2}{2}}\right\}\right\}$

Here is a plot with StreamPlot. The solution is blue:

In[·]:= StreamPlot[{1, x*y}, {x, −2, 2}, {y, 0, 2}, StreamPoints → {{{{0, 1}, Blue}, Automatic}}]

Out[·]=

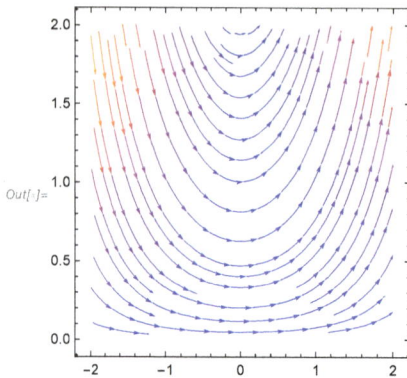

Current

For a circuit with charge q, resistance r, capacitance c and voltage v, there is the following equation:

In[·]:= r * D[q[t], t] + q/c == v[t]

Out[·]= $\dfrac{q}{c} + r\, q'[t] == v[t]$

Here t is time. Find the general solution to the equation when r, c and v are constant.

First, rearrange the equation so $q'[t]$ is by itself:

In[·]:= Solve[r * D[q[t], t] + q/c == v, D[q[t], t]]

Out[·]= $\left\{\left\{q'[t] \rightarrow \dfrac{-q + c\, v}{c\, r}\right\}\right\}$

This is a separable equation with $f[t] = 1$ and $g[q] = (-q + c*v)/(c*r)$. So rearrange and integrate:

In[·]:= $\int c*r/(-q + c*v)\, dq == \int 1\, dt$

Out[·]= $-c\, r\, Log[-q + c\, v] == t$

Solve for q:

$In[\cdot]:=$ **Normal@Solve[−c∗r Log[−q + c∗v] == t + k && c > 0, q, Reals]**

$Out[\cdot]=\left\{\left\{q \rightarrow -e^{\frac{-k-t}{cr}} + cv\right\}\right\}$

So the solution is $c*v - K\,e^{-t/(c*r)}$ for some constant K. Since $e^{-\infty} = 0$, you can see that as the time approaches ∞, the charge approaches $c*v$.

Newton's Law of Cooling

Consider an object at temperature t_0 that is placed in a medium where the temperature is held constant at s.

It can be modeled by the following equation:

$In[\cdot]:=$ **temp '[t] == k (temp[t] − s);**

Here k is a constant to be determined, and you have the initial value problem $temp[0] = t_0$.

This equation is separable because $f[t] = 1$ and $g[temp] = k\,(temp - s)$. So rearrange and integrate:

$In[\cdot]:=\int 1/(k\,(temp − s))\,d\,temp == \int 1\,d\,t$

$Out[\cdot]=\dfrac{Log[−s + temp]}{k} == t$

Then solve for the temperature (do not forget C, which is called c):

$In[\cdot]:=$ **Solve[$\dfrac{Log[−s + temp]}{k}$ == t + c, temp, Reals]**

$Out[\cdot]=\left\{\{temp \rightarrow e^{ck+kt} + s\}\right\}$

Since k and c are constants, e^{k*c} is a constant, which is again called c. Now solve for c using the initial condition:

$In[\cdot]:=$ **Solve[c ∗ Exp[k ∗ 0] + s == t0, c, Reals]**

$Out[\cdot]=\{\{c \rightarrow −s + t0\}\}$

So the solution is $(-s + t_0)\,e^{k*t} + s$. Confirm with DSolve:

$In[\cdot]:=$ **DSolve[{temp '[t] == k (temp[t] − s), temp[0] == t0}, temp[t], t]**

$Out[\cdot]=\{\{temp[t] \rightarrow s − e^{kt}\,s + e^{kt}\,t0\}\}$

Determining *k*

For Newton's law of cooling, there is a constant k. You cannot determine k just by knowing the initial temperature. If you know the temperature t_1 at **a later time** τ, then solve for k.

Recall that the solution to the equation is $(-s + t_0)\, e^{k*t} + s$.

Since there are two times for the temperature, you now have a system of equations, which you can solve with Solve:

In[ ]:= **Normal@**
    Solve[
        {(−s + t0) Exp[k ∗ 0] + s == t0 && (−s + t0) Exp[k ∗ τ] + s == t1 && 0 < s < t0 < τ},
        k, Reals]

Out[ ]= $\left\{\left\{ k \rightarrow \dfrac{\text{Log}\left[\frac{s-t1}{s-t0}\right]}{\tau} \right\}\right\}$

If you have a cup of boiling coffee at $100°C$, put it in a room with temperature $22°C$, and an hour later it has temperature $60°C$, you can model its temperature by the following equation:

In[ ]:= **coffee[*t*_] :=**
        (100 − 22) Exp[Log[(22 − 60)/(22 − 100)] *t* / 1] + 22

Here is a plot of the function over 10 hours:

In[ ]:= **Plot[coffee[t], {t, 0, 10}, PlotRange → {0, 100}]**

Out[ ]=

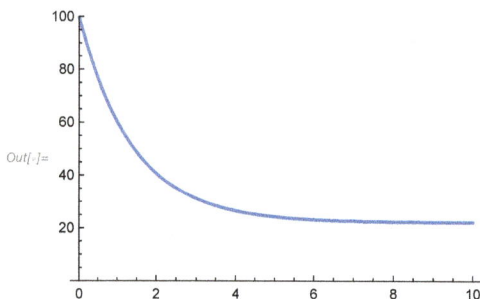

You can see that the limit as time approaches ∞ is $22°C$:

In[ ]:= **Limit[coffee[t], t → Infinity]**

Out[ ]= **22**

Summary

Separable differential equations are rather easy-to-solve differential equations.

They take the form $y' = f[x] * g[y]$ and can be rearranged with the help of differentials to solve the equation.

A circuit with constant voltage, a resistor and a capacitor can have its charge modeled by the solution to a separable differential equation.

A hot object placed into a colder environment can have its temperature over time modeled by the solution to a separable differential equation.

The next lesson will show how to use the Wolfram Language to make programs that can help solve calculus problems.

Exercises

Exercise 1—General Solution

Find the general solution to the following differential equation:

In[-]:= y' == -2 x*y^3;

Solution

Since it is separable with $f[x] = -2\,x$ and $g[y] = y^3$, rearrange and integrate:

In[-]:= $\int 1/(y\wedge 3)\,dy == \int -2\,x\,dx$

Out[-]= $-\dfrac{1}{2\,y^2} == -x^2$

Solve for y:

In[-]:= **Solve**$\left[-\dfrac{1}{2\,y^2} == -x^2 + c,\ y\right]$

Out[-]= $\left\{\left\{y \to -\dfrac{1}{\sqrt{2}\ \sqrt{-c+x^2}}\right\}, \left\{y \to \dfrac{1}{\sqrt{2}\ \sqrt{-c+x^2}}\right\}\right\}$

Here are two plots of the solution when $c = 1$:

In[-]:= **Plot**$\left[\left\{-\dfrac{1}{\sqrt{2}\ \sqrt{-1+x^2}},\ \dfrac{1}{\sqrt{2}\ \sqrt{-1+x^2}}\right\}, \{x, -5, 5\},\ \text{PlotLegends} \to \text{"Expressions"}\right]$

Out[-]=

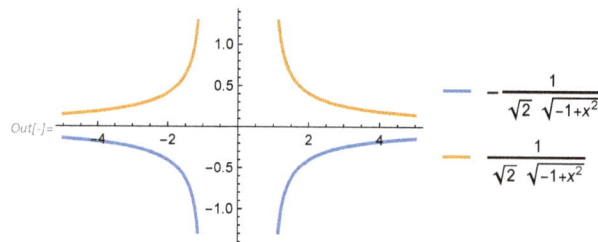

Legend:
- $-\dfrac{1}{\sqrt{2}\ \sqrt{-1+x^2}}$
- $\dfrac{1}{\sqrt{2}\ \sqrt{-1+x^2}}$

Exercise 2—Initial Value Problem

Find the equation of the curve that goes through the point (π, e) and has slope $\text{Sin}[x]/\text{Log}[y]$.

Solution

You are essentially solving the following differential equation with initial value $y[\pi] = e$:

In[ ]:= **y' == Sin[x]/Log[y];**

Since this is a separable equation $(f[x] = \mathrm{Sin}[x], g[y] = 1/\mathrm{Log}[y])$, rearrange and integrate:

In[ ]:= $\int$**Log[y]** *d***y ==** $\int$**Sin[x]** *d***x**

Out[ ]= **−y + y Log[y] == −Cos[x]**

With the initial condition, solve for c:

In[ ]:= **Solve[−y + y Log[y] == −Cos[x] + c /. {x → π, y → E}, c]**

Out[ ]= **{{c → −1}}**

So the curve has equation $y * \mathrm{Log}[y] - y = -\mathrm{Cos}[x] - 1$.

Here is its plot:

In[ ]:= **ContourPlot[−y + y Log[y] == −Cos[x] − 1, {x, −4 π, 4 π}, {y, −5, 5}, Axes → True]**

Out[ ]=

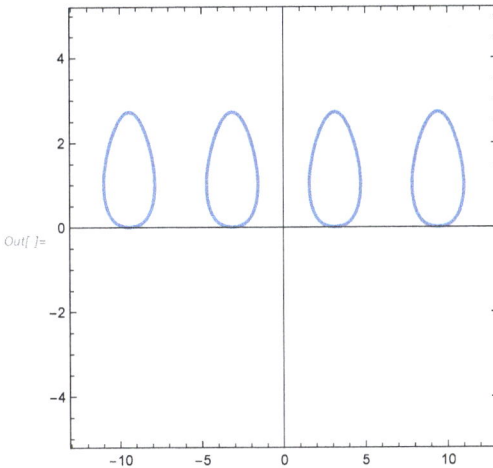

Exercise 3—Current

A circuit with inductance l, current i, resistance r and constant voltage v has the following differential equation:

In[ ]:= **l * D[i[t], t] + r * i == v**

Out[ ]= **i r + l i′[t] == v**

Find the general solution to the equation when $l = 5\ H$, $r = 6\ \Omega$ and $v = 100\ V$.

Solution

The equation can be rearranged by solving for the derivative of the current:

In[·]:= Solve[l * D[i[t], t] + r * i == v, D[i[t], t]] /. {r → 6, v → 100, l → 5}

Out[·]= $\left\{\left\{i'[t] \to \frac{1}{5}(100 - 6\,i)\right\}\right\}$

The equation is separable with $f[t] = 1$ and $g[i] = (100 - 6\,i)/5$. So rearrange and integrate:

In[·]:= $\int 5/(100 - 6\,i)\,di == \int 1\,dt$

Out[·]= $-\frac{5}{6} \text{Log}[100 - 6\,i] == t$

Then solve for i:

In[·]:= Solve$\left[-\frac{5}{6} \text{Log}[100 - 6\,i] == t + c, i, \text{Reals}\right]$

Out[·]= $\left\{\left\{i \to \frac{1}{6}\left(100 - e^{-\frac{6c}{5} - \frac{6t}{5}}\right)\right\}\right\}$

So the general solution is $i[t] = \frac{1}{6}\left(100 - e^{-6\,c/5 - 6\,t/5}\right) = 50/3 - C * e^{-6\,t/5}$ for some constant C.

Exercise 4—Newton's Law of Cooling

A block of molten lava (1000°C) is submerged in boiling water (100°C). After 10 minutes, the temperature of the lava is 450°C. How long does it take the lava to reach 110°C?

Solution

First, find k:

In[·]:= k = Log[(100 − 450)/(100 − 1000)]/10

Out[·]= $-\frac{1}{10} \text{Log}\left[\frac{18}{7}\right]$

Here is the function that models the temperature of the lava after t minutes:

In[·]:= lava[t_] := (1000 − 100) Exp[k * t] + 100

Solve for t when the temperature is $110°C$:

In[ ]:= **Solve[lava[t] == 110, t] // N // Quiet**

Out[ ]= **{{t → 47.6442}}**

So it takes about 47 minutes, 38.65 seconds for the lava to reach $110°C$.

Here is a plot of the temperature over an hour:

In[ ]:= **Plot[lava[t], {t, 0, 60}]**

Out[ ]=

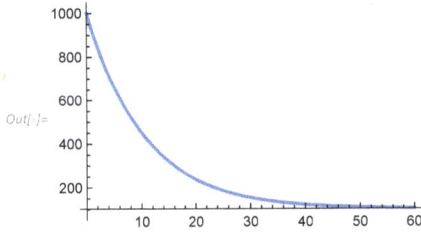

Exercise 5—Predator-Prey

The predator-prey relationship between coyotes and roadrunners can be modeled by the following differential equation:

In[ ]:= **D[c[r], r] == (−0.03 c + 0.00008 c r)/(0.1 r − 0.003 c r)**

Out[ ]= $c'[r] == \dfrac{-0.03\,c + 0.00008\,c\,r}{0.1\,r - 0.003\,c\,r}$

Solve the equation.

Solution

This is a separable equation! $f[r] = (-0.03 + 0.00008\,r)/r$ and $g[c] = c/(0.1 - 0.003\,c)$.

The equation can be rearranged and integrated as follows:

In[ ]:= $\int (0.1 - 0.003\,c)/c\ dc == \int (-0.03 + 0.00008\,r)/r\ dr + \text{const}$

Out[ ]= **−0.003 c + 0.1 Log[c] == const + 0.00008 r − 0.03 Log[r]**

Exponentiate (since **const** is a constant, e^{const} is also a constant):

In[ ]:= **Exp[−0.003 c + 0.1 Log[c]] == const * Exp[0.00008 r − 0.03 Log[r]]**

Out[ ]= $c^{0.1}\, e^{-0.003\,c} == \dfrac{\text{const}\, e^{0.00008\,r}}{r^{0.03}}$

Rewrite this as $c^{0.1}\, r^{0.03} \big/ \left(e^{0.003\, c}\, e^{0.00008\, r}\right) = $ const. Here is a plot of the predator-prey relationship when const = 1.45:

In[·]:= **ContourPlot[1.45 == (r^0.03 c^0.1)/(Exp[0.00008 r] Exp[0.003 c]), {r, 0, 1500}, {c, 0, 80}]**

Out[·]=

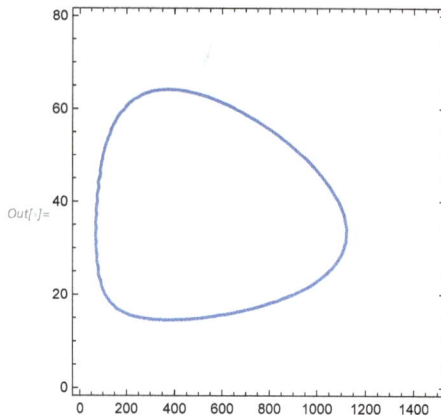

38 | Calculus and Programming

Overview

Throughout this journey into calculus, you have run into problems where you have to do a certain operation multiple times.

For example, Euler's method uses a recurrence relation, which can be tedious to do by hand.

Other times, you have found a technique that is so useful that it would be nice to have something that lets you do it very easily on a computer.

The closed interval method is helpful when finding the absolute maximum or minimum on a closed interval.

You can do both things with **programs**, and the Wolfram Language is highly suited to creating such programs.

In this lesson, you will make your own programs within the Wolfram Language.

Preliminaries

In order to make programs in the Wolfram Language, you will need a couple of tools.

You need the function If to control what to do in different situations for the programs.

The character ; will be useful. It lets you **stop** an instruction and move on to another instruction.

The function Print lets you put text on the screen and returns Null, which can be useful for debugging.

Here are some examples:

```
In[ ]:= ifexample[a_] := If[a > 0, Print["positive"], Print["not positive"]]
```

ifexample prints "positive" if the input is a positive number, and prints "not positive" otherwise:

```
In[ ]:= {ifexample[10], ifexample[0], ifexample[-4]}
```

```
positive

not positive

not positive
```

```
Out[ ]= {Null, Null, Null}
```

Now set the variable a to 5, and return the value $a + 6$:

```
In[·]:= a = 5; a + 6
```

```
Out[·]= 11
```

Comments and For

Using the notation (* *) lets you put **comments** in your code. Comments are very useful when explaining how the code works. Anything put between the two asterisks * * **will not be evaluated**.

Here is an example:

```
In[·]:= c = 4;(* c=6 *); c
```

```
Out[·]= 4
```

The For function is useful when you want to do a task repeatedly.

The following program prints the numbers from 1 to 10:

```
In[·]:= For[i = 1, (* start at 1 *)
    i < 11 , (* keep going as long as i is less than 11 *)
    i++ , (* increment i after every pass–thru *)
    Print[i] (* On each pass–thru print the number i *)]
1
2
3
4
5
6
7
8
9
10
```

Global versus Local Variables

All of the variables so far have been **global**; once they were set, they could be used anywhere and evaluate to their set value:

In[ ]:= **a**

Out[ ]= 5

You set *a* to 5 previously, and calling it here returned the same value. This can be a huge problem when making programs. You do **not** want variables from multiple programs interfering with each other.

For example, consider the following function:

In[ ]:= **f[*x_*] := *x* + a**

In[ ]:= **f[2]**

Out[ ]= 7

You may have wanted the function to return 2 + *a*, instead of 2 + 5 = 7. You could have used another variable, but instead you will use **local variables**. Their values are **local** to one instance of the program running.

The Wolfram Language uses Module to make programs with local variables:

In[ ]:= **Clear[f]**

In[ ]:= **f[*x_*] := Module[{a}, *x* + a]**

In[ ]:= **f[2]**

Out[ ]= 2 + a$27784

The extra symbols at the end of *a* indicate that it is a local variable.

exampleLimit

You now have enough to make a simple program. It will print "limit exists" and return the limit of a function if the limit exists. Otherwise, it will print "nonexistent":

In[ ]:= **exampleLimit[*function_*, *a_*] := Module[{y = *a*},**
    If[Limit[*function*[x], x → y, Direction → "FromAbove"] ==
       Limit[*function*[x], x → y, Direction → "FromBelow"],
      Print["limit exists"]; Limit[*function*[x], x → y],
      Print["nonexistent"]]]

The program first creates the local variable *y*, which equals the function input *a*. It then checks that the left-hand limit of the function at *y* equals the right-hand limit of the function at *y*.

If the limits are equal, then it prints "limit exists" and the limit is returned. Otherwise it prints "nonexistent".

Here are some example functions. The first function is continuous, so the limit exists at 3. The second function has its left and right limits equal at 0, so its limit exists at 0. The last function has its left and right limits not equal at 0, so the limit does not exist there:

```
In[ ]:= f[x_] := x^2
        g[x_] := 1/x^2
        h[x_] := 1/x
```

```
In[ ]:= exampleLimit[f, 3]
```
limit exists

Out[ ]= 9

```
In[ ]:= exampleLimit[g, 0]
```
limit exists

Out[ ]= ∞

```
In[ ]:= exampleLimit[h, 0]
```
nonexistent

limitTable

When you were estimating limits with tables in the first lesson on limits, it would have been tiresome to keep inputting the following code:

```
In[ ]:= {TraditionalForm[
          Grid[Join[{{"x", f["x"]}}, Transpose[{{-1, -0.5, -0.1, -0.05, -0.01, -0.005, -0.001},
              f[{-1, -0.5, -0.1, -0.05, -0.01, -0.005, -0.001}]}]], Frame → All]],
        TraditionalForm[
          Grid[Join[{{"x", f["x"]}}, Transpose[{{1, 0.5, 0.1, 0.05, 0.01, 0.005, 0.001},
              f[{1, 0.5, 0.1, 0.05, 0.01, 0.005, 0.001}]}]], Frame → All]]}
```

Out[ ]=

x	x^2
-1	1
-0.5	0.25
-0.1	0.01
-0.05	0.0025
-0.01	0.0001
-0.005	0.000025
-0.001	$1. \times 10^{-6}$

x	x^2
1	1
0.5	0.25
0.1	0.01
0.05	0.0025
0.01	0.0001
0.005	0.000025
0.001	$1. \times 10^{-6}$

You could copy and paste, but that takes up a lot of space. For limits at points other than zero, you also have to change the values around the point.

Here is a function that makes a table to help estimate the limit of any function around any point:

```
In[ ]:= limitTable[function_, xlimit_] := Module[{list},
          list = {1, 0.5, 0.1, 0.05, 0.01, 0.005, 0.001};
          {TraditionalForm[
              Grid[Join[{{"x", function["x"]}},
                  Transpose[{xlimit − list, function[xlimit − list] // N}]], Frame → All]],
              TraditionalForm[
                  Grid[Join[{{"x", function["x"]}},
                      Transpose[{xlimit + list, function[xlimit + list] // N}]], Frame → All]]}]
```

limitTable Explanation

The **limitTable** program does five things.

It first asks for the function (*function*) you are taking the limit of and the point (*xlimit*) at which you are taking the limit.

Then it makes the local variable *list*.

It sets the value of *list* to {1, 0.5, 0.1, 0.05, 0.01, 0.005, 0.001}.

Then it makes two tables. The first table consists of the values less than *xlimit*, made by subtracting the values of the list from *xlimit* (which makes another list), and the function values at those points.

The second table consists of the values greater than *xlimit*, which are made by adding the values of the list to *xlimit*, and the function values at those points.

Here is a table to estimate the limit of $x^3 − 2x + 9$ as x approaches 5:

```
In[ ]:= f[x_] := x^3 − 2x + 9
```

```
In[ ]:= limitTable[f, 5]
```

Out[ ]=
x	$x^3 − 2x + 9$
4	65.
4.5	91.125
4.9	116.849
4.95	120.387
4.99	123.271
4.995	123.635
4.999	123.927

,

x	$x^3 − 2x + 9$
6	213.
5.5	164.375
5.1	131.451
5.05	127.688
5.01	124.732
5.005	124.365
5.001	124.073

It appears the limit is 124, as it should be since $5^3 − 2*5 + 9 = 124$.

Closed Interval Method

The closed interval method is good for calculating the absolute maximum and minimum on a closed interval.

Make a program that does it for you:

```
In[ ]:= closedinterval[function_, a_, b_] :=
    Module[{min, fmin, max, fmax,
        sol = Solve[function'[x] == 0 && a < x < b, x],
        list = {a, b}}, (* these are our local variables *)
      For[i = 1, i ≤ Length[sol], i++,
        AppendTo[list, sol[[i, 1, 2]]]
      ]; (* we add the critical numbers to list *)
      fmin = Min[function[list]];
      fmax = Max[function[list]];
      (* we find the maximum and minimum function values *)
      For[i = 1, i ≤ Length[list], i++,
        If[fmin == function[list[[i]]], min = list[[i]]]];
      For[i = 1, i ≤ Length[list], i++,
        If[fmax == function[list[[i]]], max = list[[i]]]];
      (* we find the x-
        values corresponding to the maximum and minimum function values *)
      {{min, fmin}, {max, fmax}}
      (* finally we return the absolute max and min as a list of ordered pairs *)]
```

Find the absolute max and min for the function $f[x] = x^2$ on the interval $[-1, 2]$:

```
In[ ]:= f[x_] := x^2
```

```
In[ ]:= closedinterval[f, -1, 2]
```

```
Out[ ]= {{0, 0}, {2, 4}}
```

Try it for the same function on the interval $[3, 5]$:

```
In[ ]:= closedinterval[f, 3, 5]
```

```
Out[ ]= {{3, 9}, {5, 25}}
```

Caveats

When using the program **closedinterval**, you have to assume that the function you are inputting is differentiable on the open interval (a, b).

The program cannot find points where the slope is undefined. So it would not find the absolute minimum for $|x|$ on the interval $[-1, 1]$:

```
In[ ]:= f[x_] := RealAbs[x]
```

```
In[ ]:= closedinterval[f, -1, 1]
```

```
Out[ ]= {{1, 1}, {1, 1}}
```

As should be expected, it cannot find the (nonexistent) absolute maximum for $1/x$ on the interval $[-1, 1]$, since it has a discontinuity at 0:

```
In[ ]:= g[x_] := 1/x
```

```
In[ ]:= closedinterval[g, -1, 1]
```

```
Out[ ]= {{-1, -1}, {1, 1}}
```

$x^{2/3}$ has a cusp and a minimum at 0, but the minimum cannot be found with this program:

```
In[ ]:= h[x_] := CubeRoot[x]^2
```

```
In[ ]:= closedinterval[h, -1, 1]
```

```
Out[ ]= {{1, 1}, {1, 1}}
```

Recursion

Make a function that makes points for Euler's method:

```
In[ ]:= eulermethod[function_, start_, end_, initvalue_, stepsize_] :=
    Module[{m, xvalues, yvalues},
        m = (end - start)/stepsize; (* the number of steps to be used *)
        xvalues = Table[start + stepsize * i, {i, 0, m}];
        (* a list of our x-values *)
        yvalues = Table[initvalue, {i, 0, m}];
        (* a list of what will be our future y-values *)
        For[i = 1, i < m + 1, i++,
            yvalues[[i + 1]] = yvalues[[i]] + stepsize * function[xvalues[[i]], yvalues[[i]]]];
        (* Euler's Method being used *)
        Transpose[{xvalues, yvalues}(* the list of points obtained from Euler's Method *)]
    ]
```

Use **eulermethod** to estimate $y[2]$ for the differential equation $y' = y$ with $y[0] = 1$. Use step size 0.1:

```
In[ ]:= f[x_, y_] := y
```

```
In[ ]:= Last@eulermethod[f, 0, 2, 1, 0.1]
```

```
Out[ ]= {2., 6.7275}
```

The last ordered pair {2, 6.7275} gives the estimate $y[2] = 6.7275$. Compare it to e^2 (the actual value):

In[·]:= **RealAbs[Exp[2] – 6.7275]**

Out[·]= 0.661556

Decreasing the step size gives a better estimate:

In[·]:= **RealAbs[Exp[2] – (Last@eulermethod[f, 0, 2, 1, 0.05])〚2〛]**

Out[·]= 0.349067

Summary

The Wolfram Language is great for making programs.

Using functions like If, For, Print and Module is useful for various reasons.

Comments are **crucial** for understanding code.

It is best to use local variables in a program as much as possible, which you can do with Module.

Methods like the closed interval method and Euler's method can be programmed with the Wolfram Language.

The next lesson will give Part 1 for a sample calculus exam.

Exercises

Exercise 1—Even or Odd

Make a program that prints "even" if the input is even, and "odd" if the input is odd (hint: use Mod).

Solution

Use If, Mod and Print:

```
In[ ]:= evenodd[x_] := If[Mod[x, 2] == 0, Print["even"], Print["odd"]]
```

Check for the list of numbers 0 to 10:

```
In[ ]:= evenodd[#] & /@ Range[0, 10]
```

```
even

odd

even

odd

even

odd

even

odd

even

odd

even
```

```
Out[ ]= {Null, Null, Null, Null, Null, Null, Null, Null, Null, Null, Null}
```

Exercise 2—Sum

Sum the numbers from 0 to 956 using a for-loop.

Solution

Here is the for-loop. Note that *sum* is first set to be 0:

In[·]:= **sum = 0;**

In[·]:= **For[i = 0, i < 957, i++, sum = sum + i]**

In[·]:= **sum**

Out[·]= 457 446

Confirm with Sum:

In[·]:= **Sum[i, {i, 0, 956}]**

Out[·]= 457 446

Sum the numbers from 45 to 10203. Set *sum* to again be 0, since it is a global variable:

In[·]:= **sum = 0;**

In[·]:= **For[i = 45, i < 10 204, i++, sum = sum + i]**

In[·]:= **sum**

Out[·]= 52 054 716

Confirm with Sum:

In[·]:= **Sum[i, {i, 45, 10 203}]**

Out[·]= 52 054 716

Exercise 3—Midpoint

Make a program that uses the midpoint rule to numerically integrate any function from a to b for any number of subintervals.

Solution

The midpoint rule uses the formula $\sum_{i=1}^{n} f[(x_{i-1} + x_i)/2] \, \Delta x$ to approximate integrals, where $\Delta x = (b - a)/n$.

Set up a Module:

In[·]:= **midpoint[*function_*, *a_*, *b_*, *n_*] :=**
    Module[{width = (*b* − *a*)/*n*, sum},
      sum = Sum[*function*[(((*a* + (i − 1) width) + (*a* + i ∗ width))/2], {i, 1, *n*}] width;
      sum]

Use the program to approximate the integral of $1/x$ from 1 to e with 10 subintervals:

In[ ]:= **f[x_] := 1 / x**

In[ ]:= **midpoint[f, 1, E, 10] // N**

Out[ ]= 0.998942

The actual integral is 1:

In[ ]:= **Integrate[1 / x, {x, 1, E}]**

Out[ ]= 1

The error is rather small:

In[ ]:= **RealAbs[Integrate[1 / x, {x, 1, E}] – midpoint[f, 1, E, 10]] // N**

Out[ ]= 0.00105757

With more subintervals, it is even smaller:

In[ ]:= **RealAbs[Integrate[1 / x, {x, 1, E}] – midpoint[f, 1, E, 20]] // N**

Out[ ]= 0.00026554

Exercise 4—Mean Value Theorem

Make a program that utilizes the mean value theorem for functions with endpoints a and b.

Solution

Set up a Module:

In[ ]:= **meanvaluetheorem[*function_*, *a_*, *b_*] :=**
      Module[{sol},
         sol = Solve[D[*function*[x], x] == (*function*[*b*] – *function*[*a*]) / (*b* – *a*) && *a* < x < *b*, x];
         {sol[[1, 1, 2]], *function*[sol[[1, 1, 2]]]}]

Here is an example of Rolle's theorem for the function $f[x] = x^2$ on the interval $[-1, 1]$:

In[ ]:= **f[x_] := x^2**

In[ ]:= **meanvaluetheorem[f, –1, 1]**

Out[ ]= {0, 0}

Here is an example of the mean value theorem for the function $g[x] = x^3 - 3x$ on the interval $[2, 5]$:

```
In[•]:= g[x_] := x^3 - 3 x
```

```
In[•]:= meanvaluetheorem[g, 2, 5]
```

$$Out[•]= \left\{ \sqrt{13}, 10 \sqrt{13} \right\}$$

Here is a plot to confirm:

```
In[•]:= Plot[{g[x], 2 + (g[5] - g[2])/(5 - 2) (x - 2), 10 Sqrt[13] + g'[Sqrt[13]] (x - Sqrt[13])},
        {x, 2, 5}, PlotLegends → {"function", "secant", "tangent"}]
```

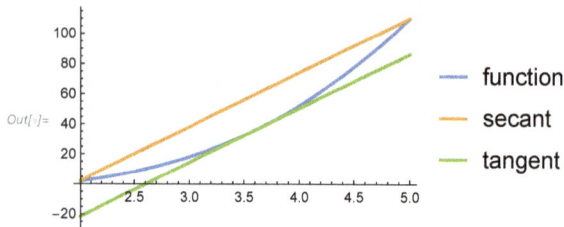

Exercise 5—Simpson's Rule

Make a program that uses Simpson's rule to numerically integrate any function from a to b for any number of subintervals. Simpson's rule uses the formula $\sum_{i=1}^{n/2} (f[x_{2i-2}] + 4 f[x_{2i-1}] + f[x_{2i}]) \Delta x / 3$ to approximate integrals, where $\Delta x = (b - a) / n$.

Solution

Set up a Module:

```
In[•]:= simpson[function_, a_, b_, n_] :=
        Module[{width = (b - a)/n, sum},
            sum = Sum[function[a + (2 i - 2) width] + 4 function[(a + (2 i - 1) width)] +
                    function[(a + 2 i * width)], {i, 1, n/2}] width/3;
            sum]
```

Use the program to again approximate the integral of $1/x$ from 1 to e with 10 subintervals:

```
In[•]:= f[x_] := 1/x
```

```
In[•]:= simpson[f, 1, E, 10] // N
```

```
Out[•]= 1.00003
```

The actual integral is 1:

In[ ]:= **Integrate[1/x, {x, 1, E}]**

Out[ ]= 1

The error is even smaller than that for the midpoint rule:

In[ ]:= **RealAbs[Integrate[1/x, {x, 1, E}] − simpson[f, 1, E, 10]] // N**

Out[ ]= 0.000026709

With more subintervals, it is even smaller:

In[ ]:= **RealAbs[Integrate[1/x, {x, 1, E}] − simpson[f, 1, E, 20]] // N**

Out[ ]= 1.75197×10^{-6}

39 | Sample Calculus Exam, Part 1

Overview

This course has gone over a lot of topics in differential calculus and integral calculus.

Now you will have a chance to discuss some exam problems to apply what you have learned.

The questions on the exam are similar to those you would see on the AP Calculus AB exam.

Question 1

Find the derivative of $f[x] = 5\,x / (9\,x^3 - 7\,x + 2)$ at 2.

Solution

The derivative can be found with the quotient rule.

The derivative is $\left((9\,x^3 - 7\,x + 2) * (5\,x)' - 5\,x * (9\,x^3 - 7\,x + 2)'\right) \big/ (9\,x^3 - 7\,x + 2)^2$.

Evaluate it at 2:

In[ ]:= `(((9 x^3 − 7 x + 2) * D[5 x, x] − 5 x * D[9 x^3 − 7 x + 2, x]) / ((9 x^3 − 7 x + 2)^2)) /. x → 2`

Out[ ]= $-\dfrac{71}{360}$

Confirm with D:

In[ ]:= `D[5 x / (9 x^3 − 7 x + 2), x] /. x → 2`

Out[ ]= $-\dfrac{71}{360}$

Question 2

Calculate the limit of $x^4 - 3x - 6\operatorname{Log}[x]$ at 5.

Solution

x^4, $3x$ and $\operatorname{Log}[x]$ are continuous at 5. So by the limit laws, the limit is just $5^4 - 3*5 - 6\operatorname{Log}[5]$:

In[·]:= **5^4 − 3 ∗ 5 − 6 Log[5]**

Out[·]= 610 − 6 Log[5]

Confirm with Limit:

In[·]:= **Limit[x^4 − 3 ∗ x − 6 Log[x], x → 5]**

Out[·]= 610 − 6 Log[5]

Here is a plot of the function around 5:

In[·]:= **Plot[x^4 − 3 x − 6 Log[x], {x, 4, 6}, Epilog → {PointSize[Large], Point[{5, 610 − 6 Log[5]}]}]**

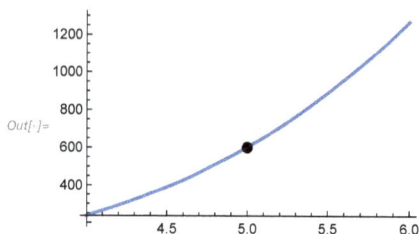

Question 3

Calculate the integral $\int_\pi^5 (\operatorname{Cos}[x] + 1/x)\,dx$.

Solution

By the fundamental theorem of calculus, the integral is $\operatorname{Sin}[5] + \operatorname{Log}[5] - (\operatorname{Sin}[\pi] + \operatorname{Log}[\pi]) = \operatorname{Sin}[5] + \operatorname{Log}[5/\pi]$.

Confirm with Integrate:

In[·]:= **Integrate[Cos[x] + 1/x, {x, π, 5}]**

Out[·]= $\operatorname{Log}\left[\dfrac{5}{\pi}\right] + \operatorname{Sin}[5]$

Here is a plot of the region:

In[ ]:= **Plot[Cos[x] + 1 / x, {x, π, 5}, Filling → Axis]**

Out[ ]=

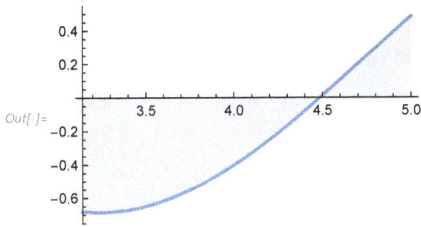

Question 4

Find the values of c that satisfy the mean value theorem for $f[x] = 2\,x^3 - x$ on the interval $[-1, 3]$.

Solution

Find the slope of the line going through $(-1, f[-1])$ and $(3, f[3])$:

In[ ]:= **f[x_] := 2 x^3 - x**

In[ ]:= **slope = (f[3] - f[-1]) / (3 - (-1))**

Out[ ]= 13

Using Solve, then find the points in the interval $[-1, 3]$ with slope 13:

In[ ]:= **Solve[D[f[x], x] == 13 && -1 < x < 3, x]**

Out[ ]= $\left\{\left\{x \to \sqrt{\dfrac{7}{3}}\right\}\right\}$

In[ ]:= **f[Sqrt[7 / 3]]**

Out[ ]= $\dfrac{11\sqrt{\dfrac{7}{3}}}{3}$

So when $c = \sqrt{7/3}$, i.e. at the point $\left(\sqrt{7/3}, 11\sqrt{7/3}/3\right)$, there is a point where the slope equals the slope of the line going through the endpoints of the interval. Here is a plot to illustrate:

```
In[·]:= Plot[{f[x], f[-1] + slope (x - (-1)), f[Sqrt[7/3]] + f'[Sqrt[7/3]] (x - Sqrt[7/3])},
       {x, -1, 3}, PlotLegends → {"f[x]", "secant", "tangent"}]
```

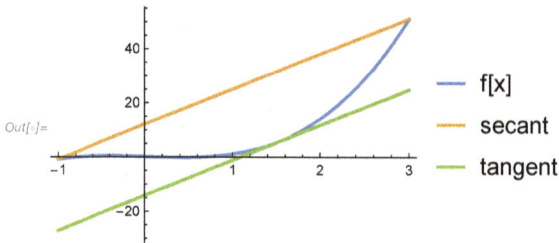

Question 5

Find the general solution to the equation $y' = 3x * (y^2 - 1)/y$.

Solution

This equation is separable. So rearrange and integrate:

```
In[·]:= ∫ y/(y^2 - 1) dy == ∫ 3 x dx + c
```

$$Out[\cdot]= \frac{1}{2} \text{Log}[-1 + y^2] == c + \frac{3x^2}{2}$$

Getting rid of the $1/2$ on both sides and exponentiating, you get:

```
In[·]:= Exp[Log[-1 + y^2]] == Exp[3 x^2 + c]
```

$$Out[\cdot]= -1 + y^2 == e^{c + 3x^2}$$

So $y^2 - 1 = c * e^{3x^2}$ describes the solutions to the equation.

Here are plots for the values $c = 0.1, 0.5, 1$ and 2:

In[ ]:= **ContourPlot[Evaluate[Table[y^2 – 1 == c Exp[3 x^2], {c, {0.1`, 0.5`, 1, 2}}]],**
    {x, –1, 1}, {y, –3, 3}, PlotLegends → "Expressions"]

Out[ ]=

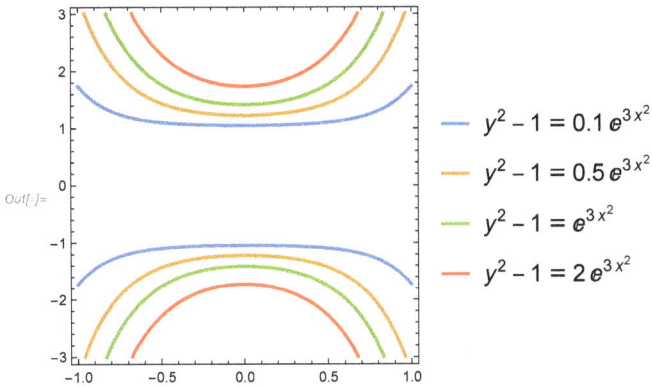

$y^2 - 1 = 0.1\,e^{3x^2}$

$y^2 - 1 = 0.5\,e^{3x^2}$

$y^2 - 1 = e^{3x^2}$

$y^2 - 1 = 2\,e^{3x^2}$

Question 6

Calculate the limit of $\left(10\,x^3 - 3\,x\right)/\left(3\,x^3 - 4\right)$ at ∞.

Solution

Since the degrees of the numerator and denominator are the same, the limit at infinity is $10/3$.

Confirm with Limit:

In[ ]:= **Limit[(10 x^3 – 3 x)/(3 x^3 – 4), x → Infinity]**

Out[ ]= $\dfrac{10}{3}$

Here is a plot to better visualize:

In[ ]:= **Plot[(10 x^3 – 3 x)/(3 x^3 – 4), {x, 200, 1000}, GridLines → {None, {10/3}}]**

Out[ ]=

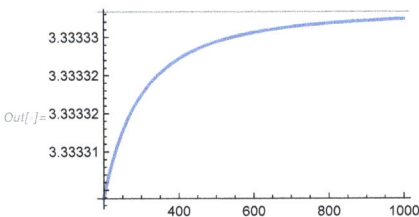

Question 7

Use the trapezoidal rule with five trapezoids to approximate $\int_2^{10} x^2 \sqrt{x+1} \; \text{Log}[x+3] \, dx$.

Solution

The width of each trapezoid is $\Delta x = (10 - 2)/5 = 8/5$.

With five trapezoids, the integral is approximately $\Sigma_{i=1}^5 (f[x_{i-1}] + f[x_i]) \, \Delta x / 2$, where $f[x_i] = x_i^2 \sqrt{x_i + 1} \; \text{Log}[x_i]$ and $x_i = 2 + i * 8 / 5$:

In[ ]:= **f[x_] := x^2 * Sqrt[x + 1] Log[x + 3]**

In[ ]:= **Sum[f[2 + (i − 1) 8 / 5] + f[2 + 8 i / 5], {i, 1, 5}] * (8 / (2 * 5)) // N**

Out[ ]= **2314.03**

Compare with Integrate:

In[ ]:= **Integrate[f[x], {x, 2, 10}] // N**

Out[ ]= **2267.1**

Here is a plot with the trapezoids:

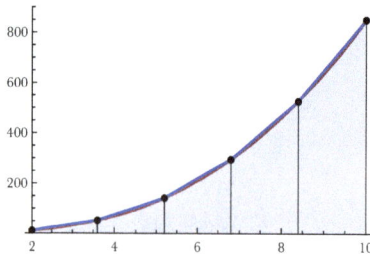

Question 8

Find the equation of the line tangent to the graph of $e^{x-1} - 3x$ at $(1, -2)$.

Solution

First, find the slope at $(1, -3)$. The derivative of $e^{x-1} - 3x$ is $e^{x-1} - 3$, so the slope is $e^0 - 3 = -2$.

Confirm with D:

In[ ]:= **D[Exp[x − 1] − 3 x, x] /. x → 1**

Out[ ]= **−2**

The point-slope formula says that the equation of the line is $y + 2 = -2 \, (x - 1)$.

Here is a plot to show the line and the graph of the function:

```
In[ ]:= Plot[{Exp[x – 1] – 3 x, –2 – 2 (x – 1)}, {x, 0, 2},
         PlotLegends → "Expressions", Epilog → {PointSize[Large], Point[{1, –2}]}]
```

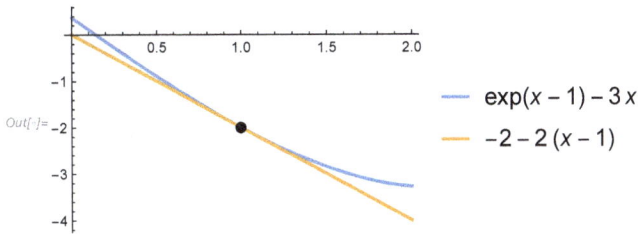

Question 9

What value of a will make the function $f[x] = \{x - 2, \ x < 2; \ a\,x^2 + 4, \ x \geq 2\}$ continuous?

Solution

The function is a piecewise function, and each piece is continuous, except possibly at 2. In order for the function to be continuous, the limit from the left at 2 must equal the limit from the right at 2.

Solve for a with Solve:

```
In[ ]:= Solve[Limit[x – 2, x → 2, Direction → "FromBelow"] ==
         Limit[a * x^2 + 4, x → 2, Direction → "FromAbove"], a]
```

```
Out[ ]= {{a → –1}}
```

So the function $f[x] = \{x - 2, \ x < 2; \ -x^2 + 4, \ x \geq 2\}$ is continuous.

Plot the function:

```
In[ ]:= Plot[ {  x – 2      x < 2
               { –x² + 4   x ≥ 2  ,  {x, –5, 5}]
```

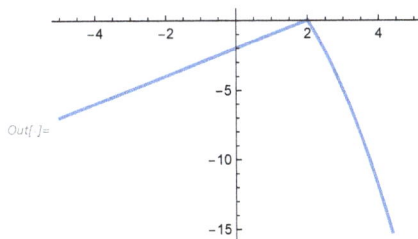

Question 10

Approximate $\sqrt{143.95}$ to four decimal places.

Solution

Since $\sqrt{144} = 12$ is close to $\sqrt{143.95}$, use a linear approximation and calculate the line tangent to $\sqrt{x}$ at $(144, 12)$.

The slope at $(144, 12)$ is $1 / (2\sqrt{144}) = 1/24$.

Confirm with D:

In[·]:= **D[Sqrt[x], x] /. x → 144**

Out[·]= $\dfrac{1}{24}$

So the tangent line has equation $y - 12 = 1/24\,(x - 144)$.

Therefore, $\sqrt{143.95}$ is approximately $12 + (143.95 - 144)/24$:

In[·]:= **12 + (143.95 − 144)/24**

Out[·]= **11.9979**

Compare it to the actual value:

In[·]:= **Sqrt[143.95]**

Out[·]= **11.9979**

In[·]:= **RealAbs[(12 + (143.95 − 144)/24) − Sqrt[143.95]]**

Out[·]= 1.80876×10^{-7}

The answers agree up to the ten millions place.

Question 11

The region bounded by the y axis, $x = 2$, the x axis and the function $f[x] = -x^3 + 5\,x$ is rotated around the y axis. Find the volume of the solid made by the rotation.

Solution

First, plot the region:

In[ ]:= **Plot[-x^3 + 5 x, {x, 0, 2}, GridLines → {{2}, None}, Filling → Axis]**

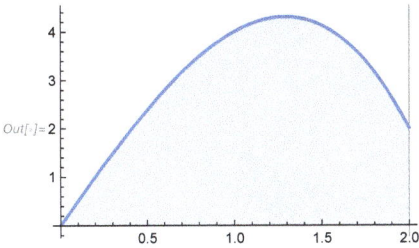

For this problem, it is best to use the cylindrical shell method. The radius is x, and the height is $-x^3 + 5x$.

So the integral you would calculate is $\int_0^2 2\pi x * (-x^3 + 5x)\, dx$:

The volume calculated with Integrate is:

In[ ]:= **Integrate[2 π x (-x^3 + 5 x), {x, 0, 2}]**

Out[ ]= $\dfrac{208\,\pi}{15}$

Question 12

Find the extrema of $f[x] = -78x + 58x^2 - 14x^3 + x^4$ on the interval $[2, 6]$. Use the closed interval method.

Solution

The function is differentiable on the interval, so the critical numbers are those with slope 0.

Use Solve and D:

In[ ]:= **Solve[D[-78 x + 58 x^2 - 14 x^3 + x^4, x] == 0 && 2 < x < 6, x]**

Out[ ]= **{{x → 3}}**

Now find the absolute maximum and minimum on the interval. They can only be at 2, 3 or 6:

In[ ]:= **f[x_] := -78 x + 58 x^2 - 14 x^3 + x^4**

In[ ]:= **f /@ {2, 3, 6}**

Out[ ]= **{-20, -9, -108}**

You can see that there is a local and absolute maximum at $(3, -9)$ and an absolute minimum at $(6, -108)$.

Here is a plot with the maximum and minimum highlighted:

In[·]:= **Plot[f[x], {x, 2, 6}, Epilog → {PointSize[Large], Point[{{3, f[3]}, {6, f[6]}}]}]**

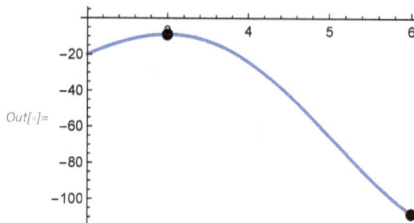

Out[·]=

Question 13

Calculate the derivative of $2\, x\, \mathrm{Tan}[x] - 5\, x^2\, \mathrm{Log}_4[x]$.

Solution

By the product rule, the derivative of the first term is $2\, x * \mathrm{Tan}\,'[x] + \mathrm{Tan}[x] * (2\, x)'$:

In[·]:= **2 x * D[Tan[x], x] + Tan[x] * D[2 x, x]**

Out[·]= $2 \, x \, \mathrm{Sec}[x]^2 + 2 \, \mathrm{Tan}[x]$

The derivative of the second term is $5\, x^2\, \mathrm{Log}_4\,'[x] + \mathrm{Log}_4[x] * (5\, x^2)'$, again by the product rule:

In[·]:= **5 x^2 * D[Log[4, x], x] + Log[4, x] * D[5 x^2, x]**

Out[·]= $\dfrac{5\,x}{\mathrm{Log}[4]} + \dfrac{10\,x\,\mathrm{Log}[x]}{\mathrm{Log}[4]}$

So by the difference rule, the derivative of $2\, x\, \mathrm{Tan}[x] - 5\, x^2\, \mathrm{Log}_4[x]$ is $2\, x * \mathrm{Tan}\,'[x] + \mathrm{Tan}[x] * (2\, x)' - (5\, x^2\, \mathrm{Log}_4\,'[x] + \mathrm{Log}_4[x] * (5\, x^2)')$.

Confirm with **D**:

In[·]:= **D[2 x * Tan[x] − 5 x^2 * Log[4, x], x]**

Out[·]= $-\dfrac{5\,x}{\mathrm{Log}[4]} - \dfrac{10\,x\,\mathrm{Log}[x]}{\mathrm{Log}[4]} + 2\,x\,\mathrm{Sec}[x]^2 + 2\,\mathrm{Tan}[x]$

40 | Sample Calculus Exam, Part 2

Overview

Here is the second part of the calculus exam.

As in Part 1, the questions are similar to those you would see on the AP Calculus AB exam.

However, the questions in this part are somewhat more difficult than those in the first part of the exam.

Question 14

Calculate the limit of $(\text{Sec}[\pi/3 + h] - \text{Sec}[\pi/3])/h$ at $h = 0$.

Solution

You could calculate the limit, but this could also be interpreted as the derivative of $\text{Sec}[x]$ at $\pi/3$, which is $\text{Sec}[\pi/3]\,\text{Tan}[\pi/3] = 2\sqrt{3}$.

Confirm with D:

In[ ]:= **D[Sec[x], x] /. x → π/3**

Out[ ]= $2\sqrt{3}$

Confirm with Limit:

In[ ]:= **Limit[(Sec[π/3 + h] − Sec[π/3])/h, h → 0]**

Out[ ]= $2\sqrt{3}$

Here is a plot around 0 as well:

In[ ]:= **Plot[(Sec[π/3 + h] − Sec[π/3])/h, {h, −1, 1},**
       Epilog → {Black, PointSize[Large], Point[{0, 2 Sqrt[3]}]}]

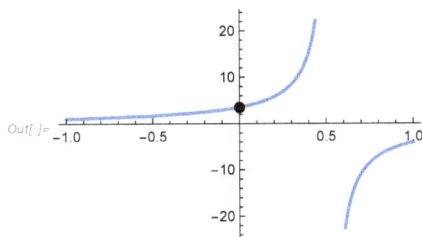

Question 15

A disk was measured to have radius 9 cm. The error was ±0.04 cm. Find the error in the area.

Solution

Since $A = \pi r^2$ for a disk of radius r and area A, you have the equation $dA = 2\pi r\,dr$:

$$A = \pi r^2$$

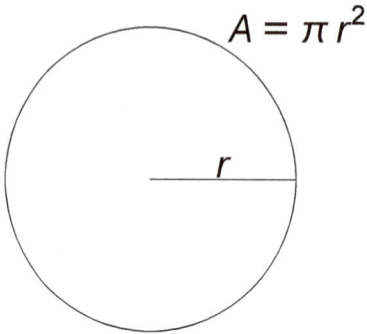

Since $r = 9$ cm and $dr = 0.04$ cm, the error in the area should be $2\pi(9 * 0.04)$ cm^2:

In[·]:= **2 π (9 ∗ 0.04)**

Out[·]= **2.26195**

Question 16

Calculate the derivative of $\text{Log}[x^2 - 3]$.

Solution

By the chain rule, the derivative of $\text{Log}[x^2 - 3]$ is $1/(x^2 - 3) * 2x = 2x/(x^2 - 3)$.

Confirm with **D**:

In[·]:= **D[Log[x^2 − 3], x]**

Out[·]= $\dfrac{2x}{-3 + x^2}$

Question 17

Find the third derivative of $3x^4 - \text{Cos}[x] + e^{2x}$.

Solution

The derivative of $3\,x^4 - \text{Cos}[x] + e^{2\,x}$ is $12\,x^3 + \text{Sin}[x] + 2\,e^{2\,x}$:

$\textit{In[ ]:=}$ **D[3 x^4 – Cos[x] + Exp[2 x], x]**

$\textit{Out[ ]=}$ $2\,e^{2\,x} + 12\,x^3 + \text{Sin[x]}$

The second derivative of $3\,x^4 - \text{Cos}[x] + e^{2\,x}$ is $36\,x^2 + \text{Cos}[x] + 4\,e^{2\,x}$:

$\textit{In[ ]:=}$ **D[3 x^4 – Cos[x] + Exp[2 x], {x, 2}]**

$\textit{Out[ ]=}$ $4\,e^{2\,x} + 36\,x^2 + \text{Cos[x]}$

The third derivative of $3\,x^4 - \text{Cos}[x] + e^{2\,x}$ is $72\,x - \text{Sin}[x] + 8\,e^{2\,x}$.

Confirm with D:

$\textit{In[ ]:=}$ **D[3 x^4 – Cos[x] + Exp[2 x], {x, 3}]**

$\textit{Out[ ]=}$ $8\,e^{2\,x} + 72\,x - \text{Sin[x]}$

Question 18

The region bounded by $x + 2$ and $-x^2 + 5\,x + 4$ is rotated around the x axis. Find the volume of the solid made by this rotation.

Solution

Plot the region:

$\textit{In[ ]:=}$ **Plot[{x + 2, –x^2 + 5 x + 4}, {x, –5, 5}]**

$\textit{Out[ ]=}$

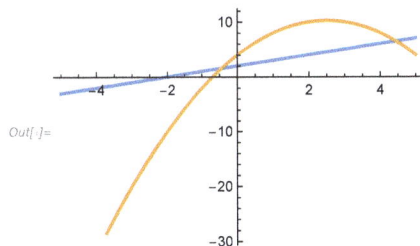

Find the places where the two curves intersect:

$\textit{In[ ]:=}$ **sol = Solve[x + 2 == –x^2 + 5 x + 4, x]**

$\textit{Out[ ]=}$ $\left\{\left\{x \to 2 - \sqrt{6}\right\}, \left\{x \to 2 + \sqrt{6}\right\}\right\}$

Let $f[x] = x + 2$ and $g[x] = -x^2 + 5\,x + 4$. f is below g.

You will use the washer method to calculate the integral. The integral is
$\int_{2-\sqrt{6}}^{2+\sqrt{6}} \pi\left(g[x]^2 - f[x]^2\right) dx.$

Compute the integral with Integrate:

In[·]:= **f[x_] := x + 2**
    g[x_] := −x^2 + 5 x + 4

In[·]:= **Integrate[π (g[x]^2 − f[x]^2), {x, 2 − Sqrt[6], 2 + Sqrt[6]}]**

Out[·]= $\dfrac{512 \sqrt{6}\ \pi}{5}$

Question 19

Calculate the limit of $\left(e^{2x} - e^{x/3}\right)/\mathrm{Tan}[x]$ at 0.

Solution

Plugging in 0 gives the indeterminate form $0/0$:

In[·]:= **{Exp[2 x] − Exp[x / 3] /. x → 0, Tan[x] /. x → 0}**

Out[·]= **{0, 0}**

So use l'Hôpital's rule:

In[·]:= **D[Exp[2 x] − Exp[x / 3], x] / D[Tan[x], x] /. x → 0**

Out[·]= $\dfrac{5}{3}$

Confirm with Limit:

In[·]:= **Limit[(Exp[2 x] − Exp[x / 3]) / Tan[x], x → 0]**

Out[·]= $\dfrac{5}{3}$

Here is a plot around 0:

In[·]:= **Plot[(Exp[2 x] − Exp[x / 3]) / Tan[x], {x, −1, 1},**
    Epilog → {Black, PointSize[Large], Point[{0, 5 / 3}]}]

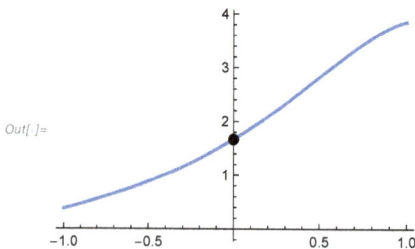

Question 20

Calculate the integral $\int_2^3 3\,x/\left(x^2 - 1\right) dx$:

Solution

Use the substitution rule and set $u = x^2 - 1$.

$du = 2\,x\,dx$ so $\frac{3}{2}\,du = \frac{3}{2} * 2\,x = 3\,x$.

The endpoints will be changed from 2 and 3 to 3 and 8.

$\int_2^3 3\,x/\left(x^2 - 1\right) dx = \int_3^8 \frac{3\,du}{2\,u} = \frac{3}{2}\,(\text{Log}[8] - \text{Log}[3]) = \frac{3}{2}\,\text{Log}\left[\frac{8}{3}\right].$

Confirm with Integrate:

In[ ]:= **Integrate[3 x/(x^2 – 1), {x, 2, 3}]**

Out[ ]:= $\dfrac{3}{2}\,\text{Log}\left[\dfrac{8}{3}\right]$

Here is a plot with the area shaded:

In[ ]:= **Plot[3 x/(x^2 – 1), {x, 2, 3}, Filling → Axis]**

Out[ ]:=

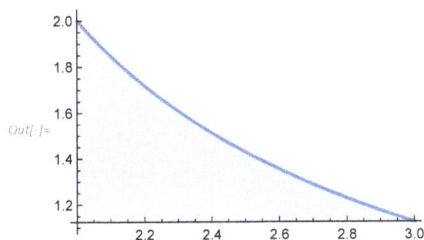

Question 21

Find y' if $y^2 + x * y = 2$.

Solution

Use implicit differentiation to find y'.

You have the equation $d\,y^2/dx + (x * d\,y/dx + y * 1) = 0$.

This is equivalent to $2\,y\,d\,y/dx + x\,d\,y/dx = -y$.

Factoring out $d\,y/dx$, you get $d\,y/dx * (2\,y + x) = -y$, so $d\,y/dx = -y/(2\,y + x)$.

Do it in the Wolfram Language:

In[ ]:= **Solve[D[y[x]^2 + x * y[x], x] == D[2, x], y'[x]]**

Out[ ]= $\left\{\left\{y'[x] \rightarrow -\dfrac{y[x]}{x + 2\,y[x]}\right\}\right\}$

Here is a plot of the curve and the tangent line at $(1, 1)$. The slope is $-1/(2*1 + 1) = -1/3$:

In[ ]:= **ContourPlot[{y^2 + x * y == 2, y == 1 − 1/3 (x − 1)},**

{x, −5, 5}, {y, −5, 5}, Epilog → {PointSize[Large], Point[{1, 1}]}]

Out[ ]=

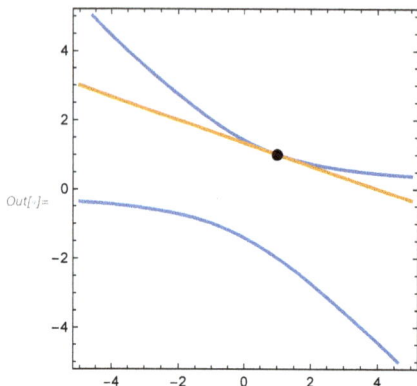

Question 22

Find the equation of the line normal to the graph of $x^2/2 + 3x$ at $(-2, -4)$.

Solution

First, find the slope at the point $(-2, -4)$.

The derivative of the curve is $x + 3$, so the slope at $(-2, -4)$ is $-2 + 3 = 1$.

Therefore, the slope of the normal line at $(-2, -4)$ is $-1/1 = -1$.

So the equation of the normal line is $y - (-4) = -1\,(x - (-2))$.

This simplifies to $y = -4 - (x + 2) = -x - 6$.

Here is a plot of the curve and the normal line:

In[ ]:= **Plot[{x^2/2 + 3 x, −x − 6}, {x, −3, −1}, AspectRatio → Full]**

Out[ ]=

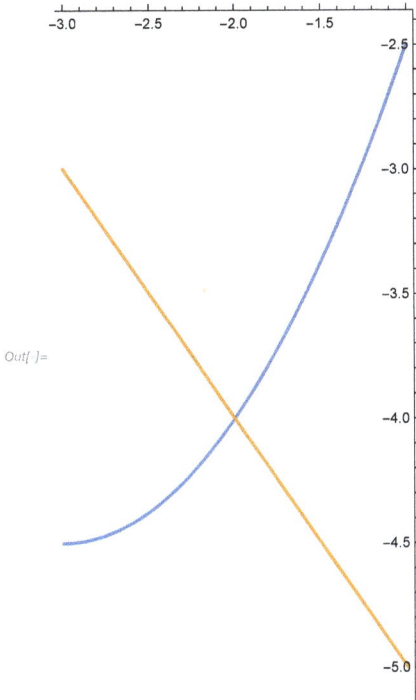

Question 23

What value of a and b will make the function
$f[x] = \{a\,x + 3,\ x < -1;\ b\,x^3 - 2\,x^2,\ x \geq -1\}$ differentiable?

Solution

You have a piecewise function, and each piece is differentiable, except maybe at -1.

In order for the function to be differentiable, the left-hand derivative at -1 must equal the right-hand derivative at -1.

The left-hand limit at -1 must also equal the right-hand limit at -1.

So you have the equations $a = 3\,b(-1)^2 - 4\,(-1) = 3\,b + 4$ and
$-a + 3 = b(-1)^3 - 2\,(-1)^2 = -b - 2$.

You can solve for a and b using Solve:

In[ ]:= **Solve[a == 3 b + 4 && −a + 3 == −b − 2, {a, b}]**

Out[ ]= $\left\{\left\{a \to \dfrac{11}{2},\, b \to \dfrac{1}{2}\right\}\right\}$

So when $a = 11/2$ and $b = 1/2$, the function is differentiable.

Here is its graph of the differentiable function:

$In[\cdot]:=$ **Plot[{** $\begin{cases} \frac{11x}{2}+3 & x<-1 \\ \text{Null} & \text{True} \end{cases}$ **,** $\begin{cases} \frac{x^3}{2}-2x^2 & x\geq-1 \\ \text{Null} & \text{True} \end{cases}$ **}, {x, -2, 0}]**

$Out[\cdot]=$

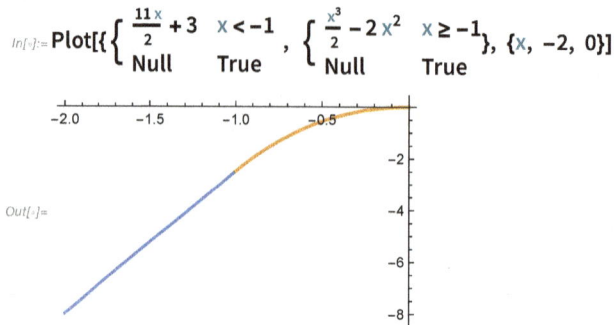

Question 24

A particle moves according to the velocity function $v[t] = 4\,\text{Cos}[t]\,\text{Sin}[t]$. Find the total distance the particle travels from $t = 0$ to $t = \pi$.

Solution

Since you want the total distance traveled and not the total displacement, you need the integral $\int_0^{6\pi} |v[t]|\,dt$.

Use Integrate and RealAbs to calculate the integral:

$In[\cdot]:=$ **Integrate[RealAbs[4 Cos[t] Sin[t]], {t, 0, π}]**

$Out[\cdot]=$ **4**

Here is a plot with the area shaded:

$In[\cdot]:=$ **Plot[4 Cos[t] Sin[t], {t, 0, π}, Filling → Axis]**

$Out[\cdot]=$

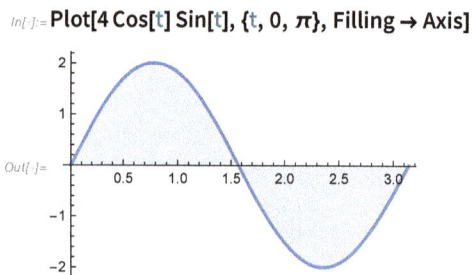

You wanted the total distance, so the negative area should count as positive.

Question 25

A spherical balloon is losing air at a rate of 2 cm^3 per sec. Find the rate at which the radius is decreasing when the radius is 10 cm.

Solution

The volume of a sphere is $V = 4\pi r^3 / 3$, where r is the radius:

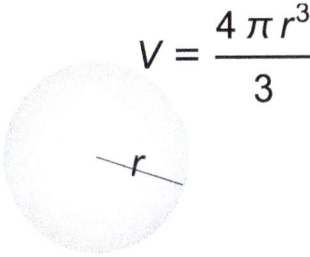

$$V = \frac{4\pi r^3}{3}$$

r

Differentiating with respect to t = time, you get $dV/dt = 4\pi r^2 \, dr/dt$.

You know $dV/dt = -2$ cm^3 and $r = 10$ cm, and want to find dr/dt.

Solve for dr/dt with Solve:

In[ ]:= **Solve[−2 == 4 π (10^2) drad, drad]**

Out[ ]:= $\left\{\left\{\text{drad} \rightarrow -\frac{1}{200\,\pi}\right\}\right\}$

So the radius is decreasing at a rate of $1/200\,\pi$ cm / sec ~ 0.00159 cm / sec.

Question 26

A cylindrical can will be produced with a volume of 216 cm^3. Find the value for the radius that will minimize the amount of material used.

Solution

To minimize the amount of material used, you need to minimize the surface area of the can.

The surface area of a cylindrical can is $2\pi r^2 + 2\pi r h$, where r is the radius and h is the height:

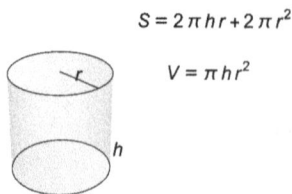

$$S = 2\pi h r + 2\pi r^2$$

$$V = \pi h r^2$$

You have the equation $\pi r^2 h = 216$, since the volume of the can is 216 cm^3.

Therefore, $h = 216/\pi r^2$, and the surface area of the can is $2\pi r^2 + 2\pi r(216/\pi r^2) = 2\pi r^2 + 432/r$.

You will minimize $2\pi r^2 + 432/r$ with respect to r when $r > 0$.

To do so, take the derivative and solve for r when the derivative is 0:

In[ ]:= **sol = Solve[D[2 π r^2 + 432/r, r] == 0 && r > 0] // N**

Out[ ]= **{{r → 3.25156}}**

So the derivative is 0 when $r = 3\sqrt[3]{4/\pi}$ cm ~ 3.25 cm.

This is a minimum, because the second derivative of $2\pi r^2 + 432/r$ is positive at $3\sqrt[3]{4/\pi}$:

In[ ]:= **D[2 π r^2 + 432/r, {r, 2}] /. sol**

Out[ ]= **{37.6991}**

So the cylindrical can with volume 216 cm^3 will use the least amount of material when the radius is $3\sqrt[3]{4/\pi}$ cm ~ 3.25 cm.

Wolfram Media

The Publishing Unit of the Wolfram Group

By Stephen Wolfram

› Predicting the Eclipse:
  A Multimillennium Tale of Computation

› The Second Law: Resolving the Mystery of the
  Second Law of Thermodynamics

› What Is ChatGPT Doing ...
  and Why Does It Work?

› Metamathematics: Foundations
  & Physicalization

› Twenty Years of A New Kind of Science

› Combinators: A Centennial View

› A Project to Find the Fundamental
  Theory of Physics

› Adventures of a Computational Explorer

› An Elementary Introduction
  to the Wolfram Language

› Idea Makers: Personal Perspectives
  on the Lives & Ideas of Some Notable People

› A New Kind of Science

By Other Authors

› The Math(s) Fix:
  An Education Blueprint for the AI Age
  Conrad Wolfram

› Query: Getting Information from Data
  with the Wolfram Language
  Seth J. Chandler

› Introduction to Machine Learning
  Etienne Bernard

› A Field Theory of Games: Introduction to
  Decision Process Engineering, Volume 1
  Gerald H. Thomas

› A Field Theory of Games: Introduction to
  Decision Process Engineering, Volume 2
  Gerald H. Thomas

› Hands-on Start to Wolfram|Alpha
  Notebook Edition
  Cliff Hastings and Kelvin Mischo

› Hands-on Start to Wolfram Mathematica
  and Programming with the Wolfram Language
  Cliff Hastings, Kelvin Mischo and Michael Morrison

› Introduction to Statistics
  with the Wolfram Language
  Juan H. Klopper

› A Numerical Approach to Real Algebraic
  Curves with the Wolfram Language
  Barry H. Dayton

www.ingramcontent.com/pod-product-compliance
Lightning Source LLC
Chambersburg PA
CBHW060954210326
41598CB00031B/4820